Modeling, Characterization, and Processing of Smart Materials

Ajay Kumar
Mechanical Engineering Department, School of Engineering and Technology, JECRC University, India

Parveen Kumar
Mechanical Engineering Department, Rawal Institute of Engineering and Technology, India

Ashish Kumar Srivastava
G.L. Bajaj Institute of Technology and Management, India

Vikas Goyat
Mechanical Engineering Department, SRM Institute of Science and Technology, India

A volume in the Advances in Chemical and Materials Engineering (ACME) Book Series

IGI Global
PUBLISHER of TIMELY KNOWLEDGE

Published in the United States of America by
IGI Global
Engineering Science Reference (an imprint of IGI Global)
701 E. Chocolate Avenue
Hershey PA, USA 17033
Tel: 717-533-8845
Fax: 717-533-8661
E-mail: cust@igi-global.com
Web site: http://www.igi-global.com

Library of Congress Cataloging-in-Publication Data

Names: Ajay, editor. | Kumar, Parveen (Professor of mechanical engineering), editor. | Srivastava, Ashish Kumar, editor. | Goyat, Vikas, 1987- editor.
Title: Modeling, characterization, and processing of smart materials / edited by Ajay, Parveen, Ashish Kumar Srivastava, and Vikas Goyat.
Description: Hershey PA : Engineering Science Reference, [2024] | Includes bibliographical references. | Summary: "This book presents a collection of studies on state-of-art techniques developed specifically for smart materials with an emphasis on using various analytical strategies, computational and simulation approach, artificial intelligence approach to develop innovative smart materials. The book offers an ideal reference guide for academic researchers and industrial engineers in the fields of different smart materials. It can also be used as a comprehensive reference source for university students in chemical engineering, material science and environmental engineering"-- Provided by publisher.
Identifiers: LCCN 2023017043 (print) | LCCN 2023017044 (ebook) | ISBN 9781668492246 (hardcover) | ISBN 9781668492253 (paperback) | ISBN 9781668492260 (ebook)
Subjects: LCSH: Smart materials.
Classification: LCC TA418.9.S62 M6235 2024 (print) | LCC TA418.9.S62 (ebook) | DDC 620.1/12--dc23/eng/20230522
LC record available at https://lccn.loc.gov/2023017043
LC ebook record available at https://lccn.loc.gov/2023017044

This book is published in the IGI Global book series Advances in Chemical and Materials Engineering (ACME) (ISSN: 2327-5448; eISSN: 2327-5456)

British Cataloguing in Publication Data
A Cataloguing in Publication record for this book is available from the British Library.

For electronic access to this publication, please contact: eresources@igi-global.com.

Advances in Chemical and Materials Engineering (ACME) Book Series

J. Paulo Davim
University of Aveiro, Portugal

ISSN:2327-5448
EISSN:2327-5456

MISSION

The cross disciplinary approach of chemical and materials engineering is rapidly growing as it applies to the study of educational, scientific and industrial research activities by solving complex chemical problems using computational techniques and statistical methods.

The **Advances in Chemical and Materials Engineering (ACME) Book Series** provides research on the recent advances throughout computational and statistical methods of analysis and modeling. This series brings together collaboration between chemists, engineers, statisticians, and computer scientists and offers a wealth of knowledge and useful tools to academics, practitioners, and professionals through high quality publications.

COVERAGE

- Fracture Mechanics
- Statistical Methods
- Computational methods
- Nanomaterials
- Toxicity of Materials
- Composites
- Industrial Chemistry
- Electrochemical and Corrosion
- Fatigue and Creep
- Artificial Intelligence Methods

IGI Global is currently accepting manuscripts for publication within this series. To submit a proposal for a volume in this series, please contact our Acquisition Editors at Acquisitions@igi-global.com or visit: http://www.igi-global.com/publish/.

The Advances in Chemical and Materials Engineering (ACME) Book Series (ISSN 2327-5448) is published by IGI Global, 701 E. Chocolate Avenue, Hershey, PA 17033-1240, USA, www.igi-global.com. This series is composed of titles available for purchase individually; each title is edited to be contextually exclusive from any other title within the series. For pricing and ordering information please visit http://www.igi-global.com/book-series/advances-chemical-materials-engineering/73687. Postmaster: Send all address changes to above address.

Titles in this Series

For a list of additional titles in this series, please visit: http://www.igi-global.com/book-series/

Handbook of Research on Advanced Functional Materials for Orthopedic Applications
R. Ranjith (SNS College of Technology, India) and J. Paulo Davim (University of Aveiro, Portugal)
Engineering Science Reference • © 2023 • 400pp • H/C (ISBN: 9781668474129) • US $325.00

Production, Properties, and Applications of Engineered Cementitious Composites
S. Praveenkumar (PSG College of Technology, India) and J. Paulo Davim (University of Aveiro, Portugal)
Engineering Science Reference • © 2023 • 300pp • H/C (ISBN: 9781668481820) • US $250.00

Innovative Multifunctional Nanomaterial for Photocatalysis, Sensing, and Imaging
Azad Kumar (Siddharth University, India)
Engineering Science Reference • © 2023 • 300pp • H/C (ISBN: 9781668487433) • US $270.00

Sustainable Utilization of Nanoparticles and Nanofluids in Engineering Applications
Sampath Boopathi (Muthayammal Engineering College, India) and J. Paulo Davim (University of Aveiro, Portugal)
Engineering Science Reference • © 2023 • 300pp • H/C (ISBN: 9781668491355) • US $270.00

Innovations in Materials Chemistry, Physics, and Engineering Research
Eugene de Silva (Virginia Research Institute, USA) and Pramudi Abeydeera (Polymeric Ltd., USA)
Engineering Science Reference • © 2023 • 300pp • H/C (ISBN: 9781668468302) • US $270.00

IGI Global
PUBLISHER of TIMELY KNOWLEDGE

701 East Chocolate Avenue, Hershey, PA 17033, USA
Tel: 717-533-8845 x100 • Fax: 717-533-8661
E-Mail: cust@igi-global.com • www.igi-global.com

Table of Contents

Detailed Table of Contents

This chapter will provide an insight into smart materials, which are a class of materials that can change their physical properties, such as shape, size, stiffness, color, or electrical conductivity, in response to external stimuli, such as temperature, pressure, light, humidity, magnetic, or electric fields. Various types of smart materials like shape memory alloys (SMAs), electroactive polymers (EAPs), piezoelectric materials, magnetostrictive materials, and thermoelectric materials were discussed in this chapter, along with their properties and applications. Nevertheless, the advantages of smart materials over the conventional materials along with the challenges were also discussed. Also, the possibility of integrating these materials with the existing systems was discussed.

Smart materials, also known as responsive or intelligent materials, are capable of responding to environmental stimuli in a predictable and controlled manner. Significant progress has been achieved in the field of smart materials over the last few decades, and these materials are now being employed in a variety of fields,

The leg bone that typically undergos the most deformation is the femur. Failures of the femur neck, which are most frequent among human bone failures due to their length and weight, are a very serious issue for people. However, before using biomaterials in implants or any orthopaedic surgery, the right biomaterials must be chosen based on their mechanical and physical properties in comparison to natural bone. The use of biomaterials in the treatment of orthopaedic-related issues is very crucial. By taking into account the mechanical and physical properties of materials, discussions and comparisons of biomaterials are conducted. In comparison to other materials, biomaterials like SS316L, Ti6Al4V, PMMA, and CoCr alloys, magnesium alloys have been found to be a better substitute.

Smart materials have made a substantial contribution to their widespread use across a number of industries. Numerous intelligent materials have the ability to alter their characteristics and behavior in response to an outside catalyst or stimulus. These materials are altering how typical engineering materials are used. To improve the desirable features of existing materials, smart materials can either be used alone or combined with them. Due to its transdisciplinary uses, smart materials are found in all human innovations. While some of these materials are in use now, some have been for a very long time. Therefore, suffice it to remark that further study is still needed on the recently produced smart materials. The development of composite materials to shape memory alloys is the subject of this review study. This literature describes the development of smart materials from traditional composites to cutting-edge shape memory alloys as well as how they are utilized.

Single point incremental forming (SPIF) is an emerging sheet forming method and has been widely accepted by the researchers for fabricating the customized items for the industrial uses and the end users. The measurement of the deforming loads

Additive manufacturing, often called 3D printing, is widely employed in all engineering sectors. Many researchers also refer to 3D printing in the biomedical field as bioprinting. Therefore, the 3D bioprinting methods are used to create patient-specific implants or devices for several tissue engineering needs. In addition, this method is also famous for drug design and targeted drug delivery systems. Nowadays, several researchers have involved smart materials with 3D printing to obtain 4D printing for biomedical applications. In this book chapter, the authors provide a quick overview of the many kinds of 3D printing. They also highlight the popular materials utilized in the technology for biomedical applications. Nanoparticles and smart materials, which are employed in biomedical applications, are also briefly explored. This technology's difficulties and future scope are also discussed. This chapter will give the readers a basic understanding of the materials used for 3D printing biomedical devices.

Manufacturing hard metals and their welding-related operations is still a challenging task for the manufacturer. In this context, many types of welding methods are developed. Among these methods, microwave welding is one of the prominent methods to weld or clad metal plates. The present work is a review chapter, and it highlights the study of microstructural and mechanical properties of the developed components through microwave welding. The present work also highlights the challenges and issues related to microwave welding technology. Finally, the future scope and its commercialization are discussed.

Ruby Jindal, Department of Physics, School of Basic and Applied Sciences, K.R. Mangalam University, Gurugram, India
Archana Tripathi, Physics Department, Gargi College, University of Delhi, New Delhi, India
Neenu Saini, Department of Physics, Pt. Jawaharlal Nehru Government College, Faridabad, India

This book chapter provides an in-depth exploration of Wilson's matrix method, a computational technique used to calculate molecular vibration frequencies and normal modes in smart materials. The chapter begins with an introduction to molecular vibrations and normal modes, emphasizing their importance in understanding material behavior at the molecular level. It then delves into the theoretical foundation and practical steps involved in Wilson's matrix method, highlighting its use of force constants and eigenvalue calculations to determine vibrational frequencies and associated modes. The chapter further examines the application of Wilson's matrix method to smart materials, emphasizing their unique vibrational characteristics and the role of molecular vibrations in shaping their functionality. The advancements in the method, including the inclusion of anharmonicity and improved force constant determination, are discussed, along with the integration of enhanced computational techniques and increased computational power.

Subhash Chandra Ram, Tula's Institute, Dehradun, India
Awani Bhushan, Vellore Institute of Technology, Chennai, India
Sunkulp Goel, University of Science and Technology, Nanjing, China

The present study provides a comprehensive state-of-the-art on currently available information about the advancement of Al-Mg2Si in-situ FG-composites for various automotive application domains, as well as their inherent benefits and drawbacks. Additionally, the chapter explored how a functionally graded material (FGMs) might serve as a suitable replacement material for automotive components. FGMs can be practically modified through the modification of their constituents to meet particular functional needs. The production of FGMs employs an extensive range of approaches. Taking into consideration Al-Si-Mg alloy as the starting materials, the fabrication methods are categorized as follows: liquid processes, gaseous processing, and solid particle methods. This chapter describes the typical centrifugal casting technique for Al-Mg2Si in-situ functionally graded lightweight composites. Optical and SEM analysis of reinforced phase Mg2Si particles, Al-Si eutectic phases, and α-Al grains size were presented in detail.

Nanowires are a class of smart materials with unique physicochemical properties owing to their electron confinement along their thickness which differs them from their bulk counterparts. Their enhanced physical, chemical, optical, and mechanical properties have heightened their suitability for a host of sensing applications. The profound sensitivity underlines their preferential applications in lighting, computing, communication, energy harvesting, and medicine for detecting and analyzing signals and molecules with a relative greater efficiency. Advanced synthesis methods have been developed to engineer the properties of nanowires for tailored sensing applications. In this chapter, captured various synthesis techniques and potential sensing applications of nanowires underpinned by their unique properties. The chapter output will enlighten recent advances in nanowires and its subsequent sensing technologies utilized in healthcare, environment, and electronic devices. Consequently, the concepts are expected to open vistas in allied nanomaterials and sensing technology development.

Different materials have been used for various purposes for the advancement of humanity ever since the beginning of human civilization. The term "smart material" encapsulates the change taking place in the field of material science. Smart materials are the key that can be used to unlock our impossible desires, such as the ability of a mobile phone screen to repair itself, the ability of some buildings to react to weather, the ability of some medical fields to release drugs into the bloodstream at the very first stage of a disease, etc. Because of its appealing and intelligent qualities, smart materials are known by other names such as responsive and intelligent materials. Overall, it can be concluded that it demonstrates a vivacious spirit and wise behavior. Temperature, pressure, tension, an electric or magnetic field, moisture, or pH can all cause these materials to react. This chapter discusses recent reviews of smart materials based on their state of the art. The authors also explore different types of smart materials and their various ways and sectors for their applications.

Preface

Smart materials are those materials which have multidirectional properties and can be changed in a controlled fashion by applying external field like temperature, pressure, stress, magnetic or electric fields and moisture or pH and examples of this behaviour defines responses to external stimuli by self-healing (in which material automatic self-repair to damage), self-sensing (allows automatic detection and quantification of external stimuli), self-diagnostic (allows automatic detection of defect), self-actuating (automatic actuation in response to external stimulus), and shape changing (in response to external stimuli material changes shape). Thus these materials provide a means of achieving 'smart' response in product and yield enhanced functionalities and capabilities that would otherwise be lacking. The main focus of smart materials are on different types of novel materials like Shape Memory Alloys (SMAs) such as Ni-Ti (Nickel-Titanium also known as Nitinol), hybrid composites, polymers, ceramics, chitosan-based gels, electroheological materials, magnetorheological fluids, sensors, nanotubes, metamaterials, quantum materials and various biomaterials etc. These smart materials highlights the multidisciplinary nature by representing how they can be utilized in different sectors like new product development, food industries, drug delivery, electronic sensors for satellite, health monitoring, biomimetic design and prosthesis, and robotics etc.

The book consists of fourteen chapters that describe perspectives of Modeling, Characterization, and Processing of Smart Materials. Chapter "Towards smart materials - enhancing the efficiency of the materials" provide an insight into smart materials, which are a class of materials that can change their physical properties, such as shape, size, stiffness, color, or electrical conductivity, in response to external stimuli, such as temperature, pressure, light, humidity, magnetic or electric fields.. Chapter "Smart Materials in Biomedicine - A Promising New Frontier" presents a thorough overview of the most recent research and development in the application of smart materials in biomedicine, including drug delivery, tissue engineering, biosensors, and imaging, as well as an explanation of the current difficulties and future prospects of this promising topic. Chapter "Smart Biodegradable and Bio-based Polymeric Biomaterials for Biomedical Applications" briefly cover smart biomaterials i.e. Smart memory polymers (SMP) which are used in biomedical applications. Chapter "Materials and technology for implant manufacturing: challenges and opportunity" discuss the mechanical and physical properties of materials, and comparisons of various biomaterials for implant manufacturing. Chapter "Review on the Development of Smart Materials, Including Shape Memory Alloys, Characterization and Recent Applications" describes the development of smart materials from traditional composites to cutting-edge shape memory alloys as well as how they are utilized. Chapter "Experimental study of impact parameters on the deforming loads in Incremental Forming" aims at investing the influence of the punch feed rate and the forming angle on the peak values of the deforming loads for AA2024 aluminum alloy sheets. Chapter "Microwave Assisted versus Conventional Hydrothermal Synthesis, Morphology, Microstructure, and Surface area analysis of g-C3N4/MoS2 Nanocomposite: Analysis of g-C3N4/MoS2 Nanocomposite" evaluated the microstructure, morphology, surface area, and pore diameters of g-C3N4/MoS2 nanocomposite produced employing hydrothermal (g-C3N4/MoS2-HT) and novel microwave methods (g-C3N4/MoS2-MW). Chapter "Metals, Polymers, Ceramics, Composites Biomaterials used in Additive Manufacturing for Biomedical Applications" provides a quick overview of the many kinds of 3D printing and give the readers a basic understanding of the materials used for 3D printing biomedical devices. Chapter "Review of microstructural and mechanical properties of microwave welding of lightweight alloys" highlights the study of microstructural and mechanical properties of the developed components through microwave welding. Chapter "Wilson's Matrix Method for Calculating Molecular Vibration Frequencies and Normal Modes for Smart Materials" delves into the theoretical foundation and practical steps involved in Wilson's matrix method, highlighting its use of force constants and eigenvalue calculations to determine vibrational frequencies and associated modes. Chapter "Industry Requirement and Future Prospects of Lightweight AlMg2Si Functionally Graded Materials for Automotive Engine Components: Review" describes the typical centrifugal casting technique for Al-Mg2Si in-situ functionally graded lightweight composites. Optical and SEM analysis of reinforced phase Mg2Si particles, Al-Si eutectic phases, and α-Al grains size were presented in detail. Chapter "Nanowires Synthesis and Application as Smart Biochemical Sensor and Energy Storage Materials" enlighten recent advances in nanowires and its subsequent sensing technologies utilized in healthcare, environment and electronic devices. Chapter "Applications and Types of Smart Materials" discusses recent reviews of smart materials based on their state of the art and explores different types of smart materials and their various ways and sectors for their applications. Chapter "Analyzing the properties of Zinc Oxide (ZnO) thin film grown on Silicon (Si) substrate, ZnO/Si using RF Magnetron sputtering approach" discusses how Zinc Oxide (ZnO) single layer of 100 nm is depoisited on silicon (Si) substrate using radio frequency magnetron sputtering approach.

Undergraduates, postgraduate students studying the materials science and its importance in different sectors like manufacturing, mechanical,, biomedical, pharmaceuticals, biotechnology, electronics engineering will find the book useful for an advance course on Smart materials and related areas like their modeling, optimization by various nature inspired algorithms, scattering techniques, nondestructive testing methods etc.

The editors acknowledge the professional support received from IGI Global and express their gratitude for this opportunity.

Reader's observations, suggestions and queries are welcome.

Ajay Kumar
Mechanical Engineering Department, School of Engineering and Technology, JECRC University, Jaipur, India

Parveen Kumar
Mechanical Engineering Department, Rawal Institute of Engineering and Technology, Faridabad, India

Ashish Kumar Srivastava
Mechanical Engineering Department, G.L. Bajaj Institute of Technology and Management, Greater Noida, India

Vikas Goyat
Mechanical Engineering Department, SRM Institute of Science and Technology, Ghaziabad, India

Acknowledgement

The editors are grateful to IGI Global for showing their interest to publish this book in the buzz area of Smart Materials. The editors express their personal adulation and gratitude to methodical staff that have helped the editors to compile and finalize the manuscript.

Editors are thankful to the researchers, experts, scientists, academicians and related professionals who generously shared their knowledge and experiences. Their contributions have enriched the content and added depth to the ideas presented in this book.

Editors would like to extend appreciation to the reviewers who diligently reviewed the manuscripts and provided valuable feedback that helped refine and improve the final version of this book.

Dr. Ajay also wishes to express his gratitude to his parents, Sh. Jagdish and Smt. Kamla, and his loving brother Sh. Parveen for their true and endless support. They have made him able to walk tall before the world regardless of sacrificing their happiness and living in a small village. He cannot close these prefatory remarks without expressing his deep sense of gratitude and reverence to his life partner Mrs. Sarita Rathee for her understanding, care, support and encouragement to keep his moral high all the time. No magnitude of words can ever quantify the love and gratitude the he feels in thanking his daughters, Sejal Rathee and Mahi Rathee, and son Kushal Rathee who are the world's best children.

Finally, the editors obligate this work to the divine creator and express their indebtedness to the "ALMIGHTY" for gifting them power to yield their ideas and concepts into substantial manifestation. The editors believe that this book would enlighten the readers about each feature of Smart Materials.

Dr. Ajay Kumar
Mr. Parveen Kumar
Dr. Ashish Kumar Srivastava
Dr. Vikas Goyat

Chapter 1
Towards Smart Materials:
Enhancing the Efficiency of the Materials

Avinash Gudimetla
Pragati Engineering College, India

Parveen Kumar
Mechanical Engineering, Rawal Institute of Engineering and Technology, India

S Sambhu Prasad
https://orcid.org/0000-0003-1331-7376
Pragati Engineering College, India

Satish Geeri
https://orcid.org/0000-0002-9039-8555
Pragati Engineering College, India

V. V. N. Sarath
Pragati Engineering College, India

ABSTRACT

This chapter will provide an insight into smart materials, which are a class of materials that can change their physical properties, such as shape, size, stiffness, color, or electrical conductivity, in response to external stimuli, such as temperature, pressure, light, humidity, magnetic, or electric fields. Various types of smart materials like shape memory alloys (SMAs), electroactive polymers (EAPs), piezoelectric materials, magnetostrictive materials, and thermoelectric materials were discussed in this chapter, along with their properties and applications. Nevertheless, the advantages of smart materials over the conventional materials along with the challenges were also discussed. Also, the possibility of integrating these materials with the existing systems was discussed.

DOI: 10.4018/978-1-6684-9224-6.ch001

1. INTRODUCTION

1.1 Definition of Smart Materials

Smart materials exhibit unique and often unusual properties and behaviours, which can be controlled or altered by an external provocation, such as temperature, light, magnetic or electric field (Ounaies, 2016). These materials are designed to respond and adapt to changes in their environment and may be employed in applications of various verticals and domains, ranging from aerospace and automotive engineering to biomedicine and consumer electronics.

Some examples of smart materials include Shape memory alloys, electrochromic materials, thermochromic materials, piezoelectric materials and magnetostrictive materials. Shape memory alloys, for example, may change shape in response to temperature changes, whereas piezoelectric materials can create an electric charge when subjected to mechanical stress (Mather, 2010).

In the recent past, there has been a enormous amount of study and innovation into the creation and use of smart materials (Dierdorf, 2002), with the goal of increasing the performance and functionality of current materials and generating new materials with unique qualities and capabilities.

1.2 Brief History on Development of Industry Important Smart Materials

Shape memory alloys (SMAs) were discovered by William Buehler and Frederick Wang at the US Naval Ordnance Laboratory in 1959. This marks the beginning of development of smart materials. SMAs have the ability to "remember" their initial shape and return to it when exposed to heat or other stimuli, making them helpful in a range of applications like as robotics and medical equipment (Bououdina, 2015).

Piezoelectric polymers, which create an electrical charge as a result of the mechanical stress (Araujo, 2015), were discovered in the 1970s by the team of researchers at the National Bureau of Standards. This property has resulted in the creation of piezoelectric sensors and actuators, which are utilised in a variety of applications such as automotive systems and medical equipment.

Electroactive polymers, which change form in reaction to an electric field, and magnetostrictive materials, which change shape in response to a magnetic field, are two further types of smart materials that have been produced throughout the years.

Nanotechnology advancements in recent past had resulted in creation of new varieties of smart materials, such as carbon nanotubes and graphene, which have unique electrical and mechanical characteristics that make them valuable in a diversified applications (J. H. Koo, 2017)(Uchino, Piezoelectric Actuators and Devices: Recent Developments and Prospects, 2006)(Xie, 2015).

1.3 Regulations and Importance of Smart Materials

Smart material regulations differ based on the use and sector. The Food and Drug Administration (FDA) in United States oversees the use of smart materials in medical devices to ensure that they are secure and effective. Similarly, to ensure compliance with safety regulations, the Federal Aviation Administration (FAA) controls the use of smart materials in aircraft (Asmatulu, 2016).

Aside from regulatory issues, there are also ethical and social implications associated with the use of smart materials (Haghniaz, 2018). Concerns have been raised, for example, pertaining to the possible influence of smart materials on employment, privacy, and security. The significance of smart materials lies in their ability to enable new technologies and applications that can improve our lives (Kandola, 2019). However, it is vital to address regulatory and ethical concerns to ensure that the benefits of these materials can be maximised while potential risks are minimised.

2. PROPERTIES OF SMART MATERIALS

2.1 Stimuli Responsiveness

Smart materials are materials with distinct characteristics that may change in reaction to external provocation such as pressure, temperature, humidity, light, and magnetic or electric fields. Smart materials' stimulus reactivity is what distinguishes them and makes them desirable for diversified applications in areas such as construction, healthcare and aerospace (Tay, 2015).

Some common types of smart materials and their corresponding stimuli are:

1. Shape memory alloys: These materials can "remember" and can retain their original shape when heated. They are responsive to changes in temperature.
2. Piezoelectric materials: These materials can generate an electrical charge when they are stressed mechanically, or deform when an electrical field is applied. They are responsive to changes in mechanical pressure or electric fields.
3. Electroactive polymers: These polymers can change shape in response to an electric field, and some can also generate an electrical charge when mechanically stressed. They are responsive to changes in electric fields or mechanical pressure.

4. Thermochromic materials: These materials can change their colour with respect to the changes in temperature.
5. Photochromic materials: These materials change colour with respect to changes in light. These respond to changes of light.

Other properties of smart materials include self-healing, self-cleaning, and self-assembling. These characteristics are the reason for making these materials to suite various applications, including biomedical devices and structural materials.

2.2 Multi-Functionality

In addition to stimulus responsiveness, another important property of smart materials is their ability to perform multiple functions at the same time. This trait, known as multi-functionality, allows smart materials to execute many functions at the same time, resulting in higher efficiency, decreased complexity, and improved performance in a variety of applications.

A smart material used in an aeroplane wing, for example, may alter form to optimise aerodynamics while also monitoring structural health and acting as a sensor for temperature and pressure changes (Mauro, 2018). This versatility can eliminate the need for multiple materials or systems, as a result of which the product is lighter, cost-effective, and more reliable product.

The electroactive polymer is a sort of a multi-functional smart material which has the ability to change its shape when an electric field is applied, create an electrical charge when mechanically strained, and operate as a pressure and deformation sensor. Self-healing materials that can repair damage while simultaneously functioning as sensors for damage detection are another example of smart materials with multi-functional characteristics, as are shape memory alloys that can alter shape while still being electrically conductive.

Overall, multi-functional smart materials have the potential to revolutionise industries ranging from aerospace to healthcare by enabling new and more efficient products and systems.

2.3 Self-Healing Ability

One of the distinguishing characteristics of smart materials is their ability to self-heal. It refers to a material's capacity to repair itself in reaction to damage or fissures without the need for external intervention. These materials have the ability to increase product longevity, reduce the costs associated with maintenance and can also increase the safety in a variety of applications (García-Moreno, 2018).

Self-healing materials are of two kinds - Non-autonomous self-healing materials require an external trigger, such as heat, light, or pressure, to commence the healing process, whereas autonomous self-healing materials may fix themselves without external intervention.

Polymers, ceramics, and metals are all examples of self-healing smart materials. Polymers having self-healing properties, for example, can be created by inserting microcapsules containing healing agents into the polymer matrix. When the material is damaged, the capsules rupture, releasing the healing agent and enabling the damage to be repaired. Ceramics that can repair cracks in reaction to changes in temperature or pressure are another example of self-healing smart materials, as are metals that can heal themselves through atom diffusion to the damaged location.

2.4 Other Unique Properties

Self-healing smart materials have the potential to revolutionize various industries, including automotive, aerospace, and construction, by reducing maintenance costs, enhancing durability and safety, and improving the sustainability of products.

Smart materials offer a unique set of features which renders them important for an extensive variety of applications. Smart materials can have qualities such as self-cleaning, self-assembling, and biocompatibility in addition to stimulus response and self-healing (Wang X. &., 2017). As self-cleaning materials repel dirt, dust, and other contaminants, they are ideal for applications such as building facades and medical devices. Superhydrophobic surfaces, which are highly water-repellent and may prevent dirt and pollutants from adhering, are an example of self-cleaning materials.

Self-assembling materials may generate complicated structures and patterns on their own, making them suitable for applications like delivering drugs and tissue engineering. Peptides and DNA molecules are examples of self-assembling materials that may form complex structures by molecular self-assembly.

Biocompatible materials are those that can interact with biological systems without harming them. They have a wide range of medicinal uses, including implants and drugs delivery systems. Hydrogels, which may reflect the qualities of human tissues, and titanium alloys, which are often used for medical implants, are examples of biocompatible materials.

3. TYPES OF SMART MATERIALS

3.1 Shape Memory Alloys

Shape memory alloys (SMAs) are a special class of materials that can revert to their initial shape after being distorted and then treated to a specific stimulus, such as heat or stress. This property correlates to a reversible phase shift that happens in the crystal structure of the material when it is subjected to the stimulus. Because of its capacity to restore their original shape, SMAs are beneficial in a broad variety of technical applications - actuators, sensors, and biomedical devices are a few to name among them.

The details of smart materials are presented in Table 1.1. NiTi alloys, being called as nitinol, are nickel-titanium alloys and are the most widely used SMAs. Because of their superior mechanical qualities, biocompatibility, and shape memory effect, these alloys have attracted substantial research (Otsuka, 1998). Copper-based alloys, iron-based alloys, and gold-cadmium alloys are examples of SMAs.

The unique properties of SMAs make them suitable for a broad spectrum of applications. They can be employed in medical devices like stents, for example, since their ability to change its shape allows them to be placed through a small incision and subsequently expand to their original shape within the body (Mohd Jani, 2014) (Ma, 2018). SMAs may also be utilised in aeronautical engineering to create self-deploying structures and morphing wings. In soft robotics, SMAs can be employed as actuators, allowing for more natural and flexible movement.

Despite the numerous potential applications of SMAs, there are some issues that must be addressed. SMAs, for example, continue to be relatively costly, and their behaviour can be complicated and difficult to forecast. Nonetheless, SMA research continues, and new applications are being investigated.

3.2 Piezo Electric Materials

A type of materials known as piezoelectric materials may convert mechanical energy into electrical energy and vice versa as shown in Figure 1.1 (Ibrahim Nazem QADER, 2019). This is due to the existence of electric dipoles in the material, which may align or reorient in response to mechanical stress or an applied electric field. Quartz, lead zirconate titanate (PZT), and barium titanate ($BaTiO_3$) are the most often used piezoelectric materials.

Piezoelectric materials are being used widely in various fields including sensing, actuation, energy harvesting, and medical devices. Piezoelectric sensors, for example, are widely employed in automotive, aerospace, and industrial applications to measure pressure, force, and acceleration. Precision positioning systems,

Table 1. Types of Smart materials along with their details

Types of smart materials	Property	Materials	Applications	Limitations
Shape Memory Alloys	Restoring their original shape	Nickel- Titanium alloys, Copper-based alloys, iron-based alloys and gold-cadmium alloys	Medical devices, aeronautical engineering, actuators etc.	Cost, Complicated behaviour-difficult to predict
Piezo electric materials	Energy convesrion	Quartz, lead zirconate titanate, and barium titanate	including sensing, actuation, energy harvesting, and medical devices	low energy density, sensitive to temperature changes
Electrochromoc materials	Colour chage	Tungsten oxide, nickel oxide, and iridium oxide, Polythiophenes, polyanilines, and viologens	smart windows, displays, and sensors	
Thermochromic materials	Colour chage	Liquid crystals, Vanadium dioxide, etc.	Temperature sensors, smart windows, and textiles	
Magnetostrictive materials	Dimension change	Terfenol-D, Galfenol	Sensing, energy harvesting, non-destructive testing and medical imaging	

microelectromechanical systems (MEMS), and robots all make use of piezoelectric actuators. Piezoelectric materials may also be utilised for collecting energy by converting mechanical vibrations into electrical energy that can be utilised to power small devices (Safari, 2019). Piezoelectric materials can be employed in ultrasonic imaging and transducers in medical equipment (Yao, 2019).

However, there are still some issues with piezoelectric materials. One significant disadvantage is their low energy density, which limits their effectiveness in some applications (Cross, 1999). Furthermore, piezoelectric materials can be sensitive to changes in temperature and humidity, which can affect their performance (Uchino, Piezoelectric actuators and ultrasonic motors, 2019). Nonetheless, ongoing research is aimed at creating new piezoelectric materials and improving their performance to suite for a wide range of applications (Zhang X. W., 2018).

3.3 Electrochromic Materials

Electrochromic materials change colour or transparency when exposed to an electric field or voltage. These materials have a wide range of uses, including smart windows, displays, and sensors. Inorganic, organic, and hybrid materials are the types of electrochromic materials.

Figure 1. Piezoelectric effects are depicted schematically – (a) Mechanical force producing electrical charge (b) Electrical charge developing mechanical force and (c) the development of shear force as a result of electrical charge applied in the direction perpendicular to the polarization of the crystal

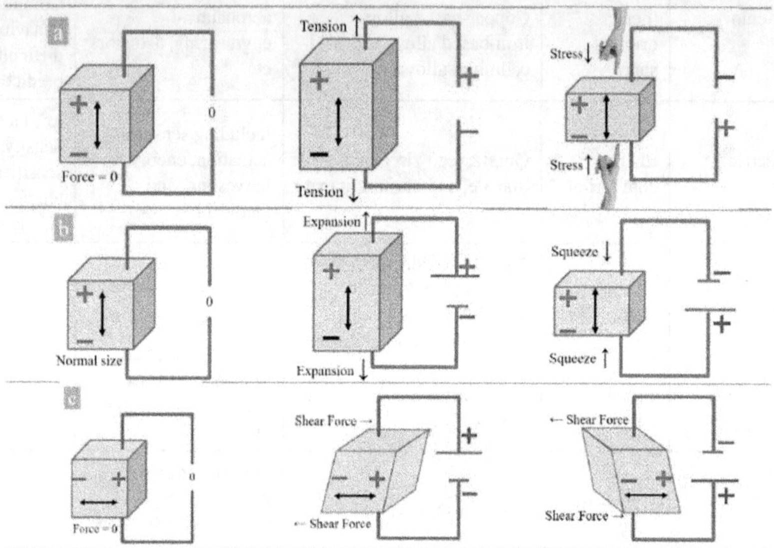

Inorganic electrochromic materials include include tungsten oxide (WO_3), nickel oxide (NiO), and iridium oxide (IrO_2). Because of its excellent optical contrast, outstanding electrochromic performance, and stability (Granqvist, 1995), WO_3 is one of the most thoroughly researched electrochromic materials. Because of its low cost and high switching speed, NiO has also showed promise as an electrochromic material. Another inorganic electrochromic substance with good optical contrast and stability is IrO_2.

Organic electrochromic materials are composed of polymers or tiny molecules that, when exposed to an electric field, undergo reversible electrochemical oxidation or reduction. Because of their low cost, versatility, and simplicity of processing, these materials have received a lot of attention (Deb, 2014). Polythiophenes, polyanilines, and viologens are examples of organic electrochromic materials.

Hybrid electrochromic materials are a combination of inorganic and organic materials that can outperform and strengthen each of their individual components (Sharma, 2018). A WO_3/polyaniline composite, for example, has demonstrated enhanced electrochromic characteristics and stability when compared to separate components.

New materials are being developed to increase the performance, stability, and utility of electrochromic materials, which is an active field of study (Zhang X. S., 2020).

The use of nanostructured materials, such as nanoparticles and nanowires, to enhance the optical and electrochromic characteristics of electrochromic materials is one of the most recent advances. Another area of study is the creation of electrochromic materials that may respond to a variety of stimuli, including light and temperature.

3.4 Thermochromic Materials

Thermochromic materials are those that change colour in response to temperature variations. Temperature sensors, smart windows, and textiles are just a few of the uses for these materials. The colour change of thermochromic materials is usually reversible, which means that it may be reversed when the temperature changes.

Liquid crystals are a typical form of thermochromic material. These materials have seen widespread application in temperature-sensitive displays such as thermometers and mood rings (Zeng, 2019). Because liquid crystals' optical properties change in response to temperature changes, they are an ideal material for temperature sensing and color-changing applications (Liu J. &., 2021).

Metal oxides, such as vanadium dioxide (VO_2), are used in another form of thermochromic material. At a specific temperature, certain materials experience a phase transition, resulting in a colour shift. Above 68°C, VO_2, for example, changes from a transparent or metallic state to a deep blue or purple colour. Because the phase transition is reversible, VO_2 is an excellent material for temperature sensing and smart window applications.

Thermochromic materials may potentially find use in the textile sector. It is feasible for creating clothing that changes colour dependent on the wearer's body temperature by incorporating thermochromic elements into textiles (Nirmalraj, 2017) (Sehrawat S. K., 2022). This might be especially helpful for athletes, who could use the colour change to track their exertion levels while exercising. Thermochromic materials, among other things, have the potential to revolutionise sensing, smart windows, and textiles.

3.5 Magnetochromic Materials

A magnetic field can be used as a stimulus to change the colour of certain materials, referred to as magnetochromic materials. Ferrofluids are fluidic variants of this substance that contain magnetic particles ranging in size from 5 to 10 nm. Whenever a magnetic field is placed on them, the particles align in a predictable fashion, causing incident rays to be dispersed from several semi-parallel planes of suspended particles. Figure 1.2 depicts a magnetochromic film that has been exposed to varying magnetic field strengths (Ibrahim Nazem QADER, 2019).

Figure 2. Diffracted light from films producing different colours because of changing magnetic field

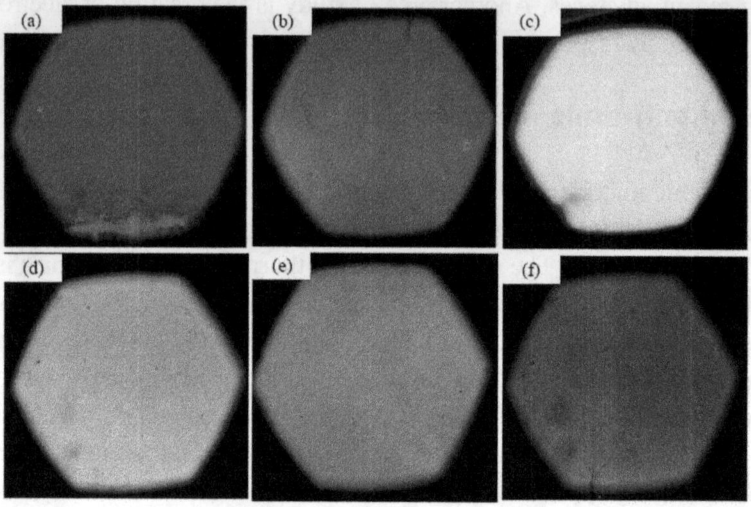

3.6 Magnetostrictive Materials

Magnetostrictive materials can change their dimensions when an magnetic field is applied. Magnetostriction is a phenomenon generated by the reorientation of magnetic domains inside the material. Sensors, actuators, and energy harvesting devices are just a few of the uses for magnetostrictive materials.

Terfenol-D, an iron-based alloy with high magnetostriction, is a common magnetostrictive material. Terfenol-D is commonly employed in high-precision positioning systems used in robotics and micromanipulation (Clark, 2004) (Sehrawat S. K., 2022). Galfenol, a form of iron-gallium alloy with high magnetostrictive qualities that can be employed in energy harvesting systems, is another commonly used magnetostrictive material.

Magnetostrictive materials have also been used in sensing applications. Highly sensitive magnetic field sensors can be created by using magnetostrictive materials as the sensing element. Magnetostrictive sensors find application in non-destructive testing and medical imaging also.

Magnetostrictive materials have potential uses in energy harvesting in addition to sensing and actuation (Snyder, 2013)(Liu X. &., 2015)(Matsumoto, 2016) (Singh V. K., 2022). It is possible to convert mechanical energy into electrical energy by utilising the magnetostrictive effect. As a result, magnetostrictive materials are a possible choice for self-powered sensors and other low-power devices. Magnetostrictive materials have the ability to convert magnetic energy into mechanical energy makes them ideal for use in sensing, actuation, and energy harvesting applications.

4. APPLICATIONS OF SMART MATERIALS

4.1 Biomedical Implants

Smart materials, also known as functional materials, may respond to environmental stimuli such as temperature, stress, and electromagnetic fields. These materials have found widespread use in a variety of industries, including biomedical implants. Biomedical implants are artificial devices inserted surgically into the body to replace or improve the function of natural organs or tissues. By providing real-time monitoring and adaptive response to changes in the body, smart materials have the capability to improve performance and functioning of biomedical implants (Shayan, 2019).

Shape memory alloys (SMAs) are an example of smart materials used in biomedical implants. When treated to a given temperature or stress, SMAs may recall their previous form and regain it. This characteristic can be utilised to develop implants that can adjust to changes in the body, such as temperature fluctuations or movement-induced stress. SMA-based implants are being employed in heart stents and orthopaedic implants.

Piezoelectric materials are another example of smart materials utilised in biomedical implants. In response to mechanical stress or pressure, piezoelectric materials can create an electrical charge. This characteristic has the potential to be exploited to construct implants that can generate electrical impulses in reaction to changes in the body, such as pressure or movement. Piezoelectric materials have been employed in a wide range of applications, such as cochlear implants and bone growth stimulators (Liu J. H., 2021) (Singh V. G., 2023). Overall, smart materials in biomedical implants have the potential to increase device performance and usefulness by offering real-time monitoring and adaptive response to changes in the body.

4.2 Aerospace Engineering

Smart materials are a kind of sophisticated material that possesses unique physical and mechanical properties that can be utilised for a variety of applications, inclusive of aerospace sector. These materials can detect and respond to changes in their surroundings, allowing them to adapt and perform efficiently under a variety of settings. Smart materials offer a variety of applications in aircraft engineering, spanning from structural monitoring to actuation and control. The details of smart materials along with their functions and applications are also presented in Table 1.2.

Table 2. Applications of Smart materials

Function of Smart Material	Example of Smart Material	Applications
Response to external stimuli - Temperature, stress, pressure, electro magnetic feld	Shape memory alloys, piezo electric materials	Bio medical Implants
Detection and respond to changes in their surroundings	Piezoelectric materials, shape-memory alloys, electroactive polymers and fibre optic sensors	Structure health monitoring, Actuators, control systems
Flexible and malleable to bend, twist and stretch	Shape memory alloys, piezoelectric materials, electroactive polymers, carbon nano tubes	Soft robots, self - healing robots
Ability to transform several types of energy	Piezoelectric and thermoelectric materials	Piezoelectric energy harvesters, thermoelectric generators
Ability to control the amount of light and heat that enters a structure	Electrochromic materials	Smart windows
Flexibility, durability, and energy-absorbing capability	Shear thickening fluids, shape-memory alloys and electroactive polymers	Impact-resistant helmets, adaptable apparel

Structure health monitoring (SHM) is one of the most important uses of smart materials in aeronautical engineering. SHM is the use of sensors to detect and monitor changes in an aircraft's structural integrity, such as fractures, deformation, and fatigue. Smart materials can be embedded into an aircraft's construction to allow for real-time monitoring and damage detection. Changes in strain, temperature, and other physical qualities may be detected by these materials and relayed to a central control system, which can notify maintenance workers of possible issues before they become critical (Aggelis, 2015) (Singh A. K., 2019). This system has the capability to save maintenance costs, improve the safety, and also can increase the service life of aircraft components.

Actuation and control systems are another important application of smart materials in aeronautical engineering. Shape-memory alloys and electroactive polymers, for example, can be used as actuators to control the motion and shape of aircraft components like wings, flaps, and control surfaces. These materials have the ability to alter shape or stiffness in response to an external input, such as temperature or voltage, allowing them to execute complicated motions and modifications. Shape-memory alloys, for example, can be used to regulate the shape of morphing wings, which can modify their geometry and lift qualities to improve aircraft performance.

Carbon nanotube (CNT) reinforced composites are one example of a smart material utilised in aeronautical engineering. Carbon nanotubes (CNTs) are high-strength, lightweight materials that may be inserted in a composite matrix (Avinash

Gudimetla, 2020) (Avinash Gudimetla D. L., 2021) (Avinash Gudimetla D. L., 2020) to improve its mechanical and electrical characteristics. CNT-reinforced composites have been employed in the manufacture of satellite antennas, which need lightweight and robust materials capable of withstanding the harsh conditions of space (Liu Y., 2018) (Ajay, 2023). The utilisation of carbon nanotubes has enabled the development of lightweight, high-performance antennas that may be deployed in space without fear of damage or failure.

From structural monitoring to actuation and control systems, smart materials have diverse applications in aeronautical engineering. These materials have distinctive properties that can enhance the performance, safety, and efficiency of aircraft components and systems. In the future years, the development of novel and innovative smart materials is expected to play an increasingly vital role in the aircraft sector.

4.3 Robotics

Smart materials have emerged as a viable class of materials in robotics because of their distinct features such as shape memory, self-healing, and responsiveness to external stimuli. These materials have the potential to transform robotics by enabling the construction of highly sophisticated and adaptive robotic systems. Here, we will look at some of the ways smart materials may be used in robotics.

The development of soft robots is one of the most vital applications of smart materials in robotics. Soft robots are a new type of robot built of flexible and malleable materials that allow them to bend, twist, and stretch in ways that standard rigid robots cannot. Shape memory alloys, piezoelectric materials, and electroactive polymers are typical materials utilised in the construction of soft robotics because they can be made to respond to particular stimuli such as changes in temperature, pressure, or electric fields. The Octobot, a soft robot built by researchers at Wyss Institute, is constructed completely of soft materials and is controlled via a pneumatic network (Rus, 2015).

Another application of smart materials in robotics is the creation of self-healing robots. Self-healing robots can fix themselves when they are broken without the need for human involvement (Gao, 2015). Shape memory polymers and carbon nanotubes are being investigated for their potential for enabling self-healing in robots. Researchers at the University of California, San Diego, for example, have created a self-healing polymer material that can restore itself in less than 24 hours after being sliced or pierced. This material might create self-healing robots that can adapt to changing situations and repair themselves when damaged.

By allowing the creation of highly sophisticated and adaptive robotic systems, smart materials have the caliber to revolutionise the field of robotics. Soft robots and self-healing robots are only two of the numerous uses of smart materials in robotics that we are likely to see in the future.

4.4 Energy Harvesting

As a result of their distinctive properties like as piezoelectricity, thermoelectricity, and photovoltaic effect, smart materials are of great interest in energy harvesting. These materials have the ability to transform several types of energy, including as mechanical, thermal, and solar energy, into electrical energy, which may then be utilised to power various devices. In this post, we will look at some of the ways smart materials may be used to capture energy.

The creation of piezoelectric energy harvesters is one of the key uses of smart materials in the harvesting of energy. As these materials were able to generate an electrical charge in response to mechanical stress, they can be used to harvest energy from a variety of sources, including vibrations, movements, and impacts. Piezoelectric energy harvesters are employed in a wide range of applications, including wireless sensor networks, wearable electronics, and structural health monitoring systems (Kim S. X., 2017). A Georgia Institute of Technology research team, for example, has created a piezoelectric energy harvester capable of producing electricity from mechanical energy produced by the human heart. The technology has the potential to power implanted medical devices like pacemakers or sensors.

The creation of thermoelectric generators is yet another application of smart materials in energy harvesting. In response to a temperature differential between two surfaces, thermoelectric materials generate an electrical voltage, which may be used to recover waste heat and transform it into useable electricity (Zhao, 2011). Thermoelectric generators finds application in space probe and satellite power generation, as well as waste heat recovery in automotive and industrial applications. A research team at the University of California, Berkeley, for example, has created a thermoelectric generator capable of collecting energy from human body heat. The technology has the potential to power wearable electronics and medical devices.

4.5 Smart Windows

Owing to its unique features such as electrochromism, photochromism, and thermochromism, smart materials have emerged as a possible alternative for the creation of smart windows. These materials have the potential to transform the area of window technology by enabling the creation of extremely energy-efficient and adaptive windows that can control the amount of light as shown in Figure 1.3, Figure 1.4 (Robert, 2014) and heat that enters a structure. Further, we will look at some of the ways smart materials may be used in smart windows.

Figure 3. Change in the density of the optical material as a function of light intensity when optical material in normal condition

Figure 4. Change in the density of the optical material as a function of light intensity when actuated condition

The creation of electrochromic windows is one of the significant uses of smart materials in smart windows. Electrochromic materials may change colour and opacity in response to an electrical signal, allowing them to be used to control the quantity of light that enters a structure (Llordés, 2013). Depending on the amount of automation necessary, electrochromic windows can be operated manually or automatically. For example, a Massachusetts Institute of Technology research team has created an electrochromic window that can transform from transparent to opaque in less than a minute while consuming only a modest amount of power (Nag, 2015). By adjusting the quantity of sunlight entering a space, the window has the ability to lower the amount of energy used for heating, cooling, and lighting in buildings.

4.6 Sports Equipment

As a result of their distinctive properties like as flexibility, durability, and energy-absorbing capability, smart materials have become more popular in the design and development of sporting equipment. These materials can enhance athlete performance and safety by enabling the development of more efficient, pleasant, and protective equipment. In this article, we will look at some of the ways smart materials may be used in sporting equipment.

The development of impact-resistant helmets is an eminent use of smart materials in sports equipment. Shear thickening fluids (STFs), for example, have been employed to increase the shock-absorbing capacity of helmets by increasing their stiffness in response to high impact forces (Lee, 2020). STFs are a form of non-Newtonian fluid that thickens and becomes more viscous in the presence of significant shear forces, such as those produced by an impact. For example, a University of Maryland research team has created a football helmet with STFs in the padding that may stiffen in reaction to impact pressures, reducing the risk of concussion.

Another use of smart materials in sports equipment is the creation of adaptable apparel. Clothing that can respond to the athlete's body temperature, wetness, and movement has been developed using smart materials like as shape-memory alloys (SMAs) and electroactive polymers (EAPs). SMAs may change form in reaction to temperature changes, whereas EAPs can change shape in response to an electrical signal (Qi, 2019). For example, a team of researchers at the Massachusetts Institute of Technology created a running suit with EAPs that can contract and expand in response to the athlete's motions, minimising drag and enhancing performance. Smart materials have the potential to revolutionise sports equipment design and development by enabling the production of more efficient, pleasant, and protective equipment.

5. ADVANTAGES OF SMART MATERIALS

5.1 Improved Performance

Smart materials provide a lot of advantages in terms of better performance in a variety of applications. These materials' characteristics may alter in response to external stimuli like temperature, light, or electric fields, which can be exploited to improve the performance of devices and systems. In this contest, we will go through some of the benefits of smart materials in terms of better performance.

One of the key benefits of smart materials is their capacity to increase device efficiency. Smart materials can be used to improve device performance by modifying their characteristics in real time in response to changing environmental conditions (Kusaka, 2016). Shape-memory alloys (SMAs), for example, can be utilised to create actuators that can change the position of a valve or other component in response to changes in temperature or pressure. This can enhance the device's efficiency by lowering the amount of energy required to run it.

Another benefit of smart materials is their capacity to increase device safety and dependability. Smart materials may be utilised to create sensors and actuators that detect and respond to environmental changes such as temperature or pressure. For example, piezoelectric materials may be utilised to create sensors that detect pressure changes and can be used to monitor a device's structural integrity. This can enhance the device's safety and dependability by recognising possible faults before they become significant.

Furthermore, smart materials can improve device performance in harsh environments such as high temperatures or radiation. By modifying their qualities in reaction to changing environmental conditions (Liu Y. S., 2018), smart materials may be utilised to build materials that can resist harsh environments. Carbon nanotubes, for example, may be utilised to create high-temperature sensors that can be used to monitor the temperature of engines in aeroplanes or other environments with elevated temperatures.

5.2 Increased Efficiency

Smart materials have significant advantages in terms of greater efficiency in a variety of applications. These materials' characteristics may vary in response to external stimuli such as temperature, light, or electric fields, allowing them to be employed to improve the performance of devices and systems. In this contest, we will go through some of the benefits of smart materials in terms of enhanced efficiency.

The capacity of smart materials in increasing the efficiency of renewable energy systems is one of its key benefits. Smart materials, such as solar panels or wind turbines, can be utilised to improve the performance of renewable energy sources by modifying their qualities in response to changing environmental circumstances. Piezoelectric materials, for example, may be utilised to create sensors that detect changes in wind direction and modify the location of wind turbines appropriately, improving system efficiency.

Smart materials can also be utilised to increase the efficiency of energy storage devices like batteries or capacitors. By changing their properties in response to changes in temperature, pressure, or other external conditions, smart materials may be utilised to build electrodes and other components that can store and release energy more effectively. For example, researchers at the University of Maryland created a battery electrode comprised of vanadium dioxide, a smart material that can flip between conductive and insulating states in response to temperature variations. This can enhance the battery's efficiency and longevity by lowering the amount of energy wasted as heat while charging and discharging.

Smart materials have significant advantages in terms of greater efficiency in a variety of applications. These materials have the potential to increase energy efficiency (Wang Z. L., 2017), the performance of renewable energy systems, and the efficiency of energy storage systems (Han, 2017). We should anticipate to see many more uses in the future as researchers continue to investigate the unique features of smart materials.

5.3 Reduced Energy Consumption

Energy consumption in heating and cooling systems may be reduced by using smart materials. Smart materials can be utilised to create materials that can automatically modify their thermal characteristics in response to changing environmental circumstances. Researchers at the University of Maryland, for example, have created a smart material that can transition between being transparent and opaque in response to temperature changes, which may assist manage a building's temperature and minimise the amount of energy required for heating and cooling (Yu, 2016). Similarly, SMAs may be utilised to develop materials which can alter their shape in with respect to change in temperature, can then be used to create smart windows that manage the amount of heat and light entering a building.

Smart materials also have the capacity to reduce the consumption of energy in transportation systems. Smart materials can helps in developing lightweight materials that can alter their characteristics with change in temperature, pressure, or other external conditions, allowing cars to be lighter and more fuel efficient. Researchers at the Massachusetts Institute of Technology (MIT), for example, have discovered a smart material called "metallic wood" that is lighter than aluminium but stronger than steel and may be used to produce lightweight components for automobiles to enhance fuel economy.

Furthermore, smart materials have the potential to reduce energy consumption in electronic devices. Smart materials can be used to create components that can modify their characteristics in reaction to changes in temperature or other external conditions, reducing the amount of energy required to operate the device (Li T.

&., 2019). Researchers at the University of California, Berkeley, for example, have created a smart material that can flip between being a conductor and an insulator in reaction to temperature variations, which may be utilised to create more energy-efficient transistors for electrical devices.

Smart materials have significant advantages in terms of reduced consumption of energy in a variety of applications. These materials can be utilised to create materials and components that can modify their characteristics in real time in response to changing environmental circumstances, reducing the amount of energy required to run devices and systems.

5.4 Enhanced Safety

Structures and infrastructure may be made safer by using smart materials. Smart materials can be used to develop materials and components that are capable of detecting changes in their surroundings, such as temperature, pressure, or humidity, and respond in real-time by changing their properties. This can be used to detect and respond to possible safety threats such as structural deterioration or environmental changes that could compromise a structure's integrity. Researchers at the University of California, San Diego, for example, have created a smart concrete material that can feel and respond to changes in stress or strain, which may be utilised to identify possible structural damage and avert catastrophic collapse (Zhang Y. L., 2016).

Smart materials also have the possibility to improve the safety of medical equipment and implants. Smart materials may be used to develop materials and components that are capable of detecting changes in their surroundings, such as temperature, pH, or the presence of specific chemicals, and can be utilised to improve the safety and efficacy of medical devices and implants. Researchers at the MIT, for example, have created a smart material that can release pharmaceuticals in response to temperature or pH changes, which may be utilised to design more effective and targeted drug delivery systems(Kim P. &., 2018).

Smart materials can also be utilised to improve the safety of transportation systems. Smart materials can be used to develop materials and components that can react to changes in their environment, such as temperature, pressure, or the presence of certain chemicals, and can be used to improve the safety and dependability of transportation systems. Researchers at the University of California, San Diego, for example, have created a smart material that can sense and respond to changes in pressure and strain, which may be used to create more dependable and safe tyres for automobiles (Kong, 2017).

Smart materials may be used to create materials and components that can detect changes in their surroundings and adapt in real time, improving the safety and dependability of devices and systems. Other advantages of smart materials are summarized in Table 1.3. (Shasshi Bahl, 2020).

Table 3. Advantages of smart materials

Type of Smart Material	Advantages
Piezoelectric	Response to high frequency, simple structure, creation of electric signals in proportion to the applied mechanical force
Magnetosatrictive	Intrinsic robostness and high energy density
Shape memory alloys	High strength, resistance to corrosion, high damping effeciency, greater life under fatigue and exhibits better elastic behaviour
Electro rheological fluid	Capacity to perform as Powefrul amplifier, stable performance of the systems
Magneto rheological fluid	Higher permiability and saturation magnetization
Optical fibers	Resistance to tapping and electro magnetic interfacing, better band width support, occupy less space and resistance to corrosive materials

6. CHALLENGES IN SMART MATERIALS DEVELOPMENT

6.1 Cost

The research and commercialization of smart materials confront a number of problems, one of which is cost. Because of the complicated manufacturing processes involved in their creation, smart materials are frequently more expensive than traditional materials. This cost barrier may limit their use in a variety of applications, particularly in large-scale businesses.

The high cost of raw materials is one of the most significant difficulties in the development of smart materials. Some smart materials, such as shape memory alloys, piezoelectric materials, and magnetostrictive materials, necessitate the use of rare earth elements, which are costly and difficult to obtain(Cui). Furthermore, the manufacturing process for smart materials is frequently complex and necessitates specialised equipment, which increases the production cost.

An additional challenge in the development of smart materials is the absence of standardisation and regulation. Because of the absence of standardisation, it is difficult to evaluate the costs of various smart materials and determine the most cost-effective materials for a certain application. Furthermore, due to a lack of regulation, some smart materials may fail to meet the necessary safety and performance standards, increasing the cost of production by necessitating additional testing and certification (Lamprecht, 2020).

To solve these issues, researchers are investigating novel production processes and alternative materials with the goal of lowering the cost of smart materials. For example, University of Maryland researchers have devised a low-cost and scalable approach for creating graphene-based smart materials using a simple 3D printing process (Li Y. Z., 2020). This technology might dramatically cut the cost of making smart materials while also opening up new applications for them.

In short, while smart materials have various benefits over traditional materials, the high cost of development and production continues to be a significant problem. Researchers are working hard to minimise the cost of smart materials by developing new production processes and using alternative materials. As the cost of smart materials continues to fall, we may expect greater acceptance and use in a variety of sectors.

6.2 Durability

Smart materials have a variety of unique features and capabilities that have the potential to revolutionise a variety of sectors. However, one of the major limitations in the development of smart materials is their durability. Smart materials must be durable in order to be successfully integrated into a variety of applications.

One of the most difficult difficulties in the creation of smart materials is the capacity to preserve their qualities over time. Smart materials are frequently subjected to a significant number of deformation cycles, and their ability to preserve their characteristics across these cycles is crucial for their longevity (Nguyen, 2018) (Kumar A. K., 2023). Furthermore, smart materials are frequently exposed to harsh environments, which can degrade their properties and reduce their lifespan.

A further challenge in the development of smart materials is their sensitivity to fatigue failure. Fatigue failure is a prevalent mechanism of failure in materials that are subjected to repeated loading and unloading cycles, which is common in smart materials (Liu Y. W., 2021). Repetitive deformation cycles can result in microstructural changes in the material, resulting in decreased strength and durability.

Researchers are researching novel materials and processing processes to increase the endurance of smart materials in order to meet these difficulties. Researchers at the University of California, Los Angeles, for example, have created a self-healing polymer composite material that can mend cracks and damage (Kaner, 2017). This self-healing material can help smart materials last longer by decreasing the impact of small damage.

Another method for increasing the durability of smart materials is to create prediction models that can identify the most significant aspects influencing their attributes and durability (Hadjiloizi, 2020). These models may be used to increase the endurance of smart materials by optimising material attributes and processing procedures.

6.3 Manufacturing Complexity

Smart materials have a variety of unique features and capabilities that have the potential to revolutionise a variety of sectors. Manufacturing complexity, on the other hand, is a key issue in the creation of smart materials. The complexity of producing smart materials is a key barrier to their mainstream adoption and commercialization.

The production of smart materials is a difficult process that needs specialised equipment and experience. Many smart materials are created through a series of phases that include raw material synthesis, processing, and post-treatment, all of which can be time-consuming and costly. Furthermore, the production of smart materials frequently necessitates the use of specialised equipment, which can be expensive and not always readily available in all manufacturing facilities.

Another hurdle in producing smart materials is establishing consistency and repeatability. To obtain the necessary qualities and functions, smart materials frequently necessitate careful control of material properties and microstructure. To guarantee that each material batch has the appropriate qualities and performance characteristics, the production process must be carefully managed.

Researchers are aiming to create new production processes that will help to simplify the manufacturing process while also improving the repeatability and homogeneity of smart materials. 3D printing technology, for example, has been utilised to create smart materials with complicated geometries and specific characteristics (Singh, 2019) (Kumar A. M., 2022). 3D printing technology has the potential to minimise production complexity while also improving the repeatability of smart materials.

A different approach to simplifying the manufacturing process is to adopt scalable processing techniques, such as roll-to-roll processing, which can enable low-cost mass production of smart materials (Cho, 2020). By lowering the variability caused during batch processing, scalable processing approaches can also improve the homogeneity and repeatability of smart materials.

6.4 Integration With Existing Systems

Owing to the variations in material qualities and control methods employed in the two systems, integrating smart materials into conventional systems can be difficult. Smart materials frequently have unique material features, such as their sensitivity to external stimuli, which can make integrating them with current systems difficult. Furthermore, smart materials frequently necessitate the use of specialised control systems that may or may not be compatible with existing control systems.

Researchers are striving to create new integration techniques that will simplify the procedure and increase the interoperability of smart materials with existing systems. Modular design principles, in which smart material components are designed to be readily incorporated into existing systems, are one method to easing the integration process (García-Vallejo, 2021). Modular design principles can aid in reducing the complexity of the integration process and improving smart material compatibility with current systems.

Another approach to improve smart material compatibility with current systems is to create new control systems that can operate both smart and traditional materials. These control systems can assist guarantee that smart materials work in tandem with current systems without requiring large changes to the control systems.

The incorporation of smart materials into current systems is a key barrier to their mainstream adoption and commercialization. As the integration of smart materials gets easier, we may anticipate growing acceptance and use in a variety of sectors.

7. CONCLUSION

This chapter will provide an overview of smart materials, their properties, types and applications. It will emphasize the potential advantages of using smart materials in various fields, as well as the challenges that must be overcome in the development of smart materials and their implementation.

8. FUTURE PROSPECTS FOR SMART MATERIALS

The future prospects for smart materials are highly promising, with significant implications for research and development across various industries. Smart materials can sense and react to external stimuli including temperature, light, pressure, or electrical fields. They are sometimes referred to as responsive or intelligent materials. These materials are extremely adaptable and advantageous for a wide range of applications since they can change their characteristics, shape, or functioning in real-time.

One of the key domains where smart materials are predicted to have a significant influence is healthcare. Smart material research gives up new opportunities for improved medical devices, medication delivery systems, and tissue engineering. Smart materials, for example, can be utilised to make implantable sensors that continuously monitor vital indicators and administer personalised therapies. They can also help with the regulated release of medications in specific areas of the body, boosting therapy efficacy and reducing side effects. Furthermore, smart materials can be developed to replicate the qualities of real tissues, allowing for the development of artificial organs or tissue scaffolds for regenerative medicine.

Another area where smart materials show significant promise is in the area of energy. As the globe seeks more sustainable and efficient energy solutions, smart materials present fresh possibilities. Smart materials, for example, might dramatically improve the energy conversion efficiency of solar panels. These materials can also be utilised to increase the performance, longevity, and safety of energy storage devices such as batteries and supercapacitors. Furthermore, smart materials can help with energy conservation by being integrated into smart windows that vary their transparency in reaction to external conditions, decreasing heating and cooling requirements in buildings.

Smart materials have the potential to revolutionise device functioning and performance in the fields of electronics and information technology. Researchers can create next-generation displays, sensors, and memory devices by using materials with unique electrical, optical, or magnetic capabilities. Flexible and transparent smart materials, for example, could enable the development of bendable or rollable electronic devices, bringing up new opportunities for wearable technology, flexible displays, and smart textiles. Smart materials can also be used to create responsive sensors capable of detecting a wide range of factors with high precision, boosting the capabilities of robotics, Internet of Things (IoT) systems, and autonomous vehicles.

The potential of smart materials for research and development are extensive. To investigate new materials, fabrication techniques, and characterisation methods, they require interdisciplinary collaboration among scientists, engineers, and material researchers. Because smart materials frequently contain nanoscale components or complicated architectures, advancements in nanotechnology and nanofabrication techniques are critical to their realisation. Furthermore, to enable widespread implementation of smart materials across industries, research efforts must focus on scalability, cost-effectiveness, and environmental sustainability.

To fully realise the potential of smart materials, more research is needed to improve their performance, durability, and reliability. Understanding their long-term behaviour, stability, and interactions with other materials is critical for successful application integration. Efforts to produce smart materials with self-healing capabilities or bio-inspired features can also improve their functioning and lifespan.

The prospects for smart materials in the future are quite intriguing, with far-reaching consequences for research and development. Their distinct features and sensitivity to environmental cues enable ground-breaking advances in healthcare, energy, electronics, and other fields. Continued research, innovation, and collaboration, on the other hand, are required to overcome hurdles and fully realise the potential of smart materials, paving the way for a more technologically sophisticated and sustainable future.

REFERENCES

Aggelis, D. G. (2015). Structural health monitoring in aerospace engineering: A review. *Journal of Intelligent Material Systems and Structures*, *26*(3), 350–374.

Ajay, P. A. (2023). Handbook of Sustainable Materials: Modelling, Characterization, and Optimization. CRC PPress.

Araujo, A. B. (2015). A Brief Review on Smart Materials. *Materials Research*, *18*(3), 529–538.

Asmatulu, R. (2016). *Smart Materials: Science and Technology*. Springer.

Avinash Gudimetla, D. L. (2020). Investigation of mechanical and tribological behavior of al 4032-SiHGM MMC. *Composites Theory and Practice*, *20*(3-4), 142–156.

Avinash Gudimetla, D. L. (2021). Modelling and optimization of wear parameters of Al4032 reinforced with coal ash using Taguchi and RSM approach. *Composites theory and practice, 21*, 3-11.

Avinash Gudimetla, S. S. (2020). Influence of RHA reinforcements on mechanical and wear behavior of Al 4032 composites. *Emerging Materials Research*, *9*(4), 1237–1249. doi:10.1680/jemmr.20.00104

Bououdina, M. (2015). Smart Materials and Their Applications: A Review. *Journal of Materials Science Research*, *4*(4), 70–89.

Cho, J. P. (2020). Scalable roll-to-roll processing of smart materials and structures for IoT devices. *Smart Materials and Structures*, *29*(8).

Clark, A. E. (2004). Magnetostrictive materials and their use in high-precision positioning systems. *Journal of Intelligent Material Systems and Structures*, *15*(10), 797–807.

Cross, L. E. (1999). *Ceramics for electro-optics, piezoelectrics, and pyroelectrics*. Springer Science & Business Media.

Cui, J. F. (2020). Development of high-performance smart materials: Progress and challenges. *Journal of Materials Chemistry. A, Materials for Energy and Sustainability*, *8*(30), 14975–15000.

Deb, S. K. (2014). Organic electrochromic materials and devices: Recent advances and future prospects. *Journal of Materials Chemistry. C, Materials for Optical and Electronic Devices*, 2(15), 2656–2675.

Dierdorf, D. S. (2002). Smart materials: Some recent developments. *Annual Review of Materials Science*, *32*(1), 285–315.

Gao, W. D. (2015). Self-Powered and Self-Healing Soft Robots. *Advanced Materials*, *27*(43), 7289–7295.

García-Moreno, F. (2018). Self-healing smart materials. *Smart Materials and Structures*, *27*(4).

García-Vallejo, D. R.-L.-L. (2021). Challenges and opportunities for smart materials in the construction industry. *Applied Sciences (Basel, Switzerland)*, *11*(1), 251.

Granqvist, C. G. (1995). Electrochromics for smart windows: Thin films of tungsten oxide and nickel oxide, and devices based on these. *Journal of Materials Chemistry*, *5*(4), 547–565.

Hadjiloizi, E. (2020). Modelling and simulation of fatigue behaviour in smart materials and structures. *Fatigue & Fracture of Engineering Materials & Structures*, *43*(7), 1636–1653.

Haghniaz, R. (2018). Smart materials: A review of applications and technologies. *Renewable & Sustainable Energy Reviews*, *91*, 1295–1307.

Han, M. D. (2017). Smart materials for energy storage systems: A review of recent advances. *Materials Today. Energy*, *6*, 169–185.

Ibrahim Nazem, QADER, M. K. (2019). A Review of Smart Materials: Researches and Applications. *El-Cezerî Journal of Science and Engineering*, *6*(3), 755–788.

Joshi, G. P., Tyagi, A., Meena, S. L., Chourasia, S., Saxena, N. S., & Kumar, A. (2021). Optical Band Gap Determination of Ni-Al Doped Polyaniline at Room Temperature and Different Annealing Temperatures. *Journal of Physics: Conference Series*, *1950*(1), 012091. doi:10.1088/1742-6596/1950/1/012091

Kandola, B. K. (2019). *Smart Textiles and Their Applications*. Woodhead Publishing.

Kaner, R. B. (2017). Nanotechnology for the energy challenge. *Nanotechnology*, *28*(11).

Kim, P. (2018). Smart hydrogels for biomedical applications: A review of recent developments. *Smart Materials and Structures*, *27*(4).

Kim, S. X. (2017). Piezoelectric energy harvesters for powering wireless electronics: State-of-the-art and future challenges. *Micro and Nano Systems Letters*, *5*(1), 1–11.

Kong, X. C. (2017). A review of smart materials in tire applications. *Smart Materials and Structures*, *26*(2).

Kumar, A. K. (2023). Materials processed by additive manufacturing techniques. In *Advances in Additive Manufacturing Artificial Intelligence, Nature-Inspired, and Biomanufacturing* (pp. 217–233). Elsevier. doi:10.1016/B978-0-323-91834-3.00014-4

Kumar, A. M. (2022). *Advances in Additive Manufacturing: Artificial Intelligence, Nature-Inspired, and Biomanufacturing*. Elsevier.

Kusaka, K. (2016). Application of smart materials for improving efficiency and reliability of mechanical systems. *Journal of Mechanical Science and Technology*, *30*(9), 3939–3946.

Lamprecht, C. P. (2020). Smart materials: Challenges and approaches in production and applications. *Journal of Materials Processing Technology*, 275.

Lee, S. J. (2020). Application of shear thickening fluid (STF) in sports equipment. *Polymers*, *12*(9).

Li, T. (2019). Metallic wood: A perspective on its synthesis, properties, and applications. *Materials Today*, *27*, 56–69.

Li, Y. Z. (2020). 3D printing of smart materials: A review on recent progresses in additive manufacturing. *Materials Horizons*, *7*(4), 1041–1065.

Liu, J. (2021). Thermochromic VO2 thin films for smart windows: A review. *Journal of Materials Chemistry. C, Materials for Optical and Electronic Devices*, *9*(4), 1102–1118.

Liu, J. H. (2021). Advances in piezoelectric materials for biomedical applications. *Journal of Materials Chemistry. B, Materials for Biology and Medicine*, *9*(25), 4957–4975. PMID:34008692

Liu, X. (2015). Magnetostrictive sensors: A review of materials and devices. *Sensors (Basel)*, *15*(3), 4655–4686.

Liu, Y. (2018). A review of smart materials in aerospace. nternational. *Journal of Aerospace Engineering*.

Liu, Y. S. (2018). Smart materials and devices for harsh environment sensing and control. *Smart Materials and Structures*, *27*(1).

Liu, Y. W. (2021). Fatigue properties of Ni–Mn–Ga/CNTs composite wire for smart actuator applications. *Journal of Alloys and Compounds*, 873.

Llordés, A. G. (2013). Electrochromic materials for smart windows: Progress and prospects. *Chemical Society Reviews*, *42*(20), 8299–8311.

Ma, N. (2018). Recent advances in shape memory polymers and composites: A review. *Journal of Materials Science*, *53*(3), 1743–1775.

Mather, P. T. (2010). Shape memory polymer research. *Annual Review of Materials Research*, *40*, 179–210.

Matsumoto, K. (2016). Magnetostrictive energy harvesting using laminated Terfenol-D/PVDF composite. *Journal of Intelligent Material Systems and Structures*, *27*(10), 1249–1255.

Mauro, M. (2018). Multifunctional Smart Materials for Future Aerospace Applications. In Handbook of Smart Materials in Analytical Chemistry, 13-34.

Mohd Jani, J. L., Leary, M., Subic, A., & Gibson, M. A. (2014). A review of shape memory alloy research, applications and opportunities. *Materials & Design*, *56*, 1078–1113. doi:10.1016/j.matdes.2013.11.084

Nag, A. (2015). Smart windows with an environment-sensitive material: A review. *Journal of Materials Chemistry. A, Materials for Energy and Sustainability*, *3*(45), 22416–22429.

Nguyen, T. H. (2018). A review of fatigue behavior in smart materials and structures. *Smart Materials and Structures*, *27*(11).

Nirmalraj, P. N. (2017). Smart textiles: A review on thermochromic materials and their applications. *Journal of Industrial Textiles*, *47*(8), 1745–1776.

Otsuka, K. &. (1998). *Shape memory materials.*

Ounaies, Z. P. (2016). *Smart composite material.* CRC Press.

Qi, Z. W. (2019). Recent progress in smart materials-based sports textiles. *Journal of Textile Research*, *40*(6), 107–116.

J., H. & Y., H. (2017). Recent Progress in Smart Materials Research. *Journal of Materials Science*, *52*(22), 12947–12966.

Robert, B. (2014). Smart materials: A review of capabilities. *Assembly Automation*, *34*(1), 16–22. doi:10.1108/AA-10-2013-094

Rus, D., & Tolley, M. T. (2015). Design, fabrication and control of soft robots. *Nature*, *521*(7553), 467–475. doi:10.1038/nature14543 PMID:26017446

Safari, A. (2019). *Piezoelectric energy harvesting: Fundamentals, materials, and devices*. Springer.

Sehrawat, S. K., Kumar, A., Grover, S., Dogra, N., Nindra, J., Rathee, S., Dahiya, M., & Kumar, A. (2022). Study of 3D scanning technologies and scanners in orthodontics. *Materials Today: Proceedings*, *56*, 186–193. doi:10.1016/j.matpr.2022.01.064

Sehrawat, S. K., Kumar, A., Prabhakar, M., & Nindra, J. (2022). The expanding domains of 3D printing pertaining to the speciality of orthodontics. *Materials Today: Proceedings*, *50*, 1611–1618. doi:10.1016/j.matpr.2021.09.124

Sharma, G. D. (2018). Hybrid electrochromic materials and devices: A review. *Journal of Materials Chemistry. C, Materials for Optical and Electronic Devices*, *6*(31), 8217–8232.

Shasshi Bahl, H. N. (2020). Smart materials types, properties and applications: A review. *Materials Today: Proceedings*, *28*(3), 1302–1306. doi:10.1016/j.matpr.2020.04.505

Shayan, M. M. (2019). Shape memory alloys for biomedical applications: A review. *Journal of Materials Science. Materials in Medicine*, *30*(7), 79. PMID:31240399

Singh, A. K. (2019). 3D printing of smart materials: A review on recent progresses in the field. *Materials Today: Proceedings*, *18*, 5017–5022.

Singh, V. G. (2023). Experimental investigations of thermo physical properties and convective heat transfer of Al2O3 and CuO nanofluids in a copper tube: Proposing New Correlations. *Biointerface Research in Applied Chemistry*, 229.

Singh, V. G., Gupta, M., & Kumar, A. (2019). Investigating the morphology and thermophysical properties of Al2O3 and CuO nanofluids. *IOP Conference Series. Materials Science and Engineering*, *577*(1), 012052. doi:10.1088/1757-899X/577/1/012052

Singh, V. K., Kumar, A., Alam, M., Kumar, A., Kumar, P., & Goyat, V. (2022). A study of morphology, UV measurements and zeta potential of zinc ferrite and Al2O3 nanofluids. *Materials Today: Proceedings*, *59*, 1034–1039. doi:10.1016/j.matpr.2022.02.371

Snyder, J. E. (2013). Magnetostrictive materials for energy harvesting. *Journal of Applied Physics*, *113*(17).

Tay, F. E. (2015). Smart materials: Recent advances and future directions. *Progress in Materials Science*, *74*, 163–239.

Uchino, K. (2006). Piezoelectric Actuators and Devices: Recent Developments and Prospects. *Japanese Journal of Applied Physics*, *45*(10A), 7571–7579.

Uchino, K. (2019). *Piezoelectric actuators and ultrasonic motors*. Springer.

Wang, X. (2017). Smart Materials for Advanced Biomedical Applications. *Smart Materials for Advanced Environmental Applications,* 313-346.

Wang, Z. L. (2017). Smart materials for energy efficiency in buildings: A review. *Renewable & Sustainable Energy Reviews*, *68*, 1074–1089.

Xie, C. Z. (2015). Recent Advances in Shape Memory Polymers: Structure, Mechanism, Functionality, Modeling and Applications. *Progress in Polymer Science*, *49-50*, 79–120.

Yao, K. S. (2019). Piezoelectric materials for biomedical applications: A review. *Materials Science and Engineering C*, 104.

Yu, Y. I. (2016). Smart windows based on thermochromic materials: A review. *Solar Energy Materials and Solar Cells*, *157*, 2–23.

Zeng, X. (2019). Recent progress in thermochromic materials for smart windows: A review. *Solar Energy Materials and Solar Cells*, *197*, 219–239.

Zhang, X. S. (2020). Recent developments in electrochromic materials and devices. *Journal of Materials Chemistry. C, Materials for Optical and Electronic Devices*, *8*(36), 12423–12444.

Zhang, X. W. (2018). Progress in the development of piezoelectric materials for acoustic transducers: A review. *Progress in Materials Science*, *97*, 187–259.

Zhang, Y. L. (2016). A review of recent developments in sensing and monitoring of concrete structures using smart materials. *Sensors (Basel)*, *16*(10).

Zhao, L. D. (2011). Thermoelectric materials: From history, a glance to future. *Materials Today*, *14*(7-8), 322–332.

Chapter 2
Smart Materials in Biomedicine:
A Promising New Frontier

Mangaiyarkkarasi J.
NMSS Vellaichamy Nadar College, India

Shanthalakshmi Revathy J.
https://orcid.org/0000-0003-1724-7117
Velammal College of Engineering and Technology, India

ABSTRACT

Smart materials, also known as responsive or intelligent materials, are capable of responding to environmental stimuli in a predictable and controlled manner. Significant progress has been achieved in the field of smart materials over the last few decades, and these materials are now being employed in a variety of fields, including biomedical engineering. This chapter explores the use of smart materials in biomedicine, which has opened up a new frontier in the field. The chapter covers the fundamental principles of smart materials and their types, properties, and applications in biomedicine. It presents a thorough overview of the most recent research and development in the application of smart materials in biomedicine, including drug delivery, tissue engineering, biosensors, and imaging, as well as an explanation of the current difficulties and future prospects of this promising topic. This chapter concludes that smart materials possess pronounced potential in biomedicine and could revolutionize the way medical treatments are administered in the future.

DOI: 10.4018/978-1-6684-9224-6.ch002

1. INTRODUCTION

In recent years, the growth of smart materials has fetched a new era in biomedical research and clinical practice. These materials are designed with unique properties that allow them to respond and adapt to changes in their environment. This remarkable characteristic has made them an attractive candidate for a broad range of biomedical uses which includes delivery of drugs, tissue engineering process, and medical imaging systems. Smart materials have the potential to revolutionize the biomedical industry by offering fresh approaches to many of the problems that face the medical community. A family of materials known as "smart materials" can adapt their characteristics or behavior in response to external stimuli such changes in temperature, light, or magnetic fields. (Y.-C. Li et al., 2016). These materials have found numerous applications in the field of biomedicine due to their exceptional properties and abilities to mimic.

Biomedical applications of smart materials are diverse, ranging from drug delivery and tissue engineering to diagnostic tools and wearable devices (Bahl et al., 2020). Smart materials are used as scaffolds in the tissue engineering to offer mechanical support to grow cells and tissues, while also facilitate their growth and differentiation. Smart materials are used in a variety of biomedical devices, where they can perform precise functions due to their unique properties. The mechanisms behind these devices are often rely on the response of the material to specific stimuli.

Smart materials have been found application in drug delivery systems to increase the release of drugs and decrease side effects (Vashist & Ahmad, 2013). These materials can react to specific stimuli like pH or temperature, and release drugs in a controlled manner. For example, smart hydrogels can swell or shrink in response to changes in specific external stimuli, and this response can be used to control drug release. Tissue engineering involves the use of biomaterials to produce functional tissues or organs. Smart materials can be utilized in tissue engineering to generate structures that mimic natural tissues or provide specific functions (Khan & Tanaka, 2017). For instance, scaffolds that can alter shape in response to a particular stimulus can be made using shape memory polymers, enabling them to better mimic the shape of the surrounding tissue.

Smart materials found applications in implants and prosthetics to enhance their durability (Badami & Ahuja, 2014). For instance, stents made of shape memory alloys can expand and contract in response to temperature or stress changes, allowing them to adapt to the surrounding tissue. Self-healing materials are helpful to extend the lifespan of implants and prosthetics by repairing any damage that may occur over time. In sensors and actuators, smart materials are used to detect and respond to specific stimuli. For example, piezoelectric materials can create an electric charge related to mechanical stress, and this response can be used to produce sensors that detect pressure, strain, or vibration (Song et al., 2021). Shape memory alloys can be employed as actuators that can respond to changes in magnetic or thermal fields in order to carry out particular tasks. Diagnostic instruments like biosensors, which can identify specific compounds or analyses in biological samples, frequently make use of smart materials. (Arab Hassani et al., 2020). They can also be used in wearable devices, such as smart fabrics and sensors, to monitor vital signs, track activity levels, and even detect changes in mood or emotions.

This chapter will explore the current state of research in smart materials and their application in biomedicine, as well as the future directions and potential impact of this promising new frontier.

1.1 Classification of Smart Materials

Based on their mode of operation and the characteristics of their reaction in the biomedical area, smart materials may be divided into a number of types (Kök et al., 2019). The main classification of smart materials are, (i) stimuli-responsive materials, (ii) self-healing materials, (iii) biomimetic materials, (iv) nano materials and (v) hybrid materials. Figure 1 explains classification of smart materials used for biomedical applications.

Figure 1. Classification of smart materials

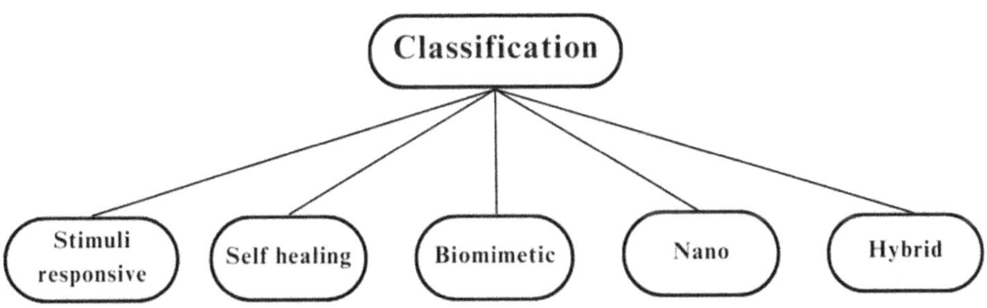

Stimuli-responsive materials respond to a specific stimulus, such as temperature, pH, light, or electric field. Examples: shape-memory alloys, hydrogels, and liquid crystals. Self-healing materials have the capability to renovate themselves after damage or wear. Examples: self-healing polymers and ceramics. Materials that mirror the characteristics and operations of naturally occurring biological materials are known as biomimicry. Examples: synthetic peptides, spider silk, and bio mineralized materials. Nano materials have unique properties at the nanoscale, such as high surface area, high reactivity, and quantum confinement effects. Examples: carbon nanotubes, quantum dots, and metallic nanoparticles. Hybrid materials combine two or more different types of smart materials to create a new material with enhanced properties. Examples: nano composites, smart coatings, and multifunctional nanoparticles.

The classification of smart biomaterials is not mutually exclusive, and some materials may belong to more than one category depending on their properties and applications.

1.1.1 Exclusive Properties of Smart Materials

Biomaterials are synthetic or natural materials that are used to replace, repair, or augment biological tissues or organs. They have exclusive properties that allow them to interact with living cells and tissues, and to perform their intended functions without causing adverse reactions or harm (Dauda et al., 2021). Figure 2 discusses important properties of smart materials which are appropriate for biomedical field.

Biocompatibility is considered as a chief property of biomaterials because they must be compatible with living tissues and not cause any toxic or immunogenic reactions. This property is essential for the success of any biomaterial-based medical device or implant. In order to stimulate cellular or tissue responses, bioactivity is crucial for engineering tissues and re-forming medicine applications. It does this by encouraging cell adhesion, proliferation, and differentiation. Then, biomaterials must also have appropriate mechanical properties, such as strength, stiffness and elasticity to withstand the forces and stresses of the body. The mechanical characteristics of biomaterials can be customized to match those of the surrounding tissues.

Biomaterials can degrade over time through various mechanisms, such as enzymatic degradation, hydrolysis, or oxidation and hence degradation is utmost important for temporary implants or drug delivery systems that do not require long-term stability. Sterilization is another key property of biomaterials because biomaterials must be sterilized to prevent infections and ensure patient safety. Some biomaterials may be sensitive to certain sterilization methods, such as heat or radiation. Biomaterials can be designed with specific surface properties, such as roughness, hydrophilicity, or charge, to control cell adhesion and proliferation. Surface modification can also improve the biocompatibility of biomaterials. Biomaterials' characteristics are determined by their intended uses and the biological environment in which they will be applied. Successful biomaterials must possess the right mix of qualities in order to carry out their activities efficiently and securely.

Figure 2. Properties of smart materials

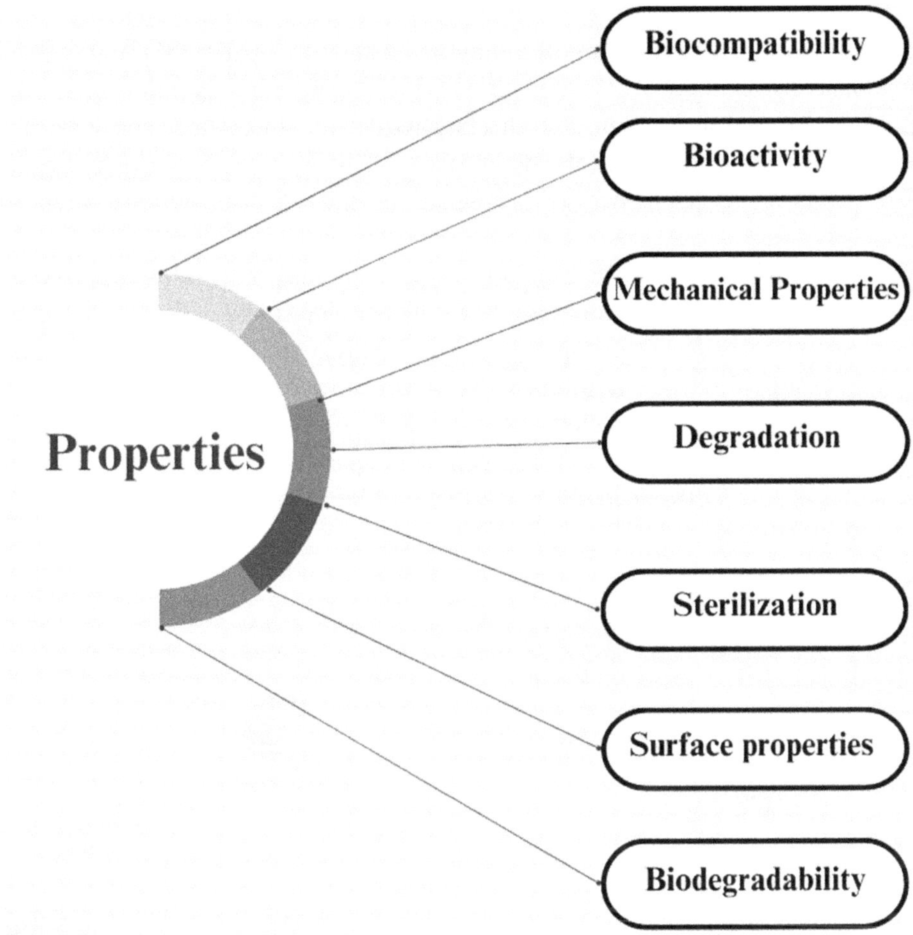

2. ESSENTIAL APPLICATIONS OF SMART MATERIALS IN BIOMEDICINE

Smart materials have numerous applications in biomedical field. Five chief applications are discussed here. (i) Drug delivery systems, (ii) tissue engineering, (iii) diagnostic tools, (iv) wearable technology, and (v) targeted therapy for cancer treatment.

2.1 Drug Delivery Systems

Drug delivery is the procedure of giving medication to a patient through various methods and techniques. These methods may include oral administration, injections, inhalation, topical application, and other targeted delivery methods. Some of the most common types of smart materials adopted for the delivery of drugs are, (i) hydrogels, (ii) liposomes, (iii) dendrimers, (iv) nanoparticles and (v) shape memory polymers.

(i) ***Hydrogels*:** One of the most important types of biomaterials are hydrogels, which have three-dimensional crosslinked polymer chains and can store enormous amounts of water. This property allows them to expand and contract in response to changes in their environment. They can be used to encapsulate drugs and release them in a controlled manner (Mahinroosta et al., 2018). Hydrogels are biocompatible, stimuli responsible like temperature, pH and light. Figure 3 shows how a drug loaded hydrogel is used in various biological systems for drug delivery.

Figure 3. Hydrogel in drug delivery

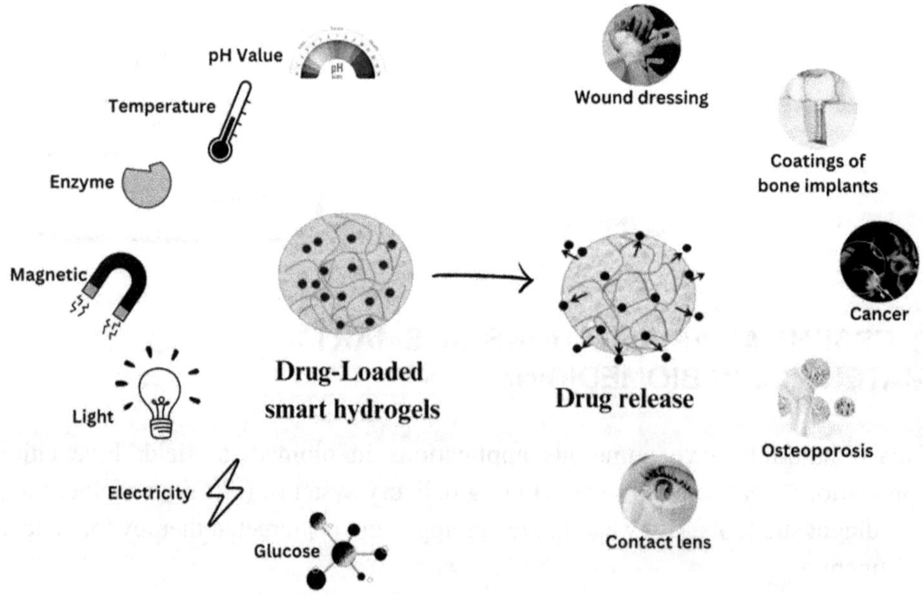

The hydrogel network can either physically enclose the medications or chemically attach them to the polymer chains. Drug delivery systems with specific targets can also be made using hydrogels. By altering the hydrogel with ligands, such as antibodies or peptides which recognize and bind to specific receptors on the target cells, drugs can be delivered to wounded tissues with greater precision. Drug stability and protection against deterioration are two additional benefits of hydrogels. Hydrogels can act as a barrier that prevents the drug from interacting with the surrounding environment, such as enzymes or other chemicals that could degrade the drug. This can help to increase the drug's half-life and lessen the frequency of dosing. Hydrogels are a flexible and promising material for the delivery of drugs that offers controlled and targeted drug release, improved drug stability, and decreased adverse effects. They may completely alter the way we handle a variety of illnesses and disorders.

(ii) ***Liposomes*:** Liposomes are spherical vesicles comprised of a lipid bilayer which encapsulate drugs on specific tissues or cells. They are modified with ligands like antibodies and peptides, to recognize and bind to specific receptors on the target cells, improving drug delivery and reducing toxicity. In drug delivery applications, liposomes can be used to encapsulate drugs and protect them from degradation. Figure 4 shows various parts present in the liposome structure.

Figure 4. Structure of liposomes

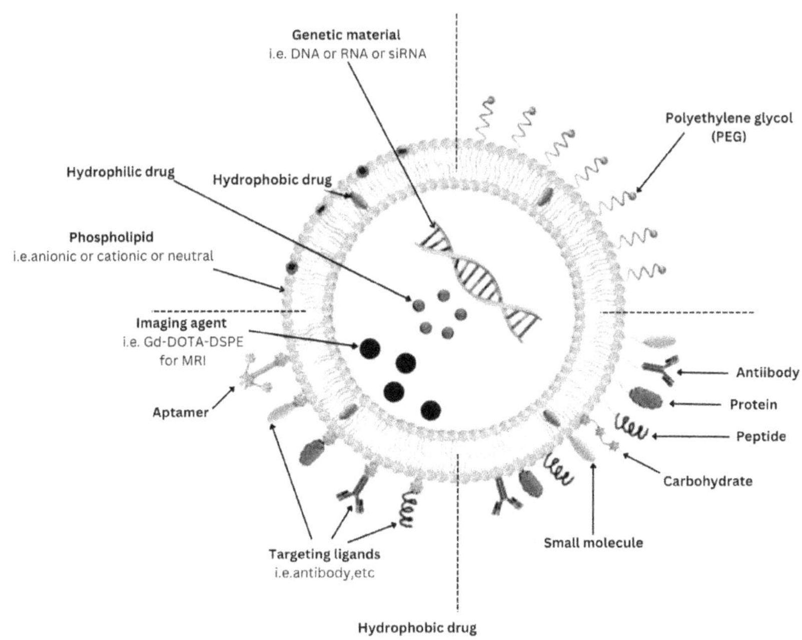

The drug can be encapsulated within the liposome membrane, which acts as a barrier that prevents the drug from interacting with the surrounding environment. This can help to increase the drug's half-life and reduce the frequency of dosing. Liposomes can also be engineered to target specific tissues or cells. The liposome membrane can be altered with ligands which recognize and bind to specific receptors on the target cells, allowing for the selective delivery of drugs to diseased or damaged tissues. This targeted drug delivery approach can reduce side effects and improve the efficacy of the drug (Nakhaei et al., 2021). Another advantage of liposomes is their ability to deliver hydrophilic and hydrophobic type drugs. Hydrophobic medications can be enclosed within the lipid bilayer, whereas hydrophilic pharmaceuticals can be encapsulated within the aqueous interior of the liposome. Since a wider variety of drugs can now be delivered using liposomes, they are a promising material for drug delivery systems. They also offer targeted drug delivery, improved drug stability, and decreased toxicity. They have the ability to completely alter how we handle a variety of illnesses and disorders.

(iii) ***Dendrimers*:** Dendrimers are extremely branched, nanoscale molecules that can be used to encapsulate drugs and deliver them to certain cells or tissues. They are altered to react to various stimuli allowing for targeted and controlled drug release. Dendrimers can be utilized to encapsulate pharmaceuticals and shield them from deterioration in medication delivery applications. Drug delivery to the target tissue or cell is possible when the dendrimer and drug combine to create a stable combination. By adding ligands that recognize and bind to particular receptors on the target cells, the dendrimer can be altered to target particular cells or tissues, such as cancer cells. (B Undre et al., 2016). Figure 5 clearly depicts how dendrimers are used in different drug delivery systems.

Dendrimers can also be engineered to respond to different stimuli, allowing for targeted and controlled drug release. For example, the dendrimer can be designed to issue the drug in response to changes in various stimuli over a period of time. Another benefit of dendrimers is their ability to carry multiple drugs simultaneously. This can be useful in situations where multiple drugs are needed to treat a disease or condition. The dendrimer can form a stable complex with each drug, which can be delivered to the target tissue or cell. Dendrimers are a promising substance for drug delivery systems because they allow for multiple drug delivery, tailored drug administration, and better drug stability. They have the ability to completely alter how we handle a variety of illnesses and disorders. To completely understand their safety and effectiveness in clinical settings, additional research is necessary.

Figure 5. Dendrimers in various drug delivery systems

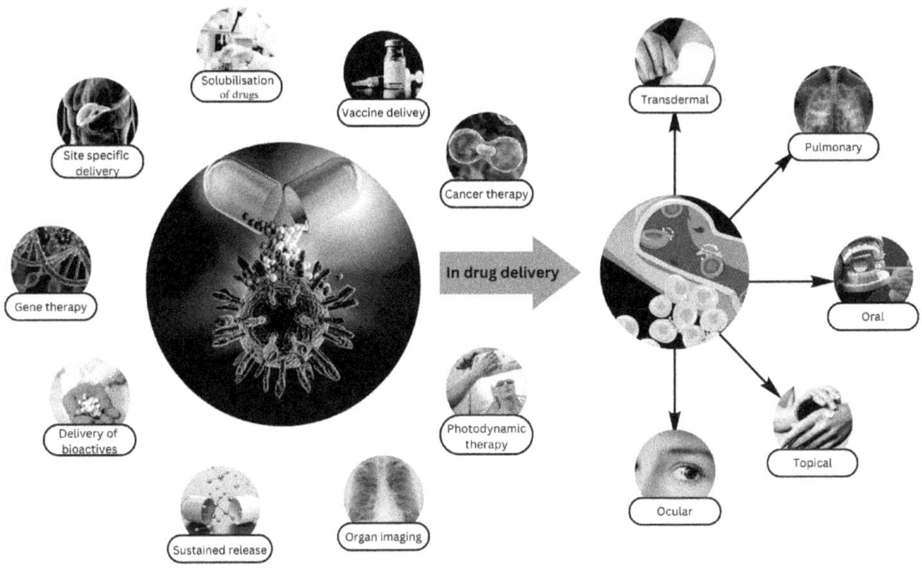

(iv) ***Nanoparticles*:** Nanoparticles are particles with dimensions in the nanometer range that can be used to encapsulate drugs and deliver them to specific cells or tissues. They can be designed to react to various stimuli and can be made from a wide spectrum of materials which comprises metals, lipids, and polymers. Nanoparticles can be utilized to encapsulate pharmaceuticals and protect them from deterioration in medication delivery applications. The medicine may be chemically or physically bound to the nanoparticle surface or physically confined inside the nanoparticle. The nanoparticle can serve as a barrier to stop the drug from interacting with nearby enzymes or other compounds that might cause the drug to break down. This can help to increase the drug's half-life and reduce the frequency of dosing (Aflori, 2021). Figure 6 shows some important biomedical applications of nanoparticles.

Nanoparticles can be created to target particular cells or regions. Drugs can be delivered selectively to diseased or damaged tissues by altering the surface of nanoparticles with ligands that recognize and bind to certain receptors on target cells. This method of focused drug delivery can reduce side effects and increase treatment efficacy. The capability of nanoparticles to deliver both hydrophilic and hydrophobic medications is another advantage. Hydrophilic drugs are encapsulated within the nanoparticle core and hydrophobic drugs are encapsulated within the nanoparticle shell. This allows for the delivery of a wider range of drugs using nanoparticles.

Figure 6. Nanoparticles in biomedicine

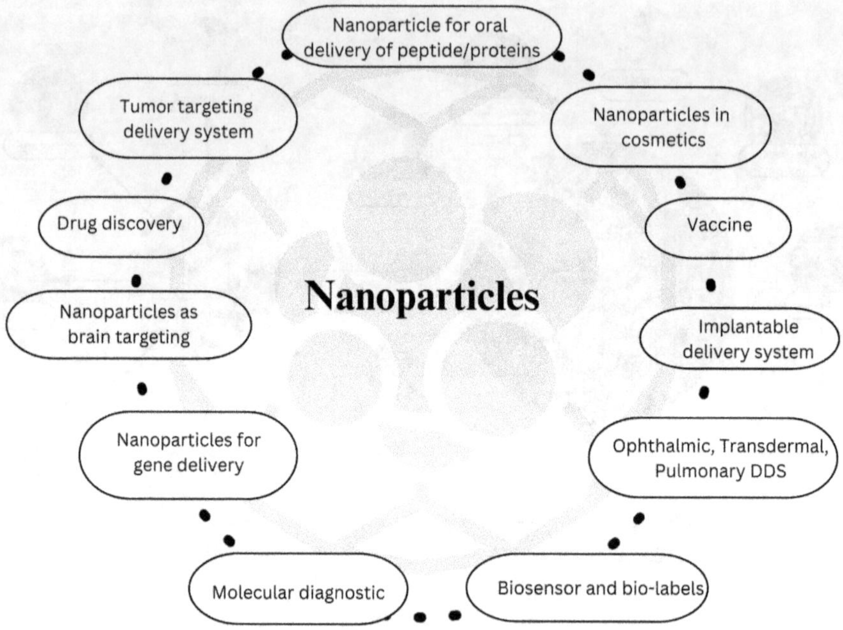

Nanoparticles can also be designed to respond to different stimuli allowing for targeted and controlled drug release. For example, the nanoparticle can be designed to discharge the drug only in the acidic environment of a tumor, minimizing the exposure of healthy tissues to the drug. The potential to transport several medications, better drug stability, and tailored drug delivery make nanoparticles an attractive component for drug delivery systems. They have the ability to completely alter how we handle a variety of illnesses and disorders. To completely understand their safety and effectiveness in clinical settings, additional research is necessary.

(v) ***Shape memory polymers (SMP)*:** Shape-memory polymers can remember and return to their unique shape following distortion. In response to changes in pH or temperature, they can be employed to make drug delivery systems that release drugs in a controlled manner. The SMP can be programmed to change form and release the medicine in response to a particular physiological triggers. This makes it possible to successfully deliver tailored drugs to particular cells or tissues. SMPs can also be used to regulate the discharge rate of drugs. This allows for sustained drug delivery, which can be beneficial in treating chronic conditions (Dhanasekaran et al., 2018). Another advantage of SMPs is their biocompatibility. SMPs can be designed to be biodegradable and non-toxic, allowing them to be safely used in drug delivery applications.

SMPs are a capable substance for drug delivery systems because they offer biocompatibility, controlled release, and targeted drug delivery. They have the ability to completely alter how we handle a variety of illnesses and disorders. To completely understand their safety and effectiveness in clinical settings, additional research is necessary. As a result, smart materials present a viable strategy for drug delivery systems, enabling targeted and controlled drug release as well as increased efficacy.

2.2 Tissue Engineering

Smart materials have significant capability for tissue engineering process, which is the development of functional tissues to substitute or repair the diseased one. To restore or enhance tissue functioning, tissue engineering tries to produce new tissues that can meld with the host tissue. The basic steps involved in tissue engineering is depicted in figure 7.

Figure 7. Process of tissue engineering

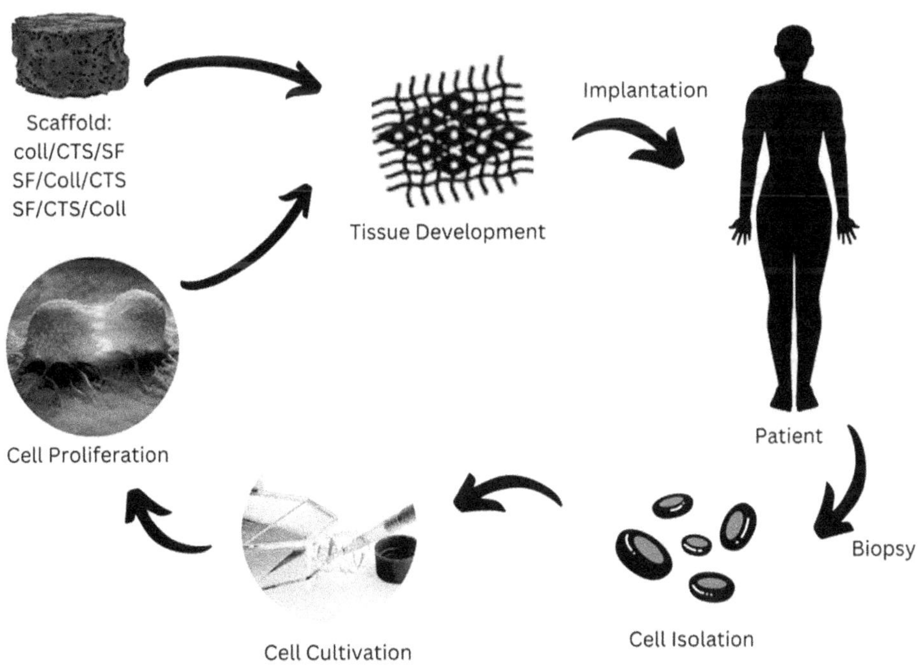

The first step in tissue engineering is cell isolation which involves the isolation of cells from the patient or a donor. Once the cells have been isolated, they are grown in a laboratory under precise conditions to form a tissue-like structure. To prepare cells for use in tissue engineering, they must initially be grown and expanded in a laboratory environment which is known as cell culture. The nutrients, growth factors, and other components required for the cells to develop and multiply are present in the culture media.

Cell scaffold is another component in tissue engineering as it offer a supportive structure for cell addition, proliferation, and distinction. Cell scaffold serves as a temporary extracellular matrix (ECM) that mimics the assembly and function of the natural ECM in vivo. A thorough understanding of cell behaviour, ECM deposition, and tissue remodelling is necessary for the complex process of tissue development in tissue engineering. Tissue engineers design specialised scaffolds that can encourage tissue formation and regeneration using a range of methods, including 3D printing, bioreactors, and microfabrication. Finally, implantation in tissue engineering is a capable approach for the treatment of various ailments and injuries, but it also poses several challenges, such as immune rejection, infection, and limited availability of donor tissues or organs. To overcome these challenges, tissue engineers are developing novel strategies, such as immune tolerance induction, biomaterial-based immunomodulation, and tissue engineering of patient-specific tissues and organs using 3D bio printing and other innovative techniques.

In tissue engineering, hydrogels, biodegradable polymers, electrospun fibres, shape memory polymers and nanoparticles play vital role (Khan & Tanaka, 2017). By offering a favourable environment for cells to attach to and multiply, hydrogels can be employed as scaffolds for cell growth. Since hydrogels are soft and pliable, they can imitate the mechanical characteristics of original tissues and their hydrophilic nature aids in the promotion of cell adhesion and migration. Drug distribution can be controlled by engineering hydrogels to react to certain stimuli, such as changes in pH or temperature. This can be beneficial in tissue engineering by delivering growing factors or other molecules that can endorse tissue regeneration. Hydrogels can be used to promote tissue regeneration by delivering growth factors or other molecules that can stimulate cell growth and differentiation. Hydrogels can also be functionalized with specific molecules or proteins that can promote cell adhesion and differentiation. Because hydrogels are biocompatible, living tissue is not harmed by them. They can thus be employed in vivo without triggering an immune response, making them a desirable material for tissue engineering. Hydrogels are considered as suitable candidate for tissue engineering because they may be created to imitate the mechanical and metabolic characteristics of natural tissues. It is feasible to create scaffolds that upkeep cell growth and differentiation and encourage tissue regeneration by manipulating the characteristics of hydrogels. Hydrogels offer significant potential for use in tissue engineering, providing a supportive scaffold for cell growth and differentiation and allowing for controlled drug delivery. With further research and development, hydrogels may be able to play a vital role in regenerative medicine by endorsing the growth and regeneration of functional tissues.

Biodegradable polymers are smart materials that can break down into non-toxic components in the body. They can be made to have particular mechanical properties that resemble those of natural tissues and can be utilized as scaffolds for tissue engineering. Biodegradable polymers can also be engineered to discharge growth factors or other molecules that can promote tissue regeneration (Colnik et al., 2020). The ability of biodegradable polymers to provide an environment that encourages cell attachment and proliferation makes them useful as scaffolds for cell growth. Biodegradable polymers can be developed to have mechanical characteristics that resemble those of natural tissues, making them appropriate in tissue engineering. The release of growth factors or other compounds that can encourage tissue regeneration can be designed into biodegradable polymers. The release of these molecules can be controlled by the rate at which the polymer degrades, allowing for controlled drug delivery. Biodegradable polymers are biocompatible, which means they are not injurious to living tissue. This makes them an attractive smart material for tissue engineering because they can be safely used in vivo without causing an immune response. By providing growth factors or other compounds that can encourage cell growth and differentiation, biodegradable polymers can be utilized to support tissue regeneration. Additionally, biodegradable polymers can be functionalized with particular molecules or proteins that can encourage cell differentiation and adhesion. Biodegradable polymers are a desirable material for tissue engineering because they may be created to match the mechanical and metabolic characteristics of natural tissues. By controlling the properties of biodegradable polymers, it is possible to create scaffolds that can support cell growth and differentiation and promote tissue regeneration. So, biodegradable polymers offer significant potential for use in tissue engineering, providing a supportive scaffold for cell growth and differentiation and allowing for controlled drug delivery. With further research and development, biodegradable polymers may be able to play a major role in regenerative medicine by stimulating the growth and regeneration of functional tissues.

An example of a smart material that is functioning as a support for the engineering tissues is electrospun fibres. They can be made to resemble the structure of natural tissues and have a greater surface area to volume ratio. Additionally, functionalizing electrospun fibres with particular molecules or proteins can help cells grow and differentiate. Shape-memory polymers can also be used to generate dynamically responsive scaffolds that can modify shape and adapt to the surrounding tissue. With dimensions ranging from nanometers to micrometres, ultrafine fibres are made using the electrospinning technology. A type of intelligent material called electrospun fibres can be utilized in tissue engineering to build scaffolds for cell growth and tissue regeneration. In this method, a polymer melt or solution is stretched and lengthened by an electrostatic field before being collected as fibre material on a grounded surface.(Parham et al., 2020). By offering an environment that is conducive to cell attachment and proliferation, electrospun fibres can be employed as scaffolds for cell growth. Electrospun fibres' high surface area to volume ratio can improve cell adhesion and development, and the fibres can be positioned in various directions to resemble the composition of natural tissues. To encourage tissue regeneration, electrospun fibres can be functionalized with medicines or growth factors. The fibres' ability to release the medications or growth factors at a controlled rate enables sustained distribution over a long period of time.

Electrospun fibers are biocompatible, which means they are not harmful to living tissue. This makes the smart materials are promising for tissue engineering because they can be safely used in vivo without causing an immune response. By providing compounds that can promote cell growth and differentiation, such as growth factors, electrospun fibres can be utilized to drive tissue regeneration. In addition to this, specific molecules or proteins that can encourage cell adhesion and differentiation can be functionalized into electrospun fibres. The mechanical and metabolic characteristics of natural tissues can be replicated in electrospun fibres, making them a suitable material for tissue engineering. By controlling the properties of electrospun fibers, it is possible to create scaffolds that can support cell growth and differentiation and promote tissue regeneration. With the ability to provide controlled doses of medication and serve as a supporting scaffold for cell development and differentiation, electrospun fibres are promising candidates for application in tissue engineering. By encouraging the creation and regeneration of functioning tissues, electrospun fibres may eventually be able to play a major part in regenerative medicine. Shape memory polymers (SMPs) are a kind of polymers that can vary their shape by responding to an external stimulus. SMPs have been used in tissue engineering to produce scaffolds that can change their shape in response to changes in the physiological environment, allowing them to adapt to the surrounding tissue and promote tissue regeneration. SMPs can be used to make 3D scaffolds with complex shapes and structures. The scaffold has the ability to be programmed to take on two different shapes. Temporary shape at high temperatures that can be readily manipulated and moulded, and a permanent shape at body temperatures that allows the scaffold to adapt to the surrounding tissue and encourage cell proliferation and tissue regeneration. In tissue engineering, SMPs are essential for drug delivery. Drug administration can be targeted and controlled using the SMPs, which is set to release medications corresponding to fluctuations in the physiological environment, such as pH, temperature, or light. This will aid in the promotion of tissue regeneration.

SMPs can also be used to create tissue engineering devices, such as stents, sutures, and other implants.

SMPs can be utilized in tissue engineering for applications including wound healing. SMPs are a promising material in tissue engineering because they can be tailored to have specific mechanical, chemical, and biological properties and react to changes in the physiological environment, allowing for better tissue regeneration and improved clinical outcomes. The SMPs can be used to create wound dressings that change their shape in response to changes in the wound environment, allowing for better wound coverage and improved healing.

Nowadays, nanoparticles are largely used in tissue engineering also to deliver growth factors or other molecules that can promote tissue regeneration. They can also be used as contrast agents for imaging, allowing researchers to monitor the growth and development of new tissues. Nanoparticles are minute particles which have 1nm to 100nm size and largely employed in tissue engineering which includes drug delivery, cell labeling, tissue imaging and in medication delivery systems. They can be designed to have certain characteristics, such as surface charge, size, and functional groups, enabling selective drug delivery to a particular cell or tissue. Nanoparticles can also shield pharmaceuticals from deterioration and clearance, enabling controlled drug release and increased therapeutic efficacy. For imaging systems like computed tomography (CT) and magnetic resonance imaging (MRI), nanoparticles are utilized as contrast agents. They can also be used to label cells and track their distribution in the body. Nanoparticles can be used to enhance tissue regeneration by improving cell adhesion, proliferation, and differentiation. They can be included into hydrogels or scaffolds to provide mechanical support and improve the mechanical and organic features of the tissue-engineered product. Nanoparticles can be used to stop bacteria from growing on tissues or implanted devices. For instance, silver nanoparticles can be used as coatings for implantable devices because they have been found to have strong antibacterial activity. Using nanoparticles, illness indicators in the body can be found. They can be created to bind specifically to proteins or nucleic acids that are associated with a particular disease, enabling early disease detection and diagnosis. For these reasons nanoparticles are a promising material in tissue engineering, as they are modified to possess particular properties and functions and can be used for wide range of applications viz., drug supply, cell labeling, tissue imaging, and tissue regeneration. However, there are still challenges to overcome, such as toxicity, biocompatibility, and long-term safety, before nanoparticles can be used in clinical applications.

2.3 Diagnostic Tools

Smart materials' distinctive qualities, including sensitivity, specificity, biocompatibility, and biodegradability, have led to an increase in their application in diagnostic equipment. Here are some instances of intelligent materials utilized in diagnostic equipment: Analytical tools called biosensors can identify and quantify biological substances or organisms. Biosensors with high sensitivity and specificity for biomarker detection can be made using smart materials including nanomaterials, quantum dots, and polymers. Biosensors can be used in point-of-care diagnostic tools for the timely finding of syndromes like cancer, cardiovascular diseases, and infectious diseases (Genchi et al., 2017). The technology of microfluidics involves controlling tiny quantities of fluid in tiny channels. Microfluidic devices can be made with smart materials including hydrogels, polymers, and nanoparticles to detect and measure biomolecules. Point-of-care diagnostics, medication screening, and personalized medicine can all be performed using microfluidic devices. Biological processes in vivo can be seen and tracked using imaging probes.

Smart materials with high sensitivity and specificity for molecular imaging include nanoparticles, quantum dots, and fluorescent dyes. Imaging probes can be used to monitor the development of diseases and to detect them early. Another technology, lab-on-a-chip which involves integrating multiple laboratory tasks in one microchip. Smart materials such as hydrogels, polymers, and nanoparticles can be used in lab-on-a-chip gadgets for synthesis of samples, detection, and analysis. Lab-on-a-chip devices can be used for point-of-care diagnostics, drug screening, and environmental monitoring. Wearable sensors are gadgets that can be worn on the body to track physiological data including blood pressure, glucose levels, and heart rate. Wearable sensors can detect and monitor biomarkers using intelligent materials including hydrogels, polymers, and nanomaterials. Wearable sensors can be used to track the development of diseases and to detect them early.

Smart materials offer significant potential for use in diagnostic tools, providing high sensitivity, specificity, and biocompatibility. With further research and development, smart materials may be capable to improve the accuracy and efficiency of disease diagnosis, monitoring, and treatment.

2.4 Wearable Devices

Several types of smart materials are used in wearable devices (Jin et al., 2018) because of their distinctive properties like flexibility, stretch ability, and biocompatibility. Some examples of smart materials used in wearable devices are explained here.

Conductive polymers: This class of intelligent material is capable of conducting electricity. They are utilized to make stretchable, flexible electrodes for wearable technology that can be incorporated into garments and other items. To assess physiological signals including heart rate, temperature, and muscle activity, conductive polymers can be utilized.

Alloys with shape memory: A class of intelligent material known as shape-memory alloys is capable of recovering from deformation. They are used in wearable devices to create flexible and shape-changing components such as springs, hinges, and wires. Nowadays shape-memory alloys are also used in smart clothing, smart eyewear, and other wearable accessories. A type of intelligent material that can absorb and hold onto water is called a hydrogel. They help produce soft, flexible sensors that can adapt to the shape of the body for wearable technology. They are capable of measuring physiological signals like pH, humidity, and temperature.

Nanomaterials: They have distinctive optical, electrical, and mechanical capabilities. They are used to make sensors, antennae, and other parts that can be incorporated into garments or other wearable accessories for wearable devices. Nanomaterials can be used to measure physiological signals such as blood oxygen levels, heart rate, and respiration rate. Electronic textiles, a kind of smart material, are made by fusing conventional textiles with electronic components such as sensors, conductive materials, and power sources. They create clothing and other wearable technology items that can track physiological signs and other traits. Electronic textiles can be used in smart clothing, smart gloves, and other wearable accessories (Zheng et al., 2021).

Smart materials offer significant potential for use in wearable devices. With further research and development, smart materials may be successful in improving the efficiency of wearable devices for health monitoring, fitness tracking, and other applications

2.5 Targeted Therapy and Imaging in Cancer Treatment

Targeted Therapy: Drugs or other therapeutic agents can be used specifically to target cancer cells while sparing healthy cells is known as targeted therapy. Cancer is a complex disease that requires personalized treatment approaches for optimal outcomes. Smart materials have emerged as promising tools for targeted therapy and imaging in cancer treatment due to their exceptional properties and ability to respond to specific stimuli (Yang et al., 2021). The use of smart materials in cancer treatment can improve drug delivery, reduce side effects, and enable non-invasive imaging of tumors. Figure 2.6 describes different technologies adopted in targeted therapy for breast cancer treatment.

Figure 8. Targeted therapy for breast cancer

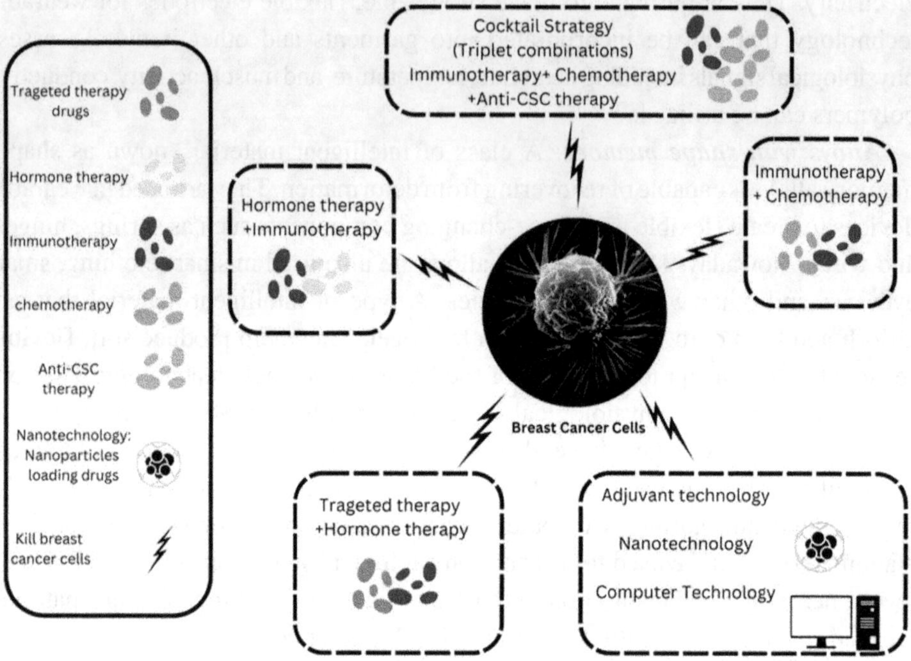

One example of a smart material used for targeted therapy is liposomes, which are in spherical structures made of lipids. Liposomes can be designed to encapsulate drugs and target specific cancer cells, allowing for effective drug delivery and reducing toxicity to healthy cells. Additionally, the effectiveness of liposomes can be increased by adding targeting ligands to their surface that precisely attach to receptors on cancer cells. The three-dimensional networks of hydrophilic polymers known as hydrogels, which have a high water-holding capacity, are another type of intelligent material employed in targeted therapy. Drug release can be triggered using hydrogels that are made to react to certain stimuli, including pH or temperature changes. For instance, a hydrogel that is pH sensitive can be created to release medication in reaction to the acidic environment of tumors.

Imaging: Smart materials can also be used for non-invasive imaging of tumors, allowing for prior detection and more precise diagnosis of cancer. Quantum dots, which are nanocrystals that release light when activated by an external light source, are one type of intelligent material utilized for imaging. Quantum dots can be modified to target specific cancer cells and emit light at specific wavelengths, allowing for the imaging of tumors with high sensitivity and specificity (J. Li et al., 2020). Another smart material used for imaging is magnetic nanoparticles, which can be used for magnetic resonance imaging (MRI) of tumors. Magnetic nanoparticles can be coated with targeting ligands to specifically bind to cancer cells and allow for more accurate imaging of tumors. Smart materials have shown great potential for targeted therapy and imaging in cancer treatment. Continued research and development of smart materials for cancer treatment can lead to more effective and personalized treatment methodologies for cancer patients.

3. REGULATORY CHALLENGES AND FUTURE PROSPECTS OF SMART MATERIALS IN BIOMEDICAL FIELD

Due to their distinctive properties, smart materials have shown pronounced assurance in biomedical field, however, there are still a number of regulatory issues that need to be resolved. These difficulties include ensuring the safety and effectiveness of smart materials as well as creating regulatory frameworks that can keep up with the field's rapid advancement (Yadav et al., 2022).

One of the main challenges for regulatory agencies is ensuring the safety of smart materials. Some smart materials, such as nanoparticles, have been shown to have potential toxicity, which could limit their usage in biomedical applications. Regulatory agencies should will need to ensure that smart materials used in biomedical applications undergo severe testing to ensure their safety. Another challenge is ensuring the effectiveness of smart materials. Due to the complexity of smart materials, a variety of variables, including alterations in the environment, may have an impact on their performance. As a result, it could be challenging to verify consistent performance across several applications. Guidelines for assessing the performance of smart materials in biomedical applications must be developed by regulatory organizations.

Regulatory bodies will also have the issue of keeping up with the rapidly developing field of smart materials. Regulatory authorities will need to be able to examine new materials swiftly and effectively because they are being created at a rapid rate. The scientific community and regulatory bodies will need to work closely together to build new regulatory frameworks that can keep up with the rapidly evolving field of smart materials. Despite these difficulties, smart materials have bright futures in biomedical applications. As new materials are created, they have the potential to completely transform the medical industry, from tissue engineering to drug delivery and beyond. The potential uses for smart materials are endless, and they are already used in many different health devices.

In conclusion, the usage of smart materials in biomedical field is a rapidly evolving field that presents both regulatory challenges and exciting opportunities. Regulating bodies will face a difficult task in ensuring the efficacy and safety of smart materials, but potential advantages of these materials make the effort worthwhile. The prospects for smart materials in biological applications are promising in the future, provided that research is conducted and regulatory bodies and the scientific community continue to work together.

4. BIOCOMPATIBILITY AND SAFETY CONSIDERATIONS FOR SMART MATERIALS

Due to their special characteristics and capacity to react to particular stimuli, smart materials have demonstrated significant potential for use in biological applications. However, there are significant safety and biocompatibility issues with their use in implants and medical devices. The term "biocompatibility" means the capability of a material to carry out its proposed role without causing a poisonous, immunological, or other negative reaction in the body. Biocompatibility is a critical consideration when designing and selecting smart materials for biomedical applications (Schmalz & Galler, 2017). Some of the main issues that affect the biocompatibility of smart materials are listed below.

(i) ***Chemical composition***: The chemical composition of a smart material can significantly affect its biocompatibility. Materials that are non-toxic and non-reactive are generally preferred for biomedical applications.
(ii) ***Degradation and stability***: The stability and degradation of smart materials are also critical factors for biocompatibility. Materials that degrade too quickly or too slowly can cause adverse reactions in the body.
(iii) ***Surface properties***: The surface properties of a smart material can affect its interactions with biological tissues and cells. Materials with uneven or hydrophobic surfaces can induce inflammation or other adverse effects.
(iv) ***Mechanicalproperties***: Mechanical properties of smart materials like stiffness and elasticity, can also affect their biocompatibility. Materials that are too rigid or too flexible may cause tissue damage or other adverse reactions
(v) ***Biofouling***: Biofouling refers to the accumulation of biological material on the surface of a material. Smart materials that are prone to biofouling can cause infections or other adverse reactions.

It is also essential to consider the safety of smart materials in terms of potential toxicity, immunogenicity, and other adverse effects. This requires thorough testing and evaluation of smart materials before they are used in medical devices or implants. The biocompatibility and safety of smart materials are critical considerations for their use in biomedical applications. To make sure that they can fulfil their intended purpose without endangering the body, these materials must be carefully designed and tested. Smart materials have the potential to revolutionize medicine and enhance patient outcomes with proper design, testing, and evaluation.

5. RECENT ADVANCES AND EMERGING TRENDS IN SMART MATERIALS FOR BIOMEDICAL APPLICATIONS

In recent decades, the practice of smart materials in the biomedical trade has been the focus of extensive research. On the basis of recent research, recent developments and emerging trends in smart materials for the biomedical area are presented below (Dai et al., 2022).

Smart hydrogels are capable to discharge pharmaceuticals in response to certain stimuli and hence, researchers have been looking into its application for drug delivery. For instance, scientists have created a smart hydrogel that can release insulin in reaction to variations in blood glucose levels, thereby opening up a new diabetes treatment option. Next, it was reported that shape-memory alloys were used for implantable devices like stents because they could be inserted compactly and then expand to take on the desired shape once inside the body. This innovation has the potential to completely change how cardiovascular diseases are treated. The usage of electroactive polymers for engineering tissues is also a hot topic of research since they can create scaffolds that replicate the electrical characteristics of real tissues. Examples include the exploration of the functionality of smart sutures, which can release medications or monitor wound healing in real-time, and the development of an electroactive polymer scaffold that can encourage the regeneration of injured nerves. For example, researchers have developed a smart suture that can release antibiotics in response to infection(Basu et al., 2022).

The function of smart materials in the biomedical field is an electrifying area of exploration with many important applications. Future developments in their applications should increase as our knowledge of these materials deepens. As a result, there are exciting possibilities for the creation of brand-new, cutting-edge medical implants and devices thanks to recent developments of smart materials in biomedical field. Smart materials have the potential to revolutionise medicine and enhance patient outcomes with sustained study and development (Ajay et al., 2023).

6. FUTURE PROSPECTS FOR SMART MATERIALS

According to recent researches smart materials have the competing potential to change a number of industries and how we live and work (Allizond et al., 2022).

Research and technological advancements are continuing to push the envelope of what is possible, the prospects for smart materials in the future are positive. The fields of healthcare, energy, aerospace, defence, building materials, and consumer electronics are among those where smart materials could be used and developed. Smart materials can be employed in the healthcare industry to create cutting-edge medical equipment including artificial organs, implantable computers, and drug delivery systems. They can also be used to make intelligent fabrics that can deliver medication or monitor vital signs. Smart materials can be used to develop new energy storing and renovation technologies like solar cells, batteries, and fuel cells. In the field of aerospace and defence, smart materials can be used to create lightweight, strong, and durable materials for aircraft and spacecraft. They can also be used to develop self-healing materials that can repair damage caused by extreme conditions (Kumar et al., 2023).

Modern building materials that are more effective, long-lasting, and ecologically friendly are created using smart materials. They can also be used to build intelligent structures that can recognise and react to alterations in their surroundings. They discovered intriguing uses for it in the area of consumer electronics, where it can be applied to the development of novel and ground-breaking goods including flexible screens, wearable electronics, and smart sensors. We may anticipate even more intriguing innovations and applications of smart materials in the future as research in this area advances. The potential benefits to society are enormous, ranging from improved healthcare and energy efficiency to more sustainable and resilient infrastructure.

7. CONCLUSION

Smart materials have emerged as a promising new frontier in the field of biomedicine, with the potential to revolutionize various aspects of healthcare. The ability to build materials that may respond to specific stimuli has opened up new opportunities for the development of revolutionary medical devices, drug delivery systems, and tissue engineering scaffolds. The need for personalised medicine, the rise in chronic illness incidence, and the growing desire for more effective and affordable therapies are just a few of the major problems that smart materials may help solve in the healthcare industry today. Despite the fact that there are still numerous obstacles to be cleared, such as those involving toxicity, biocompatibility, and regulatory approval, smart materials in biomedicine have a very bright future. Vigorous research and innovation in this field will certainly lead to new developments and advances in medical technology, ultimately improving the lives of millions of people around the world. As such, the study and progress of smart materials in biomedicine is an exciting and rapidly growing area of research, with far-reaching implications for the future of healthcare. We can only imagine the possibilities that will arise as we continue to explore this fascinating and promising new frontier.

REFERENCES

Aflori, M. (2021). Smart Nanomaterials for Biomedical Applications—A Review. *Nanomaterials (Basel, Switzerland)*, *11*(2), 396. doi:10.3390/nano11020396 PMID:33557177

Ajay, P. Ahmad, S., Sharma, J., & Gambhir, V. (2023). Handbook of Sustainable Materials: Modelling, Characterization, and Optimization. CRC Press. doi:10.1201/9781003297772

Allizond, V., Comini, S., Cuffini, A. M., & Banche, G. (2022). Current Knowledge on Biomaterials for Orthopedic Applications Modified to Reduce Bacterial Adhesive Ability. *Antibiotics (Basel, Switzerland)*, *11*(4), 529. doi:10.3390/antibiotics11040529 PMID:35453280

Arab Hassani, F., Shi, Q., Wen, F., He, T., Haroun, A., Yang, Y., Feng, Y., & Lee, C. (2020). Smart materials for smart healthcare– moving from sensors and actuators to self-sustained nanoenergy nanosystems. *Smart Materials in Medicine*, *1*, 92–124. doi:10.1016/j.smaim.2020.07.005

Badami, V., & Ahuja, B. (2014). Biosmart Materials: Breaking New Ground in Dentistry. *TheScientificWorldJournal*, *2014*, 1–7. doi:10.1155/2014/986912 PMID:24672407

Bahl, S., Nagar, H., Singh, I., & Sehgal, S. (2020). Smart materials types, properties and applications: A review. *Materials Today: Proceedings*, *28*, 1302–1306. doi:10.1016/j.matpr.2020.04.505

Basu, B., Gowtham, N. H., Xiao, Y., Kalidindi, S. R., & Leong, K. W. (2022). Biomaterialomics: Data science-driven pathways to develop fourth-generation biomaterials. *Acta Biomaterialia*, *143*, 1–25. doi:10.1016/j.actbio.2022.02.027 PMID:35202854

Colnik, M., Knez-Hrncic, M., Skerget, M., & Knez, Z. (2020). Biodegradable polymers, current trends of research and their applications, a review. *Chemical Industry & Chemical Engineering Quarterly*, *26*(4), 401–418. doi:10.2298/CICEQ191210018C

Dai, H., Fan, Q., & Wang, C. (2022). Recent applications of immunomodulatory biomaterials for disease immunotherapy. *Exploration*, *2*(6), 20210157. doi:10.1002/EXP.20210157 PMID:37324799

Dauda, K. T., Majebi, C., Ayoola, A. A., Agboola, O., Popoola, A. P., Fayomi, O. S. I., Banjo, S. O., Sojobi, J., & Ogunsanya, A. O. (2021). Biosmart Materials and Its Innovative Characteristics in Protective Composite Coating Application. *IOP Conference Series. Materials Science and Engineering*, *1107*(1), 012216. doi:10.1088/1757-899X/1107/1/012216

Dhanasekaran, R., Sreenatha Reddy, S., Girish Kumar, B., & Anirudh, A. S. (2018). Shape Memory Materials for Bio-medical and Aerospace Applications. *Materials Today: Proceedings*, *5*(10), 21427–21435. doi:10.1016/j.matpr.2018.06.551

Genchi, G. G., Marino, A., Tapeinos, C., & Ciofani, G. (2017). Smart Materials Meet Multifunctional Biomedical Devices: Current and Prospective Implications for Nanomedicine. *Frontiers in Bioengineering and Biotechnology*, *5*, 80. doi:10.3389/fbioe.2017.00080 PMID:29326928

Jin, H., Jin, Q., & Jian, J. (2018). Smart Materials for Wearable Healthcare Devices. In *Wearable Technologies*. InTech. doi:10.5772/intechopen.76604

Khan, F., & Tanaka, M. (2017). Designing Smart Biomaterials for Tissue Engineering. *International Journal of Molecular Sciences*, *19*(1), 17. doi:10.3390/ijms19010017 PMID:29267207

KÖK, M., Qader, İ. N., Dağdelen, F., & Aydoğdu, Y. (2019). Akıllı Malzemeler üzerine derleme: araştırmalar ve uygulamaları. *El-Cezeri Fen ve Mühendislik Dergisi*. doi:10.31202/ecjse.562177

Kumar, A., Kumar, P., Mittal, R. K., & Gambhir, V. (2023). Materials processed by additive manufacturing techniques. In Advances in Additive Manufacturing Artificial Intelligence, Nature-Inspired, and Biomanufacturing (pp. 217–233). Elsevier. doi:10.1016/B978-0-323-91834-3.00014-4

Li, J., Guan, X., Fan, Z., Ching, L.-M., Li, Y., Wang, X., Cao, W.-M., & Liu, D.-X. (2020). Non-Invasive Biomarkers for Early Detection of Breast Cancer. *Cancers (Basel)*, *12*(10), 2767. doi:10.3390/cancers12102767 PMID:32992445

Li, Y.-C., Zhang, Y. S., Akpek, A., Shin, S. R., & Khademhosseini, A. (2016). 4D bioprinting: The next-generation technology for biofabrication enabled by stimuli-responsive materials. *Biofabrication*, *9*(1), 012001. doi:10.1088/1758-5090/9/1/012001 PMID:27910820

Mahinroosta, M., Jomeh Farsangi, Z., Allahverdi, A., & Shakoori, Z. (2018). Hydrogels as intelligent materials: A brief review of synthesis, properties and applications. *Materials Today. Chemistry*, *8*, 42–55. doi:10.1016/j.mtchem.2018.02.004

Nakhaei, P., Margiana, R., Bokov, D. O., Abdelbasset, W. K., Jadidi Kouhbanani, M. A., Varma, R. S., Marofi, F., Jarahian, M., & Beheshtkhoo, N. (2021). Liposomes: Structure, Biomedical Applications, and Stability Parameters With Emphasis on Cholesterol. *Frontiers in Bioengineering and Biotechnology*, *9*, 705886. doi:10.3389/fbioe.2021.705886 PMID:34568298

Parham, S., Kharazi, A. Z., Bakhsheshi-Rad, H. R., Ghayour, H., Ismail, A. F., Nur, H., & Berto, F. (2020). Electrospun Nano-Fibers for Biomedical and Tissue Engineering Applications: A Comprehensive Review. *Materials (Basel)*, *13*(9), 2153. doi:10.3390/ma13092153 PMID:32384813

Schmalz, G., & Galler, K. M. (2017). Biocompatibility of biomaterials – Lessons learned and considerations for the design of novel materials. *Dental Materials*, *33*(4), 382–393. doi:10.1016/j.dental.2017.01.011 PMID:28236437

Song, M., Lin, X., Peng, Z., Xu, S., Jin, L., Zheng, X., & Luo, H. (2021). Materials and Methods of Biosensor Interfaces With Stability. *Frontiers in Materials*, *7*, 583739. doi:10.3389/fmats.2020.583739

Undre, B, S., R. Pandya, S., Kumar, V., & Singh, M. (2016). Dendrimers As Smart Materials For Developing The Various Applications In The Field Of Biomedical Sciences. *Advanced Materials Letters*, *7*(7), 502–516. doi:10.5185/amlett.2016.6153

Vashist, A., & Ahmad, S.VASHIST. (2013). Hydrogels: Smart Materials for Drug Delivery. *Oriental Journal of Chemistry*, *29*(03), 861–870. doi:10.13005/ojc/290303

Yadav, S., Sadique, M. A., Kaushik, A., Ranjan, P., Khan, R., & Srivastava, A. K. (2022). Borophene as an emerging 2D flatland for biomedical applications: Current challenges and future prospects. *Journal of Materials Chemistry. B, Materials for Biology and Medicine*, *10*(8), 1146–1175. doi:10.1039/D1TB02277F PMID:35107476

Yang, Y., Zeng, W., Huang, P., Zeng, X., & Mei, L. (2021). Smart materials for drug delivery and cancer therapy. *VIEW*, *2*(2), 20200042. doi:10.1002/VIW.20200042

Zheng, Z., Jur, J., & Cheng, W. (2021). Smart materials and devices for electronic textiles. *MRS Bulletin*, *46*(6), 488–490. doi:10.155743577-021-00120-5

Chapter 3
Smart Biodegradable and Bio-Based Polymeric Biomaterials for Biomedical Applications

Samir Das
Indian Institute of Technology, Kharagpur, India

Sudipto Datta
https://orcid.org/0000-0003-2161-6878
Indian Institute of Science, Bangalore, India

Anamika Barman
Indian Institute of Technology, Kharagpur, India

Ranjit Barua
https://orcid.org/0000-0003-2236-3876
Indian Institute of Engineering Science and Technology, Shibpur, India

ABSTRACT

Polymeric materials have gained too much attention in the past few decades because of their suitable properties that makes these materials applicable for various biomedical applications. More specifically, biopolymers have gained main attraction because of their properties like inexpensiveness, renewability, biocompatibility, and biodegradability. Additionally, these smart polymeric biomaterials are sensitive to some physical factors like magnetic field, electric field, light intensity, pH, humidity, and temperature. Moreover, this book chapter will briefly cover smart biomaterials, i.e., smart memory polymers (SMP), which are used in biomedical applications. The applications of these materials for tissue engineering and drug delivery is mainly covered. This book chapter will help to develop a brief idea for the readers on smart polymeric biomaterials.

DOI: 10.4018/978-1-6684-9224-6.ch003

1. INTRODUCTION

Biomaterials are materials that are used in medical applications to interact with biological systems, such as living tissue or organs. These materials can be natural or synthetic and are designed to be cytocompatible, meaning that they have not any adverse reactions in the body. Biomaterials have various applications in medical devices, implants, drug delivery systems which include polymers, metals, ceramics, and composites. Because of their durability and strength, metals like titanium and stainless steel are employed in orthopaedic implants. Polymers based biomaterials also known as biopolymers excel in medical applications due to their distinctive properties such as biocompatibility, biodegradability, and versatility. Biopolymers, or biologically degradable polymers, are those that are created by living things. In a nutshell, Polymer-based biomolecules consist of monomeric elements connected covalently and they are frequently bearable and renewable resources since they are arranged from alive (plant) ingredients that can be continually recreated. It play a remarkable range of roles in human body, including holding cells together to form tissues and sending the cells tiny chemical cues to control their behaviour (Kumar et al., 2022). Research on biopolymers is a great interest in both academia and industry since 1960 (Wichterle et al., 1960). Biopolymers can be produced from natural and fossil sources (Yadav et al., 2015) (Azimi et al., 2020). Several functional groups have been included into the molecular sequence to govern the chemical, physical, and natural belongings of the diversity of biopolymers for a number of biomedical uses, including the controlled administration of bioactive substances and cell-based therapies (Kwon et al., 1991) (Hoffman., 2012) (Khan et al., 2017). However, the biological applications of these material systems are determined by their cytocompatibility, relations to cells that are alive, their macromolecular structure, and how they aggregate to form three-dimensional structures (Ajay et al., 2023). If applied in settings where they increase functionality and produce additional benefits, biopolymers have unmatched advantages over standard plastics. By using these materials in live systems, it may be possible to lessen the toxicity and simulation of chronic inflammation that typically occur while a synthetic polymer method is transplanted into the host (Yates et al., 2013). Biopolymers modified for mechanical and electrical properties. (Van de Velde et al., 2002) (Kim et al., 2020). For usage in biomedical applications, biomaterial scientists have consistently sought to develop new methods, produce innovative medical equipment, and process and synthesise novel biomaterials. The majority of synthetic biomaterials now used in tissue engineering (TE) are created either from natural sources or by specifically modifying forms, sizes, porosities, and architectural structures for TE applications (Weigel et al., 2006) (Barua et al., 2020) (Khan et al., 2015) (Tsang et al., 2007).

Biomaterial scaffolds are crucial in any tissue regeneration because they not only support temporary structural exactness but also participate in cell-biomolecule interactions, cell adhesion and proliferation, and tissue formation. These conventional biomaterials, however, are unable to adjust to the fluctuating pH and body temperature brought on by disease. Hence, scientists working with polymers have been attempting to develop intelligent polymeric biomaterials (Datta et al., 2019) that resemble live tissues in the previous 20 yearsEnvironmental cues like temperature, light intensity, composition of solvent, electric field, pH, and ions cause reversible changes in smart polymer microstructure. (Dušek et al, 1968) (Barua et al., 2023) (Barua et al., 2022) (Osada etal., 1993) (DeRossi et al., 1991). At the macroscopic level, the adjustments can be seen as precipitate growth, phase separation, or, in the case of hydrogels, as magnitude changes. Reversible phenomena enable systems to return to their initial state when a trigger is removed, triggered by pH shifts, polymer additions, temperature changes, and polymer network disintegration.In particular, microfluidic devices are being developed using polymers with such sensitive properties for use in bulk engineering and microscale medicine (Barker et al., 2000), time dependent drug release systems (Kikuchi et al., 2002) (Li et al., 2001) (Kost et al., 2012) (Peppas et al., 1997), bioadhesion intermediaries (Ista et al., 1998) (Cunliffe et al., 2003) (Yamato et al., 2002) and actuators (Li et al., 2001). These polymers are valuable in a range of applications, such as drug delivery, tissue engineering, and biosensors, since they can undergo reversible changes in their structure or characteristics in response to various stimuli. It is well-known that smart polymers have certain properties (Kumar et al., 2007) that shift reversibly in response to sliding variations in system parameters like temperature, pH, ionic factor, and biological molecules, as shown in figure 1. We can anticipate that developments in cell biology and chemistry will have an increasing impact on biomaterials, and that the combination of these intelligent biomaterials with biosensors, fresh drug delivery methods, growth factors, and DNA will hasten the creation and clinical use of new medical devices. Smart biomaterials made of polymeric material have thus become cutting-edge research fields in the biomedical field, particularly as bioactive molecule delivery systems, tissue engineering scaffolds, and cellular adhesion and growth.

2. CLASSIFICATION OF SMART BIOPOLYMERS

Throughout the last 20 years, the application of functional polymers has improved significantly. These polymers react as desired to changes in temperature, magnetic or electric fields, or other conditions like pH. Smart polymeric materials expand rapidly in modern medicine and biomedical applications. For many diverse uses, various smart biopolymer kinds have been developed. Figure 2 illustrates a general classification of these polymers, which are discussed briefly in the following section.

Smart materials play a significant role in various biomedical applications due to their unique properties and responses to external stimuli. Here is a list of commonly used smart materials in biomedical applications, as mentioned in table 1.

The details applications and types of smart materials are depending on factors such as desired pH range, response kinetics, biocompatibility, and specific application requirements, that has been described in the following sessions.

Figure 1. Smart biopolymers materials for TE applications: (a) elastic modulus of various tissues during scaffold fabrication; (b) stimuli responsive system (ΔT, ΔpH, Δλ, ΔE variation of temperature, pH, wave length, and electric field respectively)

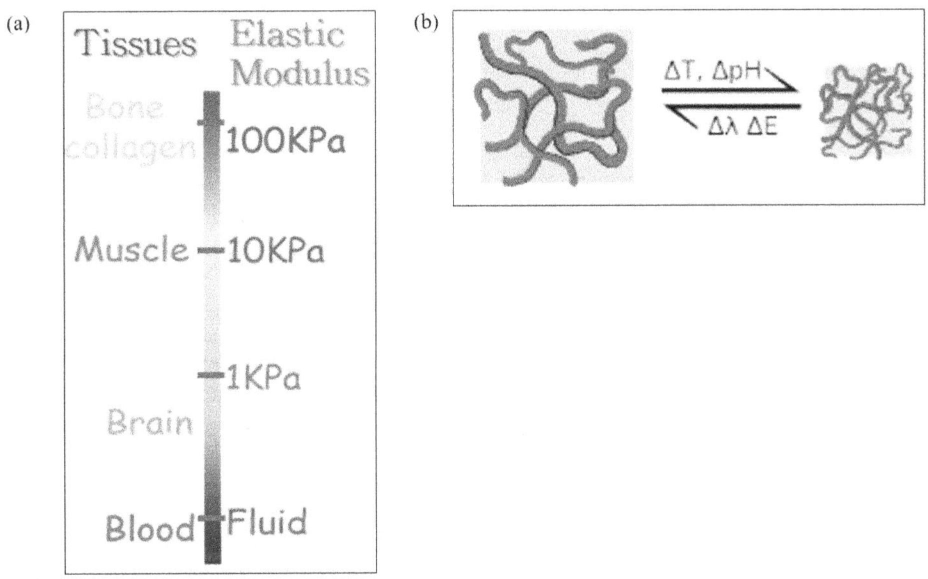

2.1. pH-Sensitive Smart Polymers

The human body manifests significant differences in pH at a particular area, tissue, or cellular portions, utilized for therapeutic purposes. Such pH variations make the gastrointestinal tract prominent for applying pH-sensitive polymers. Polymer structures of ionizable functional groups responsive to pH variation are known as pH-sensitive polymers. The functional groups can be acidic (sulphonic/carboxylic) or basic (amino salts). Thus, the polymers may accept or release protons for pH fluctuations (Qiu et al., 2001). The acidic functional groups accept (release) protons

Figure 2. Classification of smart polymers based on stimuli, functionality, and intrinsic properties

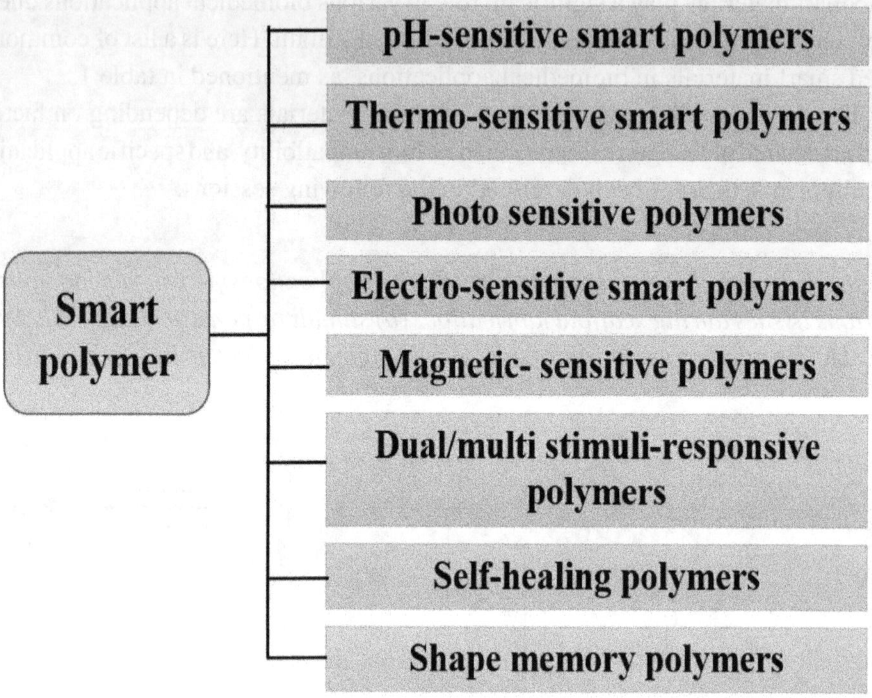

at low (high) pH values. In response to the electrostatic attraction of negatively charged groups like polyacrylic acid (PAA) and polymethacrylic acid (PMAA), polyanions cause polymer swelling. In contrast, basic functional groups accept (release) protons in an acid (basic) environment (Pattanashetti et al., 2017) and forms ion at neutral pH values. They are termed polycations configured with amino groups, like poly(vinyl amine) (PVAm), poly (2-diethyl aminoethyl methacrylate) (PDEAEMA), poly (4-vinyl pyridine), and poly (2-vinyl pyridine) (PVP).

Polymers with pH sensitivity can have the structures, such as linear, branched, or network-like. The ionizable group on the hydrophobic backbone is crucial for polymer structure, classified as polyelectrolytes. Figure 3 represents the pH-sensitive polymer form Micelle, swelling, and gelling, respectively.

Protonation/deprotonation affects the net electrostatic charge, affecting conformational arrangement and hydrodynamic volume. Elevated electrostatic charge can cause impulsion between chains, causing extension/opening, and influencing hydrophilicity. Differences in electrostatic charge can also cause self-assembly, such as micelles, gels, and vesicles.

Table 1. Name of smart materials with their biomedical applications

Type	Name	Application
Shape Memory Alloys (SMAs)	Nitinol (nickel-titanium (Ni-Ti))	Orthopedic implants, stents, drug delivery devices, etc.
Hydrogels	Poly(acrylic acid) (PAA), Poly(N-isopropylacrylamide) (PNIPAAm), Poly(3,4-ethylenedioxythiophene) (PEDOT), etc.	Wound healing and drug delivery systems
Stimuli-Responsive Polymers	Polyethylene glycol (PEG), Poly(2-hydroxyethyl methacrylate) (PHEMA), etc.	Biosensors, tissue engineering scaffolds, and controlled release systems
Magnetostrictive Materials	Terfenol-D (Terbium-Dysprosium-Iron Alloy), Terfenol-C (Terbium-Iron Alloy), etc.	Magnetic Resonance Imaging (MRI), and actuation systems
Electroactive Polymers (EAPs)	Ionic Polymer-Metal Composites (IPMCs), Des or Dielectric Elastomers, Liquid Crystal Elastomers (LCEs), etc.	Artificial muscles, biosensors, etc.
Shape Memory Polymers (SMPs)	Polyurethane, Polycaprolactone-Based SMPs, etc.	Stents, catheters, and tissue engineering scaffolds
Conducting Polymers	Polyaniline (PANI), Poly(3,4-ethylenedioxythiophene) (PEDOT), Polypyrrole (PPy), etc.	Neural interfaces, biosensors, and tissue engineering
Nanomaterials	Carbon dots, nano-bioceramics, Carbon nanotubes, etc.	Drug delivery, targeted therapy, etc
Piezoelectric Materials	Quartz, Lead Zirconate Titanate (PZT), Polyvinylidene Fluoride (PVDF), Barium Titanate (BaTiO3), etc.	Bone regeneration, biosensors, and energy harvesting
pH-Responsive Polymers	Poly(methacrylic acid) (PMAA), Poly(acrylic acid) (PAA), Chitosan, etc.	Drug delivery systems, biosensors, and wound dressings

2.2. Thermo-Responsive Smart Polymers

Temperature-sensitive smart polymers alter microstructural properties. These provide the safest option for biomaterials and drug administration systems (Nakayama et al., 2006). At a certain temperature, known as the critical solution temperature, the polymer phase may vary along with the solution depending on its composition. Lower and upper critical solution temperatures represent thermoresponsive polymers. The liquid crystal saturation temperature (LCST) is this temperature in which a solution of polymer displays a separation of phase, changing from an isotropic to anisotropic condition. The dissolution of the polymer occurs below the LCST due to the enthalpy involved with the hydrogen bonding between the water molecules and the polymer. Entropy (hydrophobic interactions) takes over when the temperature is raised over the LCST, resulting in polymer precipitation. Hydrophilic or hydrophobic moieties can be used to change the LCST of polymers in water solutions (Almeida et al., 2012). Large-scale applications in drug delivery systems exist for polymers with LCST. With the polymer, we can blend therapeutic substances like medications, cells, or proteins. This mixture could be injected into the injured area or applied topically to the human body's subcutaneous layer. As a result, a gel deposit may develop at the injection site and cause a temperature increase (Patel et al., 2021). Table 2 lists the most popular thermosensitive polymers and their uses for them.

Figure 3. Several methods of pH-sensitive polymer to form Micelle, swelling, and gelling, respectively

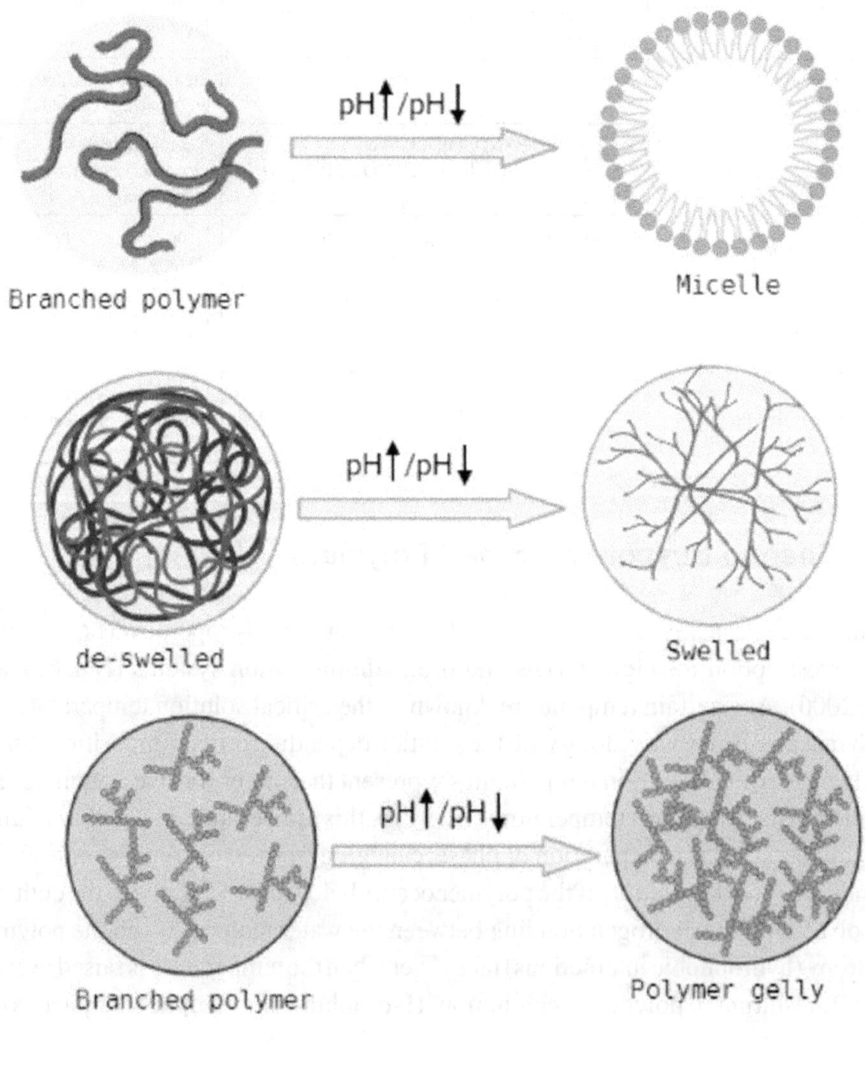

Table 2. Temperature-sensitive polymers' biomedical applications

Polymers	Treatment
F-127	Stomach cancer
Chito based glycerophosphate salt and glycerol	Depot injection therapy
Chitosan beta with glycerophosphate	Limbic ischemia
Polybenzofulvene	Tumour maligned

Temperature-sensitive polymers offer advantages like lipophilic and hydrophilic drug production, site-specific delivery, and reduced side effects, but face difficulties like high-burst issue, biocompatibility, and pH lowering (Ruel-Gariepy et al., 2004) (Schmaljohann et al., 2006).

2.3. Photo Sensitive Polymers

Photo-sensitive polymers are those that alter their phase when exposed to light. Biodegradable, water-soluble, and photo-sensitive polymers are all available. As a result, they are frequently used in engineering and biomedicine. Light-responsive polymers have gained particular attention because they control the location and timing of the release. Although these polymers can be affected by UV (Singh et al., 2022) or visible light, the latter is chosen due to its safety, affordability, availability, cleanliness, and convenience of use (Qiu et al., 2001). Table 3 depicts way of crosslinking with applications of different light-responsive polymers. A variety of photo-sensitizers, including triphenylmethane, stilbene, and azobenzene, are used in the fabrication of these polymers. Hydrogels made of poly(N-isopropyl acrylamide) and Light-sensitive chromophores, like copper chlorophyllin trisodium salt are examples of visible light-responsive hydrogels. Chemically adding photoresponsive moieties to a hydrogel is an easy way to create a photoresponsive hydrogel. This method was used to create a poly (2-hydroxyethyl methacrylate) hydrogel that was functionalized with pendant azobenzene groups. With the use of UV light, this hydrogel's swelling was reduced; visible light may now restore it.

2.4. Electro-Sensitive Smart Polymers

Electro-sensitive polymers with modified physical characteristics in response to a slight alteration in electric current. Because they transform electrical to mechanical energy, electro-responsive polymers are frequently employed in the domains of organized medication distribution, energy transductions, sound attenuation, and artificial muscle actuations. In the support chain of these polymers, there is a

Table 3. Named of light-responsive groups used in polymers and their corresponding responsiveness and applications

Polymers	Light Sensitive moiety	Types of photo crosslinking	Application
Polyester	Graphene Quantum dots	UV-light	Drug delivery
Polyacrylamide	Azobenzene acrylamide	UV/Visible light	Drug delivery
Polymethyl methacrylate(PMMA)	1,3-Dihydro-1,3,3-trimethylspiro[2H-indole-2,3′-[3H]naphth[2,1-b][1,4] oxazine]	UV light	Smart sensor
Polyethylene	4,4-Dihydroxybenzene	UV/Visible light	Actuators based optical devices
Poly ethylene co vinyl acetate	Graphene	UV/ visible	Shape memory polymer

comparatively high concentration of ionisable groups. Most of the time, polymers that are electro-responsive also show pH responsiveness. A tiny change in electric current results in a pH change, which disrupts the hydrogen bonds that hold polymer chains together. This causes the polymer chain to degrade or bend, which releases the medication. Electro sensitive hydrogels often contract or expand when exposed to an electric field, and this property has made it possible to use them as biomimetic actuators, artificial muscles, and drug delivery devices (Shiga., 1999).

2.5. Magnetic-Sensitive Biopolymers

The process of magnetic targeting involves luring magnetic nano and micro-particles to a magnetic field of an external source. A delivery system's magnetic-responsive design takes shape, field strength, and carrier characteristics into account. The carrier contains medications, which are administered intravenously or intra-arterially, and focused at tumor locations using high-gradient external magnetic fields (Bawa et al., 2009).

2.6. Multi-Stimuli Based Biopolymers

Dual- or multi-stimulus sensitive polymers react to multiple stimuli simultaneously, using hydrophobic, and ionization functional groups. They can be created by copolymerizing monomers, combining polyelectrolytes, or developing new monomers (Gil et al., 2004). A pH-responsive polymer and temperature responsive polymer (N-isopropylacrylamide-co-methacrylic acid) may be created by combining temperature- and pH-sensitive monomers, such as poly (ethylene oxide) (You et al.,

2010). Because to the advanced level of difficulty of the polymer, the inclusion of one more stimulus is very beneficial because it can increase precision, widen the switching window, or even alter the switching circumstances. PNIPAM was functionalized by Sumaru et al., 2004 using spirobenzopyran. The resulting polymer's responsiveness to temperature, light, and pH made it a logic gate, whereas the PNIPAm based polymer delivered the common thermo-responsive actions.

2.7. Self-Healing Biopolymers

Self-healing polymers, with high water levels and controlled rheological characteristics, are ideal for biomedical applications. These hydrogels replicate extracellular matrix, making them strong candidates for bioelectronics devices, strain sensors, cell/drug/ protein transporters, and wound dressings. Table 4 lists two interaction methods and their applications.

Table 4. Name of self-healing polymers based on tissue interaction, properties and applications

Interaction	Polymers	Properties	Applications
Physical	Polyaniline	Mechanically robust	Wearable Electronics
	Polyglycerylamine	Greater diffusion ability	Dermal Drug Delivery
	Poly styrene-acrylic-acid	Higher mechanical strength	Artificial skin or cartilage fabrication
Covalent bond formation	Polyurea	Highly cytocompatable	Flooring in hospital premises
	Hydroxyapatite with chitosan	Stop crack propagation at implant site	Cartilage regeneration

2.8. Shape Memory Polymers(SMPs):

SMPs are polymeric materials that respond to external stimuli like heat, chemicals, pH levels, or light to provisionally fix many shapes before returning to their initial structure (Pethe et al., 2019). Figure 4 shows a schematic of the shape memory scaffold due to external stimuli.

Supramolecular interactions or dynamic covalent connections frequently cause SMPs to exhibit shape memory behaviour. SMPs with promise in biomedical applications include those that respond to external stimuli under physiological settings, as shown in table 5.

Figure 4. Shape memory scaffold due to external stimuli

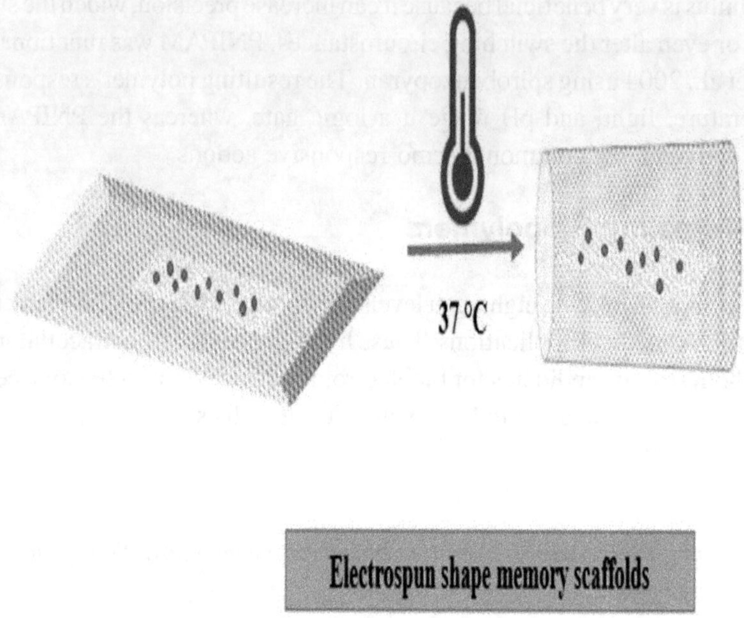

Table 5. Name of Shape healing polymers based on properties and application

External factor	Polymers	Properties	Applications
Temperature Sensitive	Polyurethane	Biodegradability	Shape memory scaffold
	N.N-dimethylacrylamide	Strong interface with the host tissue	Artifical intervertebral disk
pH sensitive	Alginate	Bone adhesive	Bioglue
Water Sensitive	Cellulose	Durable and greater mechanical stability	Pressure sensor

3. SMART BIOMATERIALS DESIGN

Designing of materials which are bio-functional is crucial for latent uses of hybrid smart biomaterials, which are made up of both synthetic and natural polymers. Those can be greatly improved by a variety of design strategies. With the use of these procedures, bio-recognition moieties can be inserted into the polymer's structure, which affects how they self-assemble into specific 3D structure forms. Designing functional monomers, feed ratio, polymerization method, molecular architecture, and crosslink networks significantly impact network structure and the properties of the smart polymer-based biomaterials. Controlling certain chemical and physical constraints throughout design, synthesis, processing, and fabrication is necessary for the development of sophisticated functional biomaterials. 3D network structure smart material properties influence polymers, monomers, oligomers, and production techniques. There are numerous approaches available for producing biomaterials of 3D networks, notably hydrogels having various characteristics and structures, including radiation-induced crosslinking, chemical crosslinking, and physical crosslinking. Covalent bonds between physical interactions or polymer chains like hydrogen bonding and ionic bonding are what create the 3D networks in hydrogels. The crosslinked network creates a condition of equilibrium between cohesive and dispersive forces between polymer chains, causing the gel to be insoluble in liquid. These materials can be used as biomaterials, delivery systems, scaffolds for functional tissue regeneration, and medicinal devices. In particular, swelling is encouraged by the positive change in thermal energy that takes place when a polymer and solvent are mixed, and it is discouraged by the negative change in thermal energy that results from the conformations of the polymer chains (Singh et al., 2019). The synthesis of new structural smart materials can also be accomplished through the physical mixing of several polymers that have been thermally or solution-processed (Singh et al., 2023), but this method still preserves the unique features of the individual polymers (Khan et al., 2009). This system has a number of benefits, one of which is how simple it is to regulate the creation and processing of smart devices. This method achieves desired qualities like porosity, pore diameters, biodegradability, swelling ratio, and mechanical properties by adjusting process variables, polymer feed ratios, and solvent choice. Copolymerizing a variety of useful building blocks while regulating their hydrophobicity, biocompatibility, biodegradability, mechanical characteristics, and receptivity to stimuli outside of their immediate environment, scientists can design complex polymeric biomaterials for biomedical and industrial applications. Future material scientists will keep looking at these systems in an effort to make advances in science on a wider scale. Multidisciplinary methodologies enable rational design and fabrication of smart biomaterials, involving polymer chemists, tissue engineers, biomaterials researchers, and surgeons.

4. APPLICATION OF SMART BIOMATERIALS

Smart polymer applications in precision medicine Smart polymers can be used to create scaffolds or stents that can accommodate various cell types for cell treatment by having tuneable mechanical strength, accurate geometries, and eco-friendly reactivity. Smart polymers with stimuli responsiveness can be created as trigger-release carriers for proteins, drugs, and cells. One fabrication technique to reduce the gap between materials and cell therapy is additive manufacturing (AM). Thus, drug delivery, medical devices, tissue engineering, 3D printing, and immunological engineering will be covered in the sections that follow as promising techniques to help make precision in biomedical application.

4.1. Smart Materials for Medical Devices and Drug Delivery

The items designed to be used for medical purposes are known as medical devices. Scaffolds have been extensively researched among them for decades. Fabricating scaffolds is mostly done to repair or replicate damaged tissues or organs or extracellular matrix (Asghari et al., 2017). In practically every tissue, major advancements have previously been made, including heart valves (Fallahiarezoudar et al., 2015), the brain (Kajte et al., 2021), the retina (Singh et al., 2018), the tracheal tissue (Hsieh et al., 2018), and the skin (Ge et al., 2016). Figure 5 lists the ways in which smart materials work in applications involving drug administration. These medical gadgets are adaptable for use in biomedical applications thanks to the development of smart polymers. For instance, drug carriers of stimuli-responsive types permitted the spiral or temporal regulation of medication discharging in the sick tissues (Ramasamy et al., 2017), and for a cardiac patch, biodegradable and flexible scaffolds of shape memory types were engaged. (Montgomery et al., 2017). In general, smart materials show promise for producing specialised medical equipment.

4.2. Tissue Engineering Using Smart Biomaterials

Tissue engineering combines medicine, chemistry, biology, engineering, and material science to improve tissue function (Vacanti et al., 1999). Figure 6 summarizes smart biomaterials in tissue engineering and biomedicine, highlighting their potential but overcoming barriers for clinical translation (Khademhosseini et al., 2016). The development of tissue engineering biomaterials are hindered significantly by the absence of a important knowledge of how the material interacts with cells or tissues. Therefore, creating materials that can control cellular behaviour is quite challenging. Intelligent biomaterials enhance tissue engineering by incorporating specific functional groups, promoting cell adhesion, growth, and migration. (Ruskowitz et al., 2018). The two pertinent tactics are nature- and bioinspired approaches (Magin et al., 2016) or click-based orthogonal methods (Truong et al., 2015), and as a result, the synthetic methods used for the creation of smart biomaterials must be carefully valuated

Figure 5. Overview of smart materials in Drug Delivery and Medical Devices applications

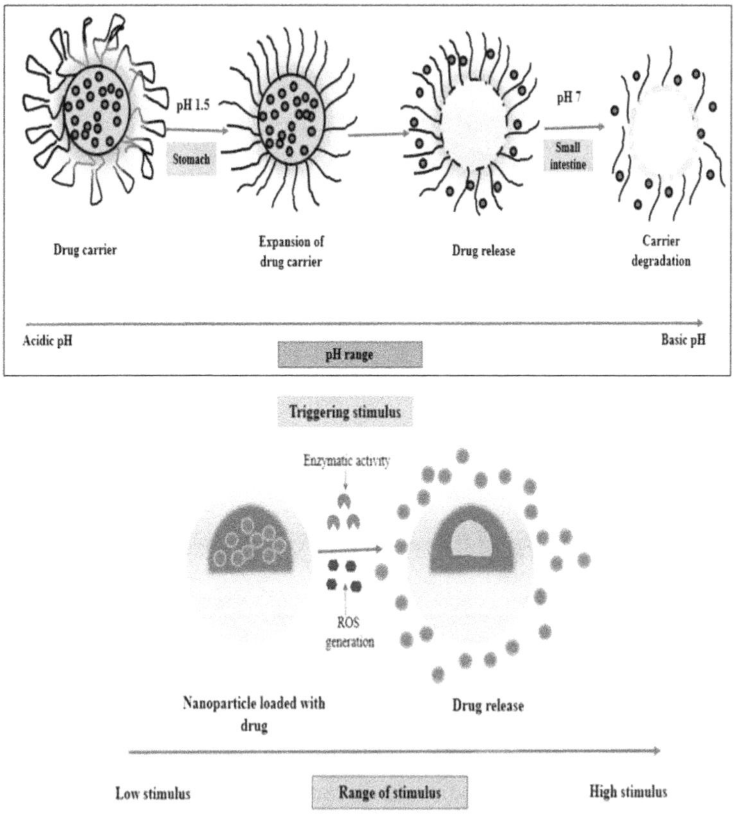

4.3. 3D Printing

The combination of 3D printing and smart biomaterials holds great promise for advancing tissue engineering, regenerative medicine, and personalized healthcare (Sehrawat et al., 2022) (Barua et al., 2020). It allows researchers and clinicians to create patient-specific tissues and organs with tailored properties, bringing us closer to the era of functional organ replacements and improved patient outcomes (Banerjee et al., 2022). 3D printing, also known as additive manufacturing can be aided by smart technologies that opens up new possibilities for creating complex functional tissues analogues and organs. Formulating smart bioink with incorporating cells, growth factors, and other bioactive molecules can be processed for native, patient specific tissue development that can possess stimuli-responsive properties, enabling controlled release of therapeutic agents or triggering specific cellular responses

Figure 6. Overview of smart biomaterials in the field of biomedical and tissue engineering application

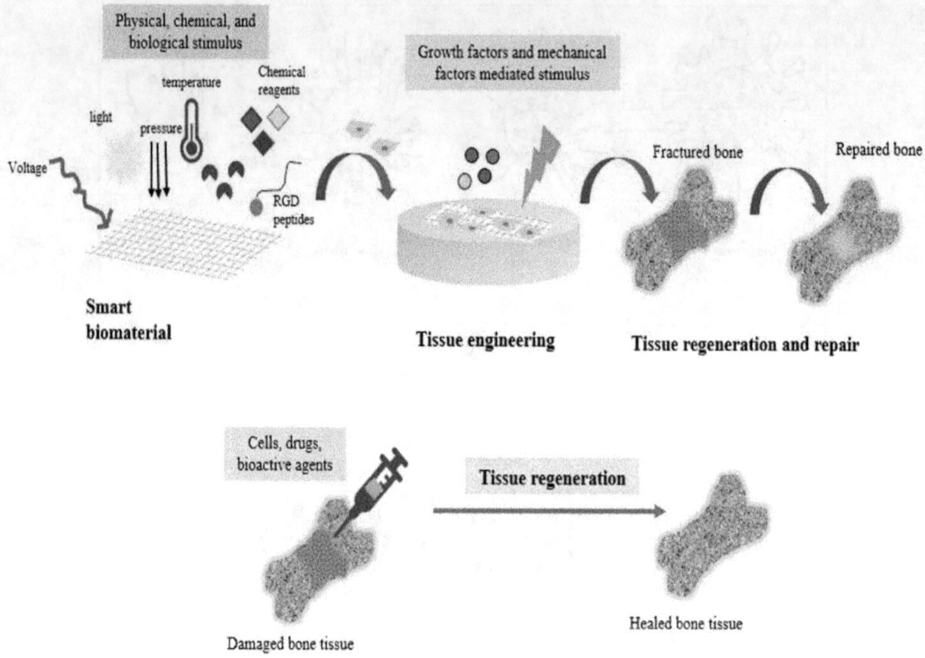

(Barua et al., 2022) (Sehrawat et al., 2022). Further, Smart materials processing by 3D printing offers numerous benefits and opens up exciting possibilities in various fields. Here are some key roles of smart materials processing through 3D printing:

1. Multi-Material 3D Printing allows the creation of complex architectures for simulating human-scale tissues with multiple cell types, vascular networks, and biomaterial gradients, closely mimicking engineered tissue environment by modulating the immune response (Lee et al., 2017).
2. 3D bioprinting of Extrusion-based is a method which enables high cell density to be incorporated in the aforementioned smart materials (Ozbolat et al., 2016), in which cell survival is greatly influenced by the crosslinking processes and mechanical characteristics of bioink (Guvendiren et al., 2016).
3. Responsive and Adaptive Structures: Smart materials possess properties that can respond to external stimuli such as temperature, light, pressure, or electric fields. By integrating these materials into 3D printing, it becomes possible to create structures and devices that can adapt, change shape, or exhibit specific behaviors based on the surrounding conditions. This capability has applications in areas like robotics, biomedical devices, and wearable technology.

4. Prototyping and Rapid Iteration: 3D printing has revolutionized the prototyping process by allowing quick and cost-effective fabrication of physical models. By incorporating smart materials into the printing process, engineers and designers can rapidly prototype and iterate on products that require functional elements. Similarly, self-sensing materials can provide real-time feedback on structural integrity and enhancing quality control in biomedical applications.
5. Biomimicry and Bioinspired Design: Smart materials processing through 3D printing allows for the creation of structures and materials that mimic natural systems to develop innovative solutions like advanced filtration systems.

It has been claimed that stem cell-seeded bioinks made of collagen and agarose sustain cell viability and promote cell differentiation (Duarte Campos et al., 2015). Combining stem cells and bioprinting allows for the customization and design of smart materials. Materials and stem cells are depicted in the schematic of transplantation and cell treatment to control cell growth or differentiation. Figure 7 elaborates on integrating stem cells and smart materials for therapeutic applications in bioprinting. For the past ten years, smart materials have had the ability to fulfil stem cell differentiation by changing their properties in response to external stimulations (Dawson et al., 2008 & Singh et al, 2008). Hence, Smart biomaterials, combined with 3D printing technology, have revolutionized the field of regenerative medicine and tissue engineering.

Overall, the role of smart materials processing by 3D printing is to expand the capabilities of additive manufacturing, enabling the creation of functional, responsive, and customizable patient specific implant with enhanced biocompatibility. This integration has the potential to transform healthcare industries to electronics and consumer products.

4.4. Immune Engineering Using Smart Biomaterials

The immune system is crucial for maintaining homeostasis and treating diseases, but biomaterials can cause adverse reactions, hindering transplants, medical devices, and medication delivery systems. Lack of information on these interactions is a major issue. Intelligent biomaterials enable immune system manipulation, crucial for material-based implants (Tang et al., 2018), showed how antigen-specific T lymphocytes may be used to create protein nanogels interleukin-15 that are reactive to changes in the surface reduction potential of the T cell that brought on by antigen which recognize in the tumour microenvironment. Figure 8 shows an overview of smart materials in immune engineering applications with the nanogel.

Figure 7. Schematic of using stem cells and smart materials together for therapeutic applications in bioprinting

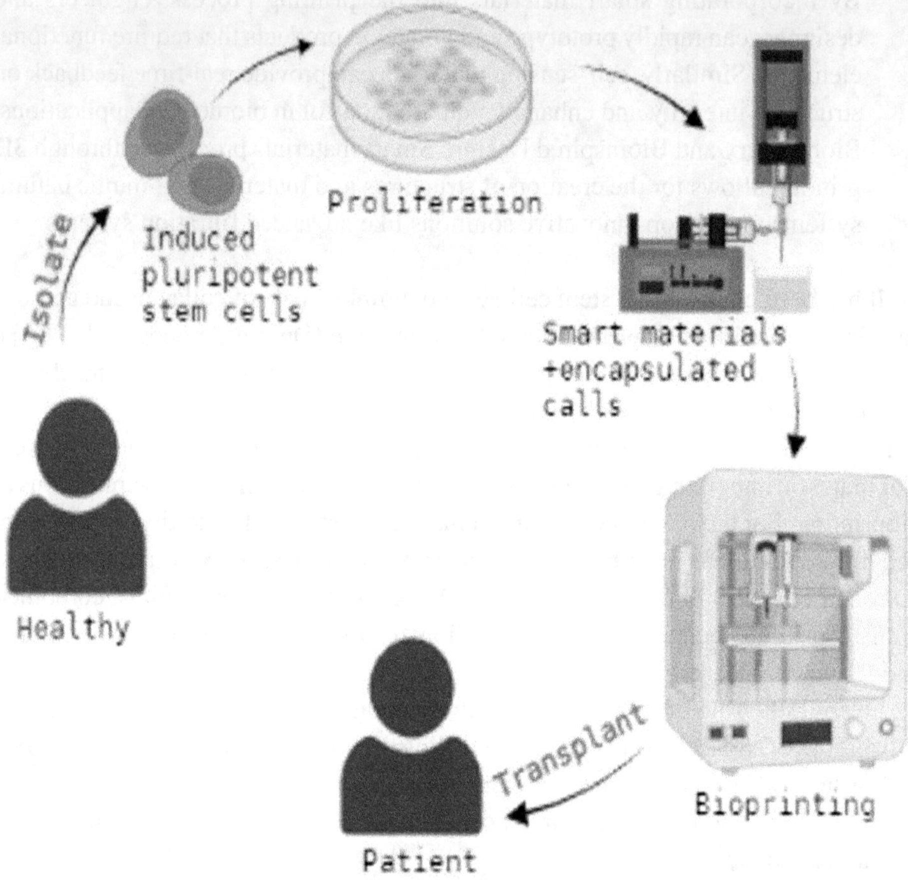

5. CHALLENGES AND FUTURE DIRECTIONS

5.1. Challenges

Smart materials aim to develop biocompatible, responsive materials for tissue engineering, medicine delivery, immunological engineering, and medical instruments. These materials should interact with tissues and cells, provide tissue-specific, stimuli-responsive delivery, and facilitate targeted immune cell distribution to prevent undesirable reactions or inflammation. It can be difficult to create smart materials that combine a number of these attributes. For instance, it can be difficult to create a smart material with the best mechanical, chemical, and biological qualities without impairing those same properties. The advantages of smart materials made from natural products include biodegradability, biocompatibility, cell recognition sites, and frequently the best physico-chemical characteristics. Yet, laborious purification processes and potential immunological reactions are difficulties related to the utilisation of biologically produced materials like collagen. Synthetic biomaterials can overcome challenges by providing desired properties, but lack biocompatibility and can have long-term negative effects. Future smart materials should understand complex signaling pathways and develop immune-protective biocompatible materials using biorthogonal and biocompatible synthesis techniques, natural building blocks, and degradability features. This will aid in rational design and economically viable manufacturing of smart biomaterials.

Figure 8. Overview of smart materials in Immune engineering applications

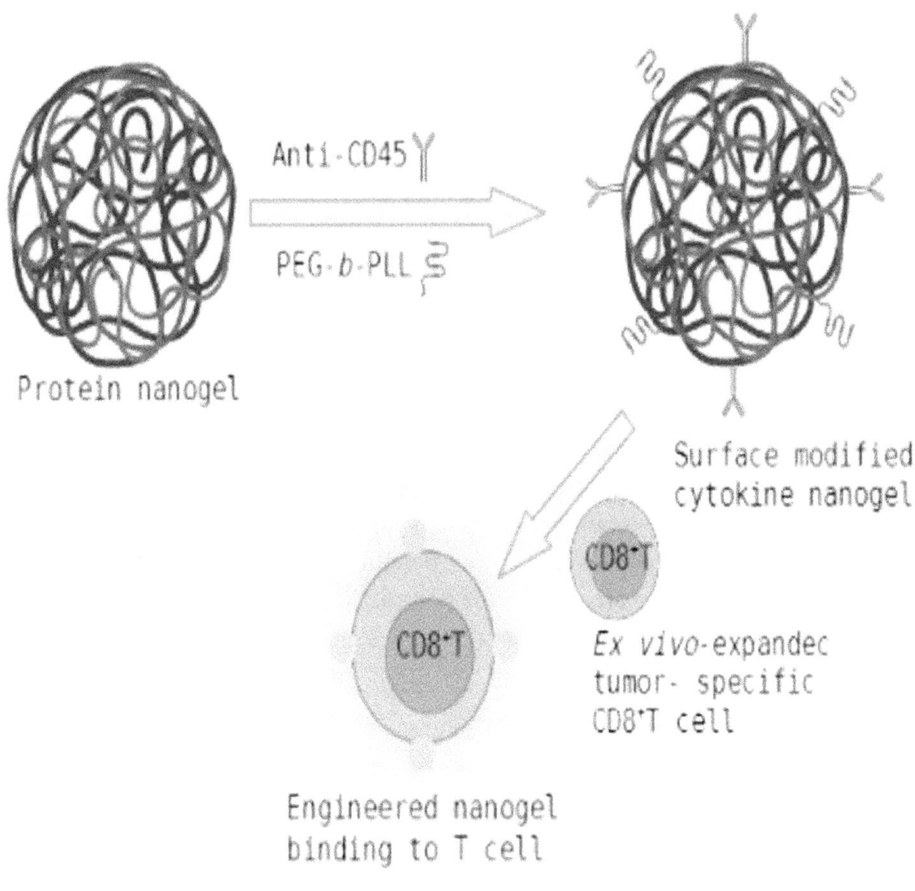

5.2. Future Scope

Smart materials have a promising future in biological applications, with a wide range of possible breakthroughs. Smart materials have special qualities and functions and are made to react to outside stimuli like temperature, pH, light, or magnetic fields. Based on these stimuli, we may take them into account for various biological applications with a variety of innovative possibilities. In order to modulate the release of bioactive molecules that can influence the healing process with the body environment, smart materials-based hydrogel, for instance, can respond to specific cues, such as enzymatic activity or mechanical stress, or it can respond to changes in pH or temperature at the target site. Further smart materials based scaffold can be mimic the natural extracellular matrix can provide mechanical support by tailoring the scaffold properties to support cell growth, differentiation, and tissue regeneration. Hence Smart materials have the potential to improve the performance and functionality of implants and prosthetics by incorporating bioactive coatings and surface treatment. In addition, with these materials can be useful for biosensor and diagnosis instrument development for continuous health monitoring. With ongoing advancements in smart biomaterials we can expect further innovations and breakthroughs in the biomedical field towards affordable heath.

6. CONCLUSION

Smart polymeric biomaterials have gained significant attention in applications in biomedical engineering because of their diverse characteristics and benefits. These materials are crucial for cell treatment, tissue engineering, and medical devices. They are perfect for control-release systems and self-healing injectable hydrogels due to their stimulus reactivity, self-healing properties and shape memory behaviour. 3D printing is essential for combining cell therapy with intelligent polymeric materials, allowing for the creation of bioink infused with patient stem cells. These three elements are essential for creating smart polymeric materials for cell therapy and precision medicine.

REFERENCES

Ajay, P. Ahmad, S., Sharma, J., & Gambhir, V. (2023). Handbook of Sustainable Materials: Modelling, Characterization, and Optimization (1st ed.). CRC Press. doi:10.1201/9781003297772

Almeida, H., Amaral, M. H., & Lobão, P. (2012). Temperature and pH stimuli-responsive polymers and their applications in controlled and selfregulated drug delivery. *Journal of Applied Pharmaceutical Science*, 01-10.

Asghari, F., Samiei, M., Adibkia, K., Akbarzadeh, A., & Davaran, S. (2017). Biodegradable and biocompatible polymers for tissue engineering application: A review. *Artificial Cells, Nanomedicine, and Biotechnology*, *45*(2), 185–192. doi:10.3109/21691401.2016.1146731 PMID:26923861

Azimi, B., Maleki, H., Zavagna, L., De la Ossa, J. G., Linari, S., Lazzeri, A., & Danti, S. (2020). Bio-based electrospun fibers for wound healing. *Journal of Functional Biomaterials*, *11*(3), 67. doi:10.3390/jfb11030067 PMID:32971968

Banerjee, A., Datta, S., Das, A., Chowdhury, A. R., & Datta, P. (2022). A Micro-Scale Non-Linear Finite Element Model to Optimize the Mechanical Behavior of Bioprinted Constructs. *3D printing and additive manufacturing, 9*(6), 490-502.

Barker, S. L., Ross, D., Tarlov, M. J., Gaitan, M., & Locascio, L. E. (2000). Control of flow direction in microfluidic devices with polyelectrolyte multilayers. *Analytical Chemistry*, *72*(24), 5925–5929. doi:10.1021/ac0008690 PMID:11140758

Barua, R., Das, S., RoyChowdhury, A., & Datta, P. (2023). Simulation and experimental investigation of the surgical needle deflection model during the rotational and steady insertion process. *The International Journal of Artificial Organs*, *46*(1), 40–51. doi:10.1177/03913988221136154 PMID:36397288

Barua, R., & Datta, S. RoyChowdhury, A., & Datta, P. (2022). Study of the surgical needle and biological soft tissue interaction phenomenon during insertion process for medical application: A Survey. *Proceedings of the Institution of Mechanical Engineers, Part H: Journal of Engineering in Medicine, 236*(10), 1465-1477.

Barua, R., Datta, S., Datta, P., & RoyChowdhury, A. (2020). Study the Bioprinting Process in Biomedical and Tissue Engineering Fields. *Proceedings of Industry Interactive Innovations in Science, Engineering & Technology (I3SET2K19).*

Barua, R., Datta, S., Datta, P., & Roychowdhury, A. (2022). Study and Application of Machine Learning Methods in Modern Additive Manufacturing Processes. In S. Salunkhe, H. Hussein, & J. Davim (Eds.), *Applications of Artificial Intelligence in Additive Manufacturing* (pp. 75–95). IGI Global. doi:10.4018/978-1-7998-8516-0.ch004

Barua, R., Giria, H., Datta, S., Roy Chowdhury, A., & Datta, P. (2020). Force modeling to develop a novel method for fabrication of hollow channels inside a gel structure. *Proceedings of the Institution of Mechanical Engineers. Part H, Journal of Engineering in Medicine*, *234*(2), 223–231. doi:10.1177/0954411919891654 PMID:31774361

Bawa, P., Pillay, V., Choonara, Y. E., & Du Toit, L. C. (2009). Stimuli-responsive polymers and their applications in drug delivery. *Biomedical Materials (Bristol, England)*, *4*(2), 022001. doi:10.1088/1748-6041/4/2/022001 PMID:19261988

Bawa, P., Pillay, V., Choonara, Y. E., & Du Toit, L. C. (2009). Stimuli-responsive polymers and their applications in drug delivery. *Biomedical Materials (Bristol, England)*, *4*(2), 022001. doi:10.1088/1748-6041/4/2/022001 PMID:19261988

Bazban-Shotorbani, S., Hasani-Sadrabadi, M. M., Karkhaneh, A., Serpooshan, V., Jacob, K. I., Moshaverinia, A., & Mahmoudi, M. (2017). Revisiting structure-property relationship of pH-responsive polymers for drug delivery applications. *Journal of Controlled Release*, *253*, 46–63. doi:10.1016/j.jconrel.2017.02.021 PMID:28242418

Bekas, D. G., Tsirka, K., Baltzis, D., & Paipetis, A. S. (2016). Self-healing materials: A review of advances in materials, evaluation, characterization and monitoring techniques. *Composites. Part B, Engineering*, *87*, 92–119. doi:10.1016/j.compositesb.2015.09.057

Cunliffe, D., de las Heras Alarcón, C., Peters, V., Smith, J. R., & Alexander, C. (2003). Thermoresponsive surface-grafted poly (N– isopropylacrylamide) copolymers: Effect of phase transitions on protein and bacterial attachment. *Langmuir*, *19*(7), 2888–2899. doi:10.1021/la026358l

Darabi, M. A., Khosrozadeh, A., Mbeleck, R., Liu, Y., Chang, Q., Jiang, J., Cai, J., Wang, Q., Luo, G., & Xing, M. (2017). Skin-inspired multifunctional autonomic-intrinsic conductive self-healing hydrogels with pressure sensitivity, stretchability, and 3D printability. *Advanced Materials*, *29*(31), 1700533. doi:10.1002/adma.201700533 PMID:28640439

Datta, S., Barua, R., & Das, J. (2020). Importance of alginate bioink for 3D bioprinting in tissue engineering and regenerative medicine. *Alginates—Recent Uses of This Natural Polymer*, 1-11.

Dawson, E., Mapili, G., Erickson, K., Taqvi, S., & Roy, K. (2008). Biomaterials for stem cell differentiation. *Advanced Drug Delivery Reviews*, *60*(2), 215–228. doi:10.1016/j.addr.2007.08.037 PMID:17997187

DeRossi, D., Kajiwara, K., Osada, Y., & Yamauchi, A. (1991). *Polymer gels*. Fundamentals and Biomedical Applications. doi:10.1007/978-1-4684-5892-3

Duarte Campos, D. F., Blaeser, A., Korsten, A., Neuss, S., Jäkel, J., Vogt, M., & Fischer, H. (2015). The stiffness and structure of three-dimensional printed hydrogels direct the differentiation of mesenchymal stromal cells toward adipogenic and osteogenic lineages. *Tissue Engineering. Part A*, *21*(3-4), 740–756. doi:10.1089/ten.tea.2014.0231 PMID:25236338

Dušek, K., & Patterson, D. (1968). Transition in swollen polymer networks induced by intramolecular condensation. *Journal of Polymer Science. Part A-2, Polymer Physics*, *6*(7), 1209–1216. doi:10.1002/pol.1968.160060701

Fallahiarezoudar, E., Ahmadipourroudposht, M., Idris, A., & Yusof, N. M. (2015). A review of: Application of synthetic scaffold in tissue engineering heart valves. *Materials Science and Engineering C*, *48*, 556–565. doi:10.1016/j.msec.2014.12.016 PMID:25579957

Ge, J., Sun, L., Zhang, F. R., Zhang, Y., Shi, L. A., Zhao, H. Y., Zhu, H.-W., Jiang, H.-L., & Yu, S. H. (2016). A stretchable electronic fabric artificial skin with pressure-, lateral strain-, and flexion-sensitive properties. *Advanced Materials*, *28*(4), 722–728. doi:10.1002/adma.201504239 PMID:26618615

Gil, E. S., & Hudson, S. M. (2004). Stimuli-reponsive polymers and their bioconjugates. *Progress in Polymer Science*, *29*(12), 1173–1222. doi:10.1016/j.progpolymsci.2004.08.003

Guvendiren, M., Molde, J., Soares, R. M., & Kohn, J. (2016). Designing biomaterials for 3D printing. *ACS Biomaterials Science & Engineering*, *2*(10), 1679–1693. doi:10.1021/acsbiomaterials.6b00121 PMID:28025653

Hoffman, A. S. (2012). Hydrogels for biomedical applications. *Advanced Drug Delivery Reviews*, *64*, 18–23. doi:10.1016/j.addr.2012.09.010 PMID:11755703

Hsieh, C. T., Liao, C. Y., Dai, N. T., Tseng, C. S., Yen, B. L., & Hsu, S. H. (2018). 3D printing of tubular scaffolds with elasticity and complex structure from multiple waterborne polyurethanes for tracheal tissue engineering. *Applied Materials Today*, *12*, 330–341. doi:10.1016/j.apmt.2018.06.004

Ista, L. K., & Lopez, G. P. (1998). Lower critical solubility temperature materials as biofouling release agents. *Journal of Industrial Microbiology & Biotechnology*, *20*(2), 121–125. doi:10.1038j.jim.2900490

Kajtez, J., Nilsson, F., Fiorenzano, A., Parmar, M., & Emnéus, J. (2021). 3D biomaterial models of human brain disease. *Neurochemistry International*, *147*, 105043. doi:10.1016/j.neuint.2021.105043 PMID:33887378

Khademhosseini, A., & Langer, R. (2016). A decade of progress in tissue engineering. *Nature Protocols*, *11*(10), 1775–1781. doi:10.1038/nprot.2016.123 PMID:27583639

Khan, F., & Tanaka, M. (2017). Designing smart biomaterials for tissue engineering. *International Journal of Molecular Sciences*, *19*(1), 17. doi:10.3390/ijms19010017 PMID:29267207

Khan, F., Tanaka, M., & Ahmad, S. R. (2015). Fabrication of polymeric biomaterials: A strategy for tissue engineering and medical devices. *Journal of Materials Chemistry. B, Materials for Biology and Medicine*, *3*(42), 8224–8249. doi:10.1039/C5TB01370D PMID:32262880

Khan, F., Tare, R. S., Oreffo, R. O., & Bradley, M. (2009). Versatile biocompatible polymer hydrogels: Scaffolds for cell growth. *Angewandte Chemie*, *121*(5), 996–1000. doi:10.1002/ange.200804096 PMID:19115339

Kikuchi, A., & Okano, T. (2002). Pulsatile drug release control using hydrogels. *Advanced Drug Delivery Reviews*, *54*(1), 53–77. doi:10.1016/S0169-409X(01)00243-5 PMID:11755706

Kim, B. S., Baez, C. E., & Atala, A. (2000). Biomaterials for tissue engineering. *World Journal of Urology*, *18*(1), 2–9. doi:10.1007003450050002 PMID:10766037

Kost, J., & Langer, R. (2012). Responsive polymeric delivery systems. *Advanced Drug Delivery Reviews*, *64*, 327–341. doi:10.1016/j.addr.2012.09.014 PMID:11259837

Kumar, A., Mittal, R. K., & Haleem, A. (Eds.). (2022). *Advances in Additive Manufacturing: Artificial Intelligence, Nature-Inspired, and Biomanufacturing*. Elsevier. doi:10.1016/C2020-0-03877-6

Kumar, A., Srivastava, A., Galaev, I. Y., & Mattiasson, B. (2007). Smart polymers: Physical forms and bioengineering applications. *Progress in Polymer Science*, *32*(10), 1205–1237. doi:10.1016/j.progpolymsci.2007.05.003

Kwon, I. C., Bae, Y. H., & Kim, S. W. (1991). Electrically erodible polymer gel for controlled release of drugs. *Nature*, *354*(6351), 291–293. doi:10.1038/354291a0 PMID:1956379

Lee, J. Y., An, J., & Chua, C. K. (2017). Fundamentals and applications of 3D printing for novel materials. *Applied Materials Today*, *7*, 120–133. doi:10.1016/j.apmt.2017.02.004

Li, S. K., & D'Emanuele, A. (2001). On–off transport through a thermoresponsive hydrogel composite membrane. *Journal of Controlled Release*, *75*(1-2), 55–67. doi:10.1016/S0168-3659(01)00365-0 PMID:11451497

Liu, Y., & Hsu, S. H. (2018). Synthesis and biomedical applications of self-healing hydrogels. *Frontiers in Chemistry*, *6*, 449. doi:10.3389/fchem.2018.00449 PMID:30333970

Magin, C. M., Alge, D. L., & Anseth, K. S. (2016). Bio-inspired 3D microenvironments: A new dimension in tissue engineering. *Biomedical Materials (Bristol, England)*, *11*(2), 022001. doi:10.1088/1748-6041/11/2/022001 PMID:26942469

Montgomery, M., Ahadian, S., Davenport Huyer, L., Lo Rito, M., Civitarese, R. A., Vanderlaan, R. D., Wu, J., Reis, L. A., Momen, A., Akbari, S., Pahnke, A., Li, R.-K., Caldarone, C. A., & Radisic, M. (2017). Flexible shape-memory scaffold for minimally invasive delivery of functional tissues. *Nature Materials*, *16*(10), 1038–1046. doi:10.1038/nmat4956 PMID:28805824

Nakayama, M., Okano, T., Miyazaki, T., Kohori, F., Sakai, K., & Yokoyama, M. (2006). Molecular design of biodegradable polymeric micelles for temperature-responsive drug release. *Journal of Controlled Release*, *115*(1), 46–56. doi:10.1016/j.jconrel.2006.07.007 PMID:16920217

Osada, Y., & Ross-Murphy, S. B. (1993). Intelligent gels. *Scientific American*, *268*(5), 82–87. doi:10.1038cientificamerican0593-82

Ozbolat, I. T., & Hospodiuk, M. (2016). Current advances and future perspectives in extrusion-based bioprinting. *Biomaterials*, *76*, 321–343. doi:10.1016/j.biomaterials.2015.10.076 PMID:26561931

Patel, D. K., & Lim, K. T. (2021). Stimuli-responsive Polymers and Their Biomedical Applications. In Functionalized Polymers: Synthesis, Characterization and Applications (pp. 250-264). CRC Press. doi:10.1201/9780367821913-13

Pattanashetti, N. A., Heggannavar, G. B., & Kariduraganavar, M. Y. (2017). Smart biopolymers and their biomedical applications. *Procedia Manufacturing*, *12*, 263–279. doi:10.1016/j.promfg.2017.08.030

Peppas, N. A. (1997). Hydrogels and drug delivery. *Current Opinion in Colloid & Interface Science*, *2*(5), 531–537. doi:10.1016/S1359-0294(97)80103-3

Pethe, A. M., & Yadav, K. S. (2019). Polymers, responsiveness and cancer therapy. *Artificial Cells, Nanomedicine, and Biotechnology*, *47*(1), 395–405. doi:10.1080/21691401.2018.1559176 PMID:30688110

Qiu, Y., & Park, K. (2001). Environment-sensitive hydrogels for drug delivery. *Advanced Drug Delivery Reviews*, *53*(3), 321–339. doi:10.1016/S0169-409X(01)00203-4 PMID:11744175

Ramasamy, T., Ruttala, H. B., Gupta, B., Poudel, B. K., Choi, H. G., Yong, C. S., & Kim, J. O. (2017). Smart chemistry-based nanosized drug delivery systems for systemic applications: A comprehensive review. *Journal of Controlled Release*, *258*, 226–253. doi:10.1016/j.jconrel.2017.04.043 PMID:28472638

Ruel-Gariepy, E., & Leroux, J. C. (2004). In situ-forming hydrogels—Review of temperature-sensitive systems. *European Journal of Pharmaceutics and Biopharmaceutics*, *58*(2), 409–426. doi:10.1016/j.ejpb.2004.03.019 PMID:15296964

Ruskowitz, E. R., & DeForest, C. A. (2018). Photoresponsive biomaterials for targeted drug delivery and 4D cell culture. *Nature Reviews. Materials*, *3*(2), 1–17. doi:10.1038/natrevmats.2017.87

Schmaljohann, D. (2006). Thermo-and pH-responsive polymers in drug delivery. *Advanced Drug Delivery Reviews*, *58*(15), 1655–1670. doi:10.1016/j.addr.2006.09.020 PMID:17125884

Sehrawat, S., Kumar, A., Grover, S., Dogra, N., Nindra, J., Rathee, S., Dahiya, M., & Kumar, A. (2022). Study of 3D scanning technologies and scanners in orthodontics. *Materials Today: Proceedings*, *56*, 186–193. doi:10.1016/j.matpr.2022.01.064

Sehrawat, S., Kumar, A., Prabhakar, M., & Nindra, J. (2022). The expanding domains of 3D printing pertaining to the speciality of orthodontics. *Materials Today: Proceedings*, *50*, 1611–1618. doi:10.1016/j.matpr.2021.09.124

Shiga, T. (1999). Deformation and viscoelastic behavior of polymer gels in electric fields. *Neutron spin echo spectroscopy viscoelasticity rheology*, 131-163.

Singh, A., & Elisseeff, J. (2010). Biomaterials for stem cell differentiation. *Journal of Materials Chemistry*, *20*(40), 8832–8847. doi:10.1039/c0jm01613f

Singh, D., Wang, S. B., Xia, T., Tainsh, L., Ghiassi-Nejad, M., Xu, T., Peng, S., Adelman, R. A., & Rizzolo, L. J. (2018). A biodegradable scaffold enhances differentiation of embryonic stem cells into a thick sheet of retinal cells. *Biomaterials*, *154*, 158–168. doi:10.1016/j.biomaterials.2017.10.052 PMID:29128844

Singh, V., Gupta, M., & Kumar, A. (2019, November). Investigating the morphology and thermophysical properties of Al2O3 and CuO nanofluids. [). IOP Publishing.]. *IOP Conference Series. Materials Science and Engineering*, *577*(1), 012052. doi:10.1088/1757-899X/577/1/012052

Singh, V., Gupta, M., Kumar, A., Luthra, S., & Kumar, A. (2023). Experimental investigations of thermo physical properties and convective heat transfer of Al2O3 and CuO nanofluids in a copper tube: Proposing New Correlations. *Biointerface Research in Applied Chemistry*, *13*(3), 229.

Singh, V., Kumar, A., Alam, M., Kumar, A., Kumar, P., & Goyat, V. (2022). A study of morphology, UV measurements and zeta potential of zinc ferrite and Al2O3 nanofluids. *Materials Today: Proceedings*, *59*, 1034–1039. doi:10.1016/j.matpr.2022.02.371

Sumaru, K., Kameda, M., Kanamori, T., & Shinbo, T. (2004). Characteristic phase transition of aqueous solution of poly (N-isopropylacrylamide) functionalized with spirobenzopyran. *Macromolecules*, *37*(13), 4949–4955. doi:10.1021/ma049661x

Tang, L., Zheng, Y., Melo, M. B., Mabardi, L., Castaño, A. P., Xie, Y. Q., Li, N., Kudchodkar, S. B., Wong, H. C., Jeng, E. K., Maus, M. V., & Irvine, D. J. (2018). Enhancing T cell therapy through TCR-signaling-responsive nanoparticle drug delivery. *Nature Biotechnology*, *36*(8), 707–716. doi:10.1038/nbt.4181 PMID:29985479

Taylor, D. L., & in het Panhuis, M. (2016). Self-healing hydrogels. *Advanced Materials*, *28*(41), 9060–9093. doi:10.1002/adma.201601613 PMID:27488822

Truong, V. X., Ablett, M. P., Richardson, S. M., Hoyland, J. A., & Dove, A. P. (2015). Simultaneous orthogonal dual-click approach to tough, in-situ-forming hydrogels for cell encapsulation. *Journal of the American Chemical Society*, *137*(4), 1618–1622. doi:10.1021/ja511681s PMID:25590670

Tsang, V. L., & Bhatia, S. N. (2007). Fabrication of three-dimensional tissues. *Tissue Engineering Ii: Basics of Tissue Engineering and Tissue Applications*, 189-205.

Vacanti, J. P., & Langer, R. (1999). Tissue engineering: The design and fabrication of living replacement devices for surgical reconstruction and transplantation. *Lancet*, *354*, S32–S34. doi:10.1016/S0140-6736(99)90247-7 PMID:10437854

Van de Velde, K., & Kiekens, P. (2002). Biopolymers: Overview of several properties and consequences on their applications. *Polymer Testing*, *21*(4), 433–442. doi:10.1016/S0142-9418(01)00107-6

Wang, S., Lee, J. M., & Yeong, W. Y. (2015). Smart hydrogels for 3D bioprinting. *International Journal of Bioprinting*, *1*(1).

Weigel, T., Schinkel, G., & Lendlein, A. (2006). Design and preparation of polymeric scaffolds for tissue engineering. *Expert Review of Medical Devices*, *3*(6), 835–851. doi:10.1586/17434440.3.6.835 PMID:17280547

Wichterle, O., & Lim, D. (1960). Hydrophilic gels for biological use. *Nature*, *185*(4706), 117–118. doi:10.1038/185117a0

Yadav, P., Yadav, H., Shah, V. G., Shah, G., & Dhaka, G. (2015). Biomedical biopolymers, their origin and evolution in biomedical sciences: A systematic review. *Journal of Clinical and Diagnostic Research : JCDR*, *9*(9), ZE21. doi:10.7860/JCDR/2015/13907.6565 PMID:26501034

Yamato, M., Konno, C., Utsumi, M., Kikuchi, A., & Okano, T. (2002). Thermally responsive polymer-grafted surfaces facilitate patterned cell seeding and co-culture. *Biomaterials*, *23*(2), 561–567. doi:10.1016/S0142-9612(01)00138-7 PMID:11761176

Yates, M. R., & Barlow, C. Y. (2013). Life cycle assessments of biodegradable, commercial biopolymers—A critical review. *Resources, Conservation and Recycling*, *78*, 54–66. doi:10.1016/j.resconrec.2013.06.010

You, J. O., Almeda, D., Ye, G. J., & Auguste, D. T. (2010). Bioresponsive matrices in drug delivery. *Journal of Biological Engineering*, *4*(1), 1–13. doi:10.1186/1754-1611-4-15 PMID:21114841

Chapter 4
Materials and Technology for Implant Manufacturing:
Challenges and Opportunity

Himanshu Kumar Tiwari
G.L. Bajaj Institute of Technology and Management, India

Ashish Kumar Srivastava
https://orcid.org/0000-0003-2834-1483
G.L. Bajaj Institute of Technology and Management, India

Parveen Kumar
Mechanical Engineering, Rawal Institute of Engineering and Technology, India

Manish Kumar Singh
G.L. Bajaj Institute of Technology and Management, India

Hritik Kumar
G.L. Bajaj Institute of Technology and Management, India

Akshit Bhadauria
G.L. Bajaj Institute of Technology and Management, India

ABSTRACT

The leg bone that typically undergos the most deformation is the femur. Failures of the femur neck, which are most frequent among human bone failures due to their length and weight, are a very serious issue for people. However, before using biomaterials in implants or any orthopaedic surgery, the right biomaterials must be chosen based on their mechanical and physical properties in comparison to natural bone. The use of biomaterials in the treatment of orthopaedic-related issues is very crucial. By taking into account the mechanical and physical properties of materials, discussions and comparisons of biomaterials are conducted. In comparison to other materials, biomaterials like SS316L, Ti6Al4V, PMMA, and CoCr alloys, magnesium alloys have been found to be a better substitute.

DOI: 10.4018/978-1-6684-9224-6.ch004

1. INTRODUCTION

In human beings the average ratio of femur to height is 25.7% for men and 25.4% for women, meaning that an individual's height is four times that of the femur (Mulenga, 2019). The longest and sturdiest bone in the skeleton, it is virtually cylindrical for the majority of its length. And it is mainly fractured due to major accidents or exerting a lot of force on the femur bone (Amalraju & Dawood, 2012)]. The components of the femur, a long bone that is made up of a shaft body, diaphysis, two epiphyses, and extremities that are different from nearby hip and knee bones are illustrated in Figure 1.

Figure 1. Femur bone anatomy

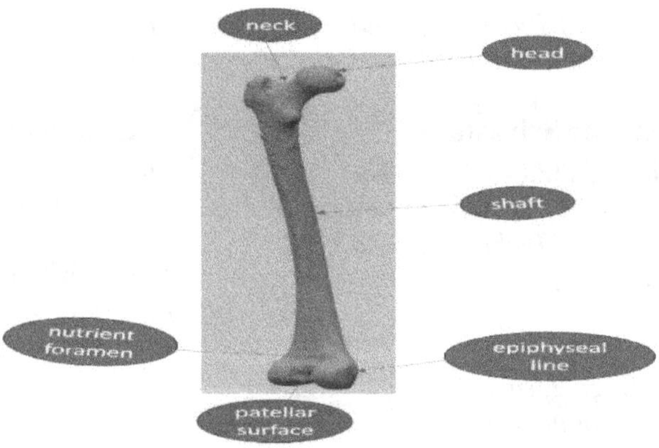

1.1. Implants

An implant is a surgically placed non-living object that serves a specific purpose in the body and is consciously inserted by a surgeon into the area where it will stay for a considerable amount of time (Amalraju & Dawood, 2012). In order to support a damaged bone, repair a missing joint, or replace a missing bone, orthopaedic implants are employed. Currently cemented and cement-less implants are generally used in the orthopaedic surgery; as far as we are known, bone cement is not used to anchor dental implants. Even though bone cement, acting as a bonding medium, can provide initial fixation, cement-less implants are growing in popularity due to the risks of cemented implant failures brought on by an accumulation of micro cracks in and around the cemented area (Dodd et al., 1990).The use of implants in different body areas are illustrated in Figure 2.

Figure 2. Implants used in various fields

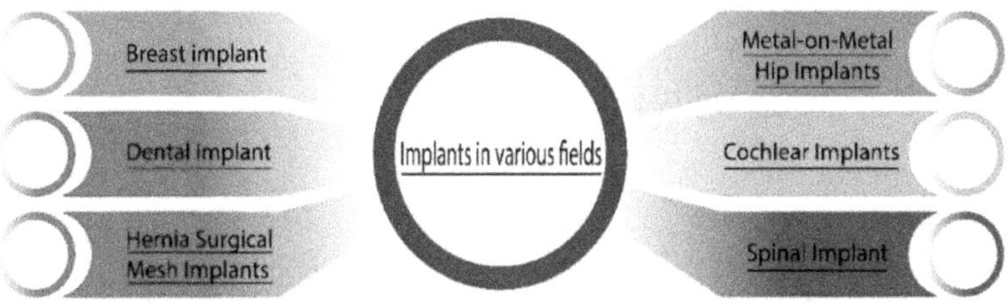

Around 70–80% of implants are made of biomedical metallic implants. In order to improve the patient's overall quality of life, metallic biomaterials are absolutely essential for the reconstruction of failed tissue, particularly failed hard tissue (Niinomi et al., 2012). As is well known, both the population and the number of accidents are increasing day by day, and this is what causes the major bone injuries. The main method of treating severely damaged bone is implantation, which required suitable biomaterials (i.e. non-toxic, non-corrosive, bio compatible, etc.). The mainly used bio materials in implants are Titanium alloys, Stainless Steel alloys, CoCr alloys, etc. niobium (Nb), Iron (Fe),tantalum (Ta), and other metallic biomaterial- like (Mg) magnesium alloys are also significant, despite having a relatively small market share in this area (Niinomi et al., 2012) .The most suitable bio materials are shown in the Figure 3.

The present work deals with the review of previously published work of several authors working in this field. It studied about the mechanical properties, types of biomaterials for implants their mechanical properties. It also discussed about different methods used for developing of implants of femur bone through different technique. At last but not least discuss the future of femur bone implants, their scope of utilization, and noble methods for the development of femur bone implants.

2. MATERIALS AND TECHNOLOGY

The aim of this review is to examine alternative biocompatible materials for implanting human femur bones. It is anticipated that these implants won't induce inflammation or refusal and will be biocompatible. Further, it is aimed a better understanding of the mechanical properties of various alternative implant materials, as well as a comparison of the results with existing scientific literature. We'll recommend the best alternative material for orthopaedic implants made of human tissue.

Figure 3. Most suitable bio material used in implants

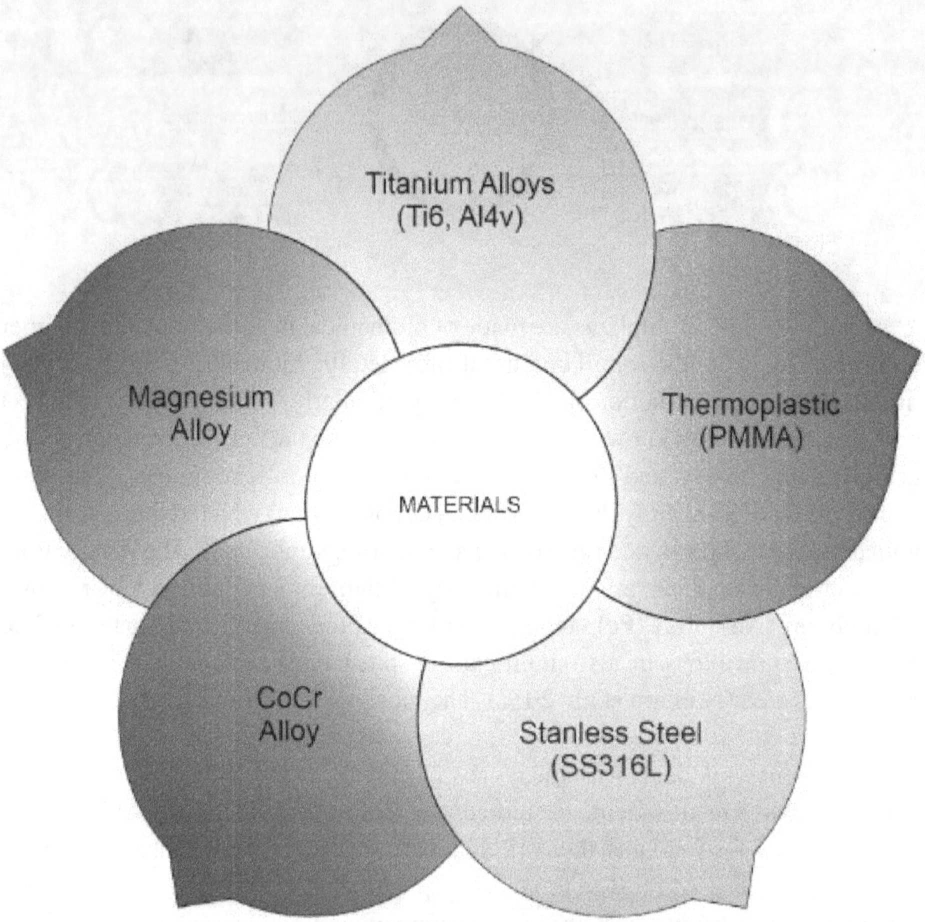

2.1. Bio Materials

A substance that has been used in biological systems for therapeutic purposes (especially in medical implants) is referred to as a "bio material." Due to its capacity as a material to perform for a given application, biomaterials, whose use had begun roughly 50 years earlier, are also in high demand today. Medical devices that interact with biological systems use biomaterials. In fact, a balancing definition of "bio compatibility" is required to comprehend the purpose of bio materials (Ratner et al., 2004; Ajay et al., 2023). The reaction of the living tissues to the disturbance caused by the foreign substances is related directly to biocompatibility of a biomaterial (Nalwa, 2007). The site, purpose, and length of the implant are all important factors in determining bio compatibility, with the amount of time needed for the material host tissue relations to develop being a major concern (Anselme, 2000).

The bio materials are classified broadly in two categories, the natural bio material and the man-made bio material (Kumar et al., 2022; Singh et al.,2023) and further the man-made materials is classified into two categories, organic and inorganic as shown in Figure 4. And in this study we are dealing basically with metallic bio materials which are a part of inorganic type bio materials.

Figure 4. Classification of bio materials

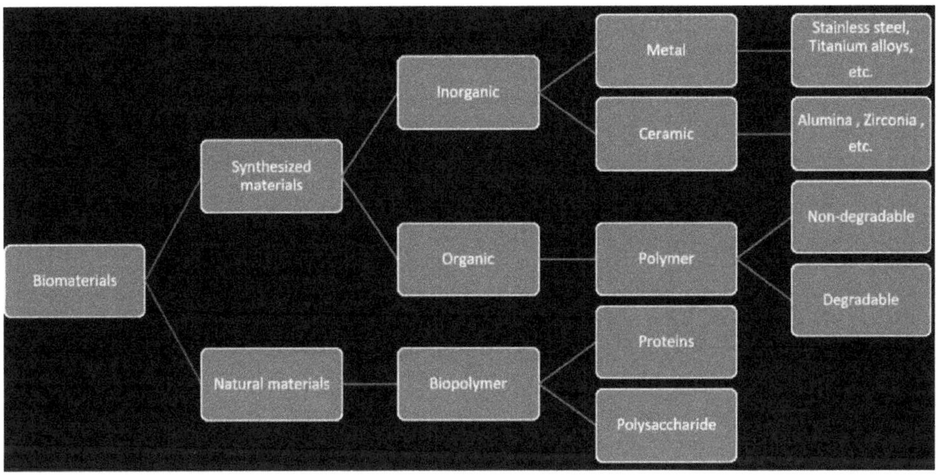

Four things are necessary for biomaterials to be successful (Patel & Gohil, 2012):

(1) Tribological, chemical, and mechanical aspects of biomaterials
(2) Biocompatibility material implanted
(3) The host's reaction to the implanted material
(4) The surgeon's knowledge of specific biomaterials

According to human anatomy, each part of the body serves a different purpose, so the type of implant material will be selected based on the purpose of the part, for example, fractured bones require metallic material, while vascular grafts and stents require polymeric material (Harun et al., 2018). Table 1 explains the applications of different biomaterials.

Applications of different bio materials are illustrated in Figure 5.

Table 1. Bio materials applications

Materials	Applications	References
Stainless Steel	For the replacement of the hip and knee joints, femoral stems, fracture plates, screws, and heads are used.	Amalraju & Dawood, 2012;, Santos, 2017
Titanium alloys (Ti6Al-4V)	Joint replacement, Dental implants devices, bone-fixing screws and pins, and trabecular bone	Long & Rack, 1998; Luo et al., 2013; Gaviria et al., 2014; Li et al., 2014
Co-Cr Alloys	dental work, a weight-bearing element of a hip replacement, and prosthetic stems.	Arden et al., 1970; Peter & Rosso, 2015
Magnesium alloys	Application in load bearing, coronary stents, bone screws, and orthopaedics is still being tested.	Li & Zheng, 2013;, Poinern wt al., 2012; Tan et al., 2018
PMMA	Ocular lens, Bone cement.	

Figure 5. Applications of bio materials in different sector

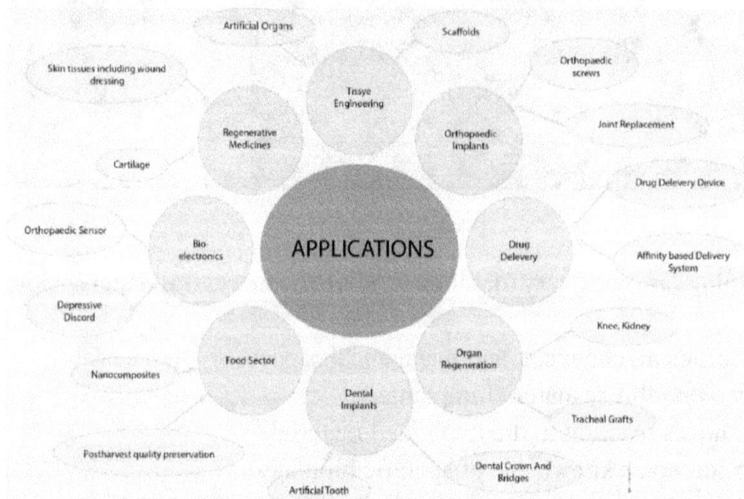

2.2. Materials for Metallic Implant

To treat fractures and bone deficiencies, it is crucial to develop medical implants. Over the past ten years, there has been a significant increase in the demand for implants. Implants need to be made more resilient. Metals are essential part of material in the fracture treatment due to their strength and ductility (Dhanopia &

Bhargava, 2017). Metals have advantage of load-bearing applications because of their fracture toughness and high mechanical strength. Since they quickly corrode and/or are not biocompatible, most of the pure metals are useless (Gu et al., 2009; Binyamin et al., 2006). Gold and silver were once thought of as inert biomaterials, but they are more expensive and have bad mechanical qualities. Metal alloys like titanium, CoCrMo, and stainless-steel alloys are frequently used (Akahori, 2002). Materials that deal with living tissues must have appropriate mechanical properties, non-toxicity, high corrosion resistance, and recognition by living tissues (Wanhill & Barter, 2011).

2.2.1. Mg Alloys

Due to its mechanical and physical characteristics, including specific strength, density, and Young's modules, a promising bio metal in the field of bio-engineering is magnesium and its alloys. It also exhibits good bio compatibility (Sanchez et al., 2015; Zheng et al., 2014; Niu et al., 2016). Magnesium and its alloys have a Young's modulus of around 45 GPa, whereas the Young's modulus of human bone is between 15 and 30 GPa. When compared to the density of human bone, which is 1.75 g/cm3, it has a density concentration of 1.79 g/cm3. In addition to having good mechanical characteristics, magnesium and its alloys can totally disintegrate in vivo, which is an appealing degradation property (Zhang et al., 2010). It is anticipated that Magnesium bone implants will serve their clinical purpose as temporary replacements flawlessly by initially providing stable mechanical support and gradually deteriorating as missing bone tissue is replaced (Yang et al., 2020).

Magnesium, the body's fourth-most abundant mineral, is essential for the growth of both bone and soft tissue. A healthy adult needs 21–28 g of magnesium to sustain normal bodily processes. A healthy adult should consume 250–350 mg of magnesium per day (Khayat et al., 2017). The human body is able to withstand high Mg concentrations, and extra ions are eliminated through urine without causing any known harm to the liver or kidneys. By increasing osteoblastic activity close to the implant, magnesium may promote the development of new bone (Ghosh et al., 2018). The use of magnesium and magnesium alloys as scaffold materials for bone tissue engineering is also thought to be promising (Yazdimamaghani et al., 2014). In the procedure known as bone tissue engineering, bone marrow stromal cells, chondrocytes, or autologous osteoblasts, are transplanted into an artificial scaffold that has excellent bio compatibility and can be gradually broken down and absorbed by the human body (Stevens, 2008; Gupta et al., 2016). The main drawback of magnesium is that it corrodes quickly in the body, which restricts its use in biocompatible applications (Chen et al., 2012; Kunchur et al., 2015; Kunchur et al., 2013; Rahmani et al., 2011)).

Due to their attractive mechanical and osteopromotive characteristics, magnesium based biomaterials have been employed in implant applications for more than a century. Additionally, their advantages over comparable standard products have been thoroughly studied, with magnesium and magnesium alloys, bioceramics, bioglasses, and polymer composite materials (Emami et al., 1991; Nabiyouni et al., 2018; Wang et al., 2020) exhibiting a special dominance in the advancement of bone growth and fracture repairing. These novel Mg and Mg alloys biomaterials are a great option to overcome the refracture constraint because their Young's modulus values are equivalent to those of human cortical bone (Wik et al., 2010; Staiger et al., 2006). Mg-based bio glasses and bio ceramics have the potential to serve as another biodegradable bone implant since they gradually deteriorate and are replaced by new bone (Nabiyouni et al., 2018).

The Figure 6 explains different aspects of Magnesium bone implants like application areas, advantages of using Magnesium bone implants, different fabrication methods and many more.

Figure 6. Application, advantages, etc. of Mg bone implants

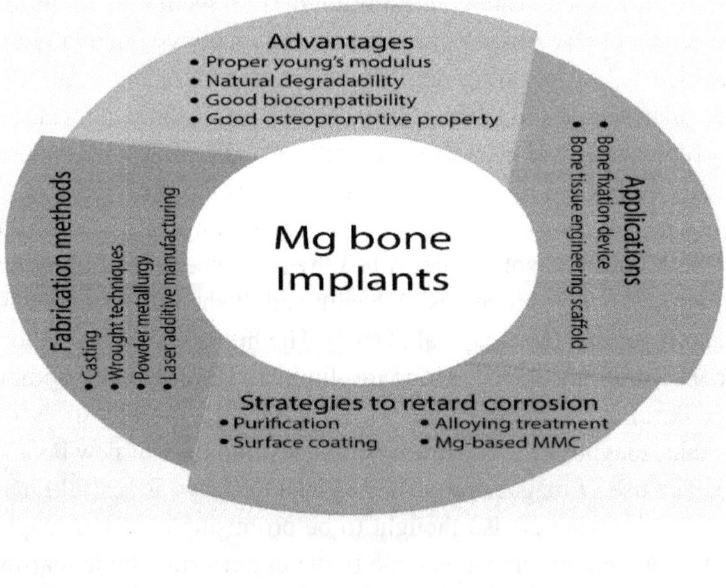

2.2.2. Titanium Alloys

In the past, biomedical applications accounted for the majority of the use of titanium alloys. Titanium (pure titanium) and its alloys are more biocompatible than the main metallic biomaterials, like Co-Cr alloys and stainless steels. Titanium alloys also have a high specific ratio and excellent corrosion resistance (Niinomi et al., 2005; Kumar et al., 2023).This is why Titanium alloys holds an extensive demand in the bio medical field. Replacement using implant devices, it is feasible to treat diseased tissue (Sehrawat et al., 2022a,b). Examples include bone plates, artificial hip joint, artificial hearts, artificial knee joints, and fracture fixation screw. Ti6Al4V (the Titanium alloy) is the most popular alloy used as a biocompatible material. Mainly Ti6Al4V alloy sometimes referred to as Ti64, is a high strength, high fracture rate, and low density titanium alloy. It has high biocompatibility, outstanding corrosion resistance, and hardness characteristics (Donachie, 2000; Cui et al., 2011). Ti6Al4V has appealing properties for applications like bridges and implants, including low density, high corrosion resistance, high strength, and biocompatibility (Cui et al., 2011; Emmelmann et al., 2011; Hao et al., 2016; Giannatsis & Dedoussis, 2009; Ciocca et al., 2011) . Because of its poor thermal conductivity (Parry et al., 2016), active chemical reaction to oxygen (Prasad et al., 2014), and propensity for strain hardening (Gupta et al., 2016; de Formanoir et al, 2017), Ti6Al4V products are difficult to manufacture despite their high demand. The isoelectric point of oxide, level of electronic conductivity, corrosion resistance, and thermodynamic condition at physiological pH value all affect how biocompatible titanium alloys are (Elias et al., 2008; Casaletto et al., 2001).

Because Titanium resists corrosion, it is occasionally etched with acids like HCl, H2SO4, and HF. According to experimental models, engraved Titanium osseotite inserts had statistically more bone in contact with them after nine months than standard implants (Abrahamsson et al., 2001). Ti6Al4V, Ti5Al2.5Fe, Ti6Al7Nb these Titanium alloys are more popular, and more used as implant materials. Combining etching and blasting is thought to be the most common way to modify titanium surfaces because etching smooth out sharp peaks and adds a high-frequency component to the implant surface that is crucial for protein adhesion while blasting creates the ideal level of roughness for mechanical fixation. Early stages of bone healing are facilitated by titanium topographical design. Improved bone-to-implant contact on blasted and etched Titanium micro implants was also noticeable in human bone after two months (Chorny et al., 2014). The bio medical applications of Titanium alloys are shown in Figure 7.

2.2.3. Stainless Steel Alloys

The most popular stainless steel is SS302, which is stronger and offers better corrosion resistance than vanadium steel. A mild amount of molybdenum, 18-8 Mo stainless steel, was produced to increase corrosion resistance in chloride environments. This alloy eventually gained notoriety under the name type 316 stainless steel. For improved corrosion resistance in chloride solution and to reduce sensitization, the

Figure 7. Application of Ti alloys in bio medical

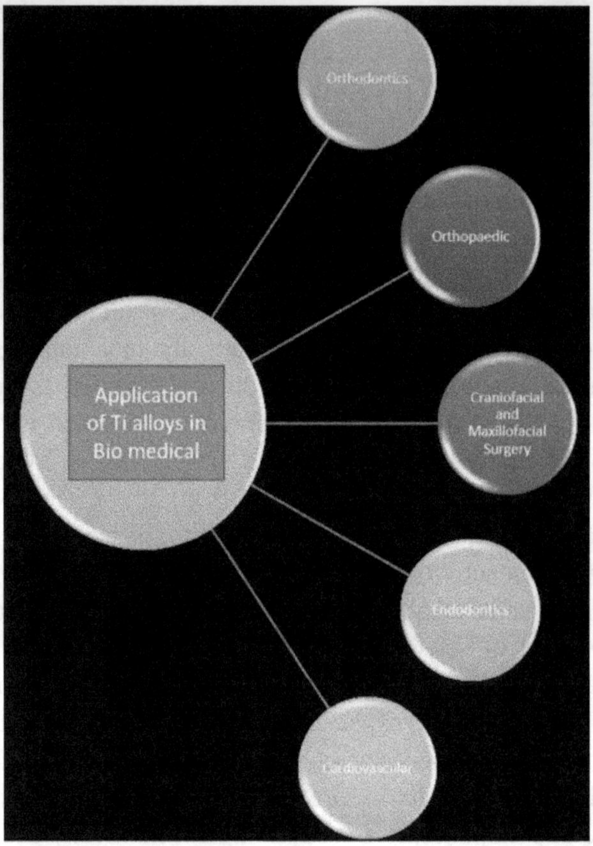

amount of carbon of SS 316 was subsequently reduced from 0.08% to a maximum level of 0.03%, and it subsequently became known as 316L stainless steel. The amount of chromium is set at 11% to provide corrosion resistance. Better corrosion resistance is achieved by passivating chromium and its alloys with 30% nitric acid (Morsiya, 2022). The most popular austenitic stainless steels used as biomaterials are 316 and 316L. They respond better to cold working than heat treatment to increase hardness. Nonmagnetic and having better corrosion resistance than other types of stainless steel, austenitic stainless steels are preferred. Molybdenum is added to increase resistance to pitting corrosion (Wong & Bronzino, 2007).

In a short period of time, the manufacturing of orthopaedic implants has gone from science fiction to a viable industry. Every year, orthopaedic techniques, tools, and products advance significantly. Orthopaedic manufacturing is crucial to enabling us to lead happier and healthier lives. Orthopaedic implants and instruments need to be designed and manufactured continuously using innovative techniques. As the cost and risks of orthopaedic surgery procedures decrease, more people who need this type of care can access the healings of modern orthopaedic medicine, dramatically improving their quality of life. Biomedical implants are made using manufacturing processes like machining, metal injection moulding, polishing, grinding, casting, forging, and rapid manufacturing. It is necessary to research the parameters that must be watched over and controlled during the production of biomedical implants because different processes used produce materials with various properties (Morsiya, 2022).

2.2.4. Co-Cr Alloys

CoCrMo alloys are well-known for use in dental and restorative applications especially in the surgical uses for implants, Co-based compounds are used as reconstruction materials for joints (stimulated) as they have excellent mechanical characteristics, biodegradability, and great resistance to corrosion. CoCrMo alloys are primarily employed in dental operations; combinations of CoCrMo further ensure biocompatibility and not having any negative effects on the body. Because high chromium is used, improved corrosion resistance is anticipated. Passive oxide has been created on the alloy's surface, protecting it from the surroundings of the body. Cast CoCrMo alloys have a reputation for having an uneven as well as a large-grained structure. Table 2 describes the function of different alloying elements in cobalt.

Notably, the fragile σ phase developed alongside the inter-dendritic area and the mechanical characteristics are broken down in this area. Regarding the amount of slip plane, phase γ (FCC) is more adequate than phase ε (HCP). Therefore, it is essential for CoCrMo alloys to prevent the development of a phase and maintain the phase's matrix. Dental implants made using a casting process similar to investment casting for the creation of artificial hip and knee joints can be made with CoCrMo alloys with Cr concentrations up to 34 weight percent that have been modified by the addition of N (Toh et al., 2017). It is required that the forged alloys (ASTM F75) outperform the cast alloys in terms of ductility and fatigue strength (ASTM F799). The coarse grain, cored matrix, and casting flaws such micro porosity are all characteristics of cast alloys. There are several Co-based medicinal alloys on the market like vitallium, MP35N, endocast, protasul-2, and protasul-21 WF. Corrosion and mechanical characteristics the chemical nature and microstructure of certain alloys influence their resistance. The most common protasul-21 WF and MP35N compounds are used in dental applications (Jurczyk, 2012).

Co-Cr alloys have lately been employed to make stents and have long been used in orthopaedic implants and dental applications (Ratner et al., 2004). Due to their high yield strength (448-648 MPa), elastic modulus (210 GPa), and tensile strength (952-1221 MPa), these alloys offer exceptional radial strength (Mani et al., 2007). Surface corrosion resistance is one of the key factors to consider when assessing biocompatibility for applications in the human body. Co-Cr alloy biomaterials' resistance to corrosion in artificial human bodily fluids has been the subject of several investigations (Kereiakes et al., 2003).

Table 2. The functions of different alloying elements in cobalt

Alloying elements	Function of alloying components
W	Reduce corrosion resistance while boosting strength.
Ni	Boost cast ability, corrosion resistance, and strength.
Mo	Strengthen and better protect against corrosion.
C	Boost casting capacity and strengthen wear resistance.
W	Reduce corrosion resistance while boosting strength.
Cr	Boost wear and corrosion protection.

2.2.5. PMMA

PMMA has a high Young's modulus, low elongation at break, and strong mechanical strength. It ruptures, but it doesn't break. One of the toughest thermoplastics, it has a high level of scratch resistance. It has capacity to absorb water and exhibit low moisture; this resulting products have better dimensional stability. Both of these characteristics become more potent as a temperature rises. Because it is comparatively less expensive, takes less time, and has good reconstruction capabilities (Dhanopia & Bhargava, 2017). PMMA has replaced previously employed denture base materials as a result of the gained qualities, such as the simplicity of manufacturing, acceptable mechanical and physical properties, inventive (Johnson, 1959), cost-effectiveness, and significantly lower toxicity (Gautam et al., 2012; Khindria et al., 2009; Kedjarune et al., 1999) . Due to differences in its mechanical and physical properties, PMMA is not the optimum material. For example, PMMA soak up water, which may affect its physical and mechanical characteristics while it is being used (Sasaki et al., 2016; Takahashi et al., 2013) and make it susceptible to failure under cyclic stress (Narva et al., 2005). Because of its benefits as a denture base biomaterial, including its beneficial aesthetic outcome, toxicological safety, higher elasticity and physical strength, heat resistance, and chemical resistance, polyamide has been gaining attention (Chang et al., 2019; Al-Thobit, 2020). The mechanical qualities of acrylic resins have been continuously improved through modifications such as the addition of butadiene styrene to PMMA to create high-impact resins and the reinforcement of denture base materials with a variety of fibres. Numerous researches have also been conducted in an effort to create a PMMA substitute (Uzun et al., 1999; Jagger et al., 2000).

3. DISCUSSION

After going over the various statistical publications, it is evident that biomaterials are being used more frequently because of their high strength, low modulus, low density, and combination of mechanical and exceptional corrosion resistance. Now days as we can see that the demand of bio-compatible has been increasing all over the world because of their numerous use in implants and medical surgery. The table 3 shows the mechanical properties of bio-compatible materials. From the table 3 it is clearly seen that Titanium alloy (Ti6Al4V) exhibits higher tensile and compressive strength compared to other bio materials. Table 4 compares the densities, Young's modulus, and poison ratios of various biomaterials, which are important for determining the mechanical characteristics of bio composites. Densities for the various bio-materials are shown in Figure 8.

Table 3. Review of tensile and compressive strength of different bio materials

Material Name	Tensile strength (Mpa)	Compressive strength (Mpa)	Author's contribution	References
Ti6Al4V	993	1086	Mechanical strength analysis of titanium alloys using ANSYS software.	Amalraju & Dawood, 2012
S.S316L	485	570	Analysis of stainless steel alloy under static loading condition using FEA techniques.	Amalraju & Dawood, 2012
PMMA	47-79	83-124	Discussed human fractured femur bone with PMMA thermoplastic using FEA technique.	Dhanopia & Bhargava, 2017
Mg alloy	180-440	100-200	Discussed about features, development and preservation of Mg bone implants.	Dhanopia & Bhargava, 2017
CoCr alloy	145-270	450-1000	Discussed mechanical properties and comparison among different Co-Cr alloys.	Ghosh et al., 2018

Table 4. Past analysis of density, young's modulus and poison's ratio of different bio materials

Materials	Density (kg/m^3)	Young's Modulus E(Gpa)	Poison Ratio
Ti6Al4V	4500	120	0.32
S.S316L	7750	193	0.31
PMMA	1180	220	0.2
Mg alloy	1800	42	0.35
CoCr alloy	10000	210	0.29
Femur cortical bone	1750	16.7	0.3
Alumina Al2o3	8500	240	0.31
Natural bone	1800-2100	14.8	0.3-0.49

Figure 8. Bar chart of density of bio-materials

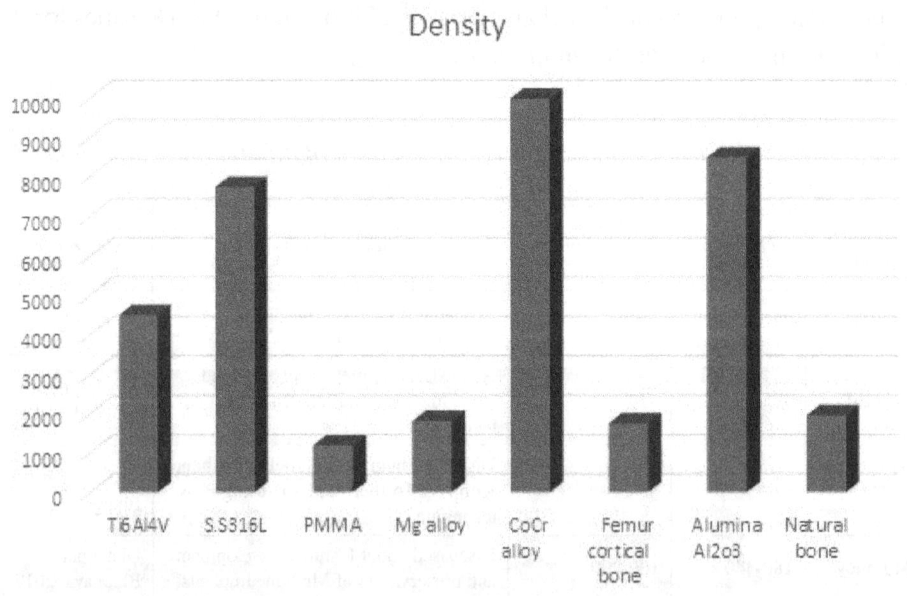

The Young's modulus of different bio materials are shown in the Figure 9.

Table 5 demonstrates that stainless steel is easily replaced by Ti6Al4V alloy due to the alloy's lower equivalent stress, maximum principal stress values, and deformation. Speaking with many doctors revealed that they would adore using Ti6Al4V alloy in place of stainless steel. Final substances have higher values for the same parameters but less application potential. Along with the latest developments

Figure 9. Bar chart of Young's modulus of bio-materials

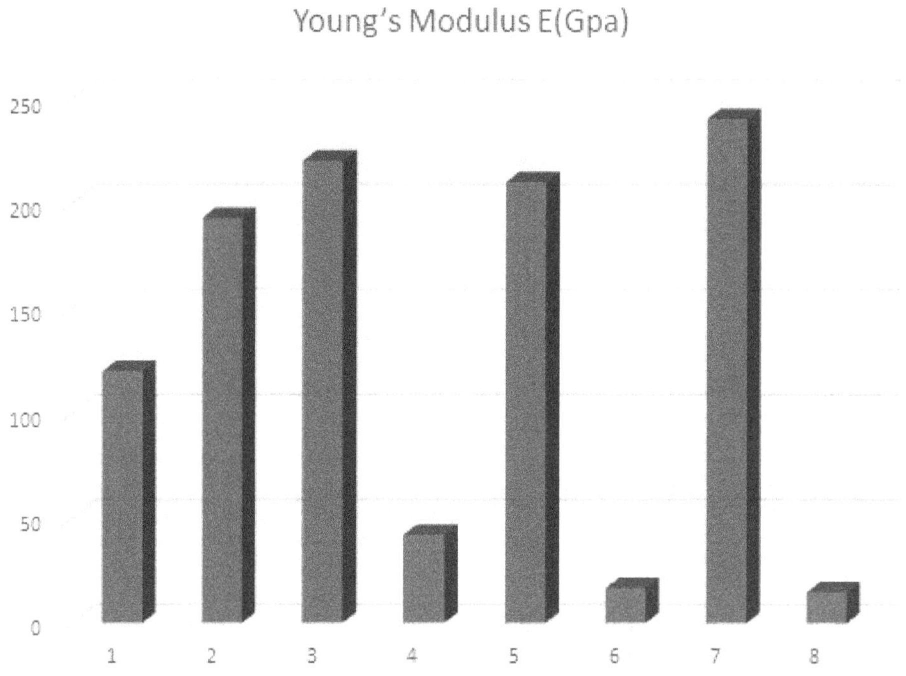

in biomaterials, consideration should also be given to composite materials, such as polymethyl methacrylate (PMMA), which has shown to be less susceptible to deformation, equivalent stress, and major stresses than metals. Composites may represent significant innovations in the field of orthopaedics.

4. CONCLUSION

Selection of bio materials for orthopaedic implants plays an essential role in the replacement of femur bone. Since a long time ago, bio-mechanics has mostly utilised stainless steel in surgical implants. We come to the conclusion that magnesium and titanium alloys might be the superior replacement for medical implants after comparing the mechanical and physical characteristics of various biomaterials. Additionally, compared to other biomaterials, alloys made of titanium and magnesium last inside the body for a very long time. Because titanium is very expensive and has physical characteristics like density and young's modulus that are very similar to those of natural bone, magnesium alloys may be a better choice than titanium alloys in the field of biomechanics.

Table 5. Comparison of stresses of different biomaterials

Materials	Total Deformation (mm)		Equivalent Stress (Mpa)		Maximum Principal Stress (Mpa)		Manimum Principal Stress (Mpa)	
	Max	**Min**	**Max**	**Min**	**Max**	**Min**	**Max**	**Min**
Ti6Al4V	2.69	0	332.27	0.00125	363.33	-22.571	40.45	-308.16
S.S316L	2.57	0	405.71	0.00125	383.77	-30.574	61.111	-385.65
PMMA	3.95	0	59.48	0.00131	46.56	-3.190	3.864	-36.575
Al2O3 (Ceramic)	2.52	0	467.12	0.00125	402.4	-39.64	69.53	-42.86
Nylon6/6 (polymer)	2.47	0	601.3	0.00124	493.88	-37.10	56.28	-463.38

Materials	Total Deformation (mm)		Equivalent Stress (Mpa)		Maximum Principal Stress (Mpa)		Manimum Principal Stress (Mpa)	
	Max	**Min**	**Max**	**Min**	**Max**	**Min**	**Max**	**Min**
Ti6Al4V	2.69	0	332.27	0.00125	363.33	-22.571	40.45	-308.16
S.S316L	2.57	0	405.71	0.00125	383.77	-30.574	61.111	-385.65
PMMA	3.95	0	59.48	0.00131	46.56	-3.190	3.864	-36.575
Al2O3 (Ceramic)	2.52	0	467.12	0.00125	402.4	-39.64	69.53	-42.86
Nylon6/6 (polymer)	2.47	0	601.3	0.00124	493.88	-37.10	56.28	-463.38

Conflict of interest statement: Authors are confirming that there is **no conflict of interest** in this work.

REFERENCES

Abrahamsson, I., Zitzmann, N. U., Berglundh, T., Wennerberg, A., & Lindhe, J. (2001). Bone and soft tissue integration to titanium implants with different surface topography: An experimental study in the dog. *The International Journal of Oral & Maxillofacial Implants*, *16*(3). PMID:11432652

Ajay, P. Ahmad, S., Sharma, J., & Gambhir, V. (Eds.). (2023). Handbook of Sustainable Materials: Modelling, Characterization, and Optimization (1st ed.). CRC Press. doi:10.1201/9781003297772

Akahori, T. (2002). Biomaterials: Journal of Japanese Society for Biomaterials 28 (3), 155-156, 2010-07-30. *Metallurgical and Materials Transactions. A, Physical Metallurgy and Materials Science*, *33*, 477–486.

Al-Thobity, A. M. (2020). The impact of polymerization technique and glass-fiber reinforcement on the flexural properties of denture base resin material. *European journal of dentistry, 14*(01), 092-099.

Amalraju, D., & Dawood, A. S. (2012). Mechanical strength evaluation analysis of stainless steel and titanium locking plate for femur bone fracture. *Engineering Science and Technology: An International Journal*, *2*(3), 381–388.

Anselme, K. (2000). Osteoblast adhesion on biomaterials. *Biomaterials*, *21*(7), 667–681. doi:10.1016/S0142-9612(99)00242-2 PMID:10711964

Arden, G. P., Taylor, A. R., & Ansell, B. M. (1970). Total hip replacement using the McKee-Farrar prosthesis. In rheumatoid arthritis, Still's disease, and ankylosing spondylitis. *Annals of the Rheumatic Diseases*, *29*(1), 1–5. doi:10.1136/ard.29.1.1 PMID:5416097

Binyamin, G., Shafi, B. M., & Mery, C. M. (2006, November). Biomaterials: A primer for surgeons. [). WB Saunders.]. *Seminars in Pediatric Surgery*, *15*(4), 276–283. doi:10.1053/j.sempedsurg.2006.07.007 PMID:17055958

Casaletto, M. P., Ingo, G. M., Kaciulis, S., Mattogno, G., Pandolfi, L., & Scavia, G. (2001). Surface studies of in vitro biocompatibility of titanium oxide coatings. *Applied Surface Science*, *172*(1-2), 167–177. doi:10.1016/S0169-4332(00)00844-8

Chang, M. C., Hung, C. C., Chen, W. C., Tseng, S. C., Chen, Y. C., & Wang, J. C. (2019). Effects of pontic span and fiber reinforcement on fracture strength of multi-unit provisional fixed partial dentures. *Journal of Dental Sciences*, *14*(3), 309–317. doi:10.1016/j.jds.2018.11.008 PMID:31528260

Chen, S., Guan, S., Li, W., Wang, H., Chen, J., Wang, Y., & Wang, H. (2012). In vivo degradation and bone response of a composite coating on Mg–Zn–Ca alloy prepared by microarc oxidation and electrochemical deposition. *Journal of Biomedical Materials Research. Part B, Applied Biomaterials*, *100*(2), 533–543. doi:10.1002/jbm.b.31982 PMID:22120974

Chorny, V., Yatsun, Y., & Golovakha, M. (2014). Historical aspects of using of biodegradable magnesium-based alloys for osteosynthesis (literature review). *Orthopaedics. Ortopediia Travmatologiia i Protezirovanie*, *0*(1), 105–109. doi:10.15674/0030-598720141105-109

Ciocca, L., Fantini, M., De Crescenzio, F., Corinaldesi, G., & Scotti, R. (2011). Direct metal laser sintering (DMLS) of a customized titanium mesh for prosthetically guided bone regeneration of atrophic maxillary arches. *Medical & Biological Engineering & Computing*, *49*(11), 1347–1352. doi:10.100711517-011-0813-4 PMID:21779902

Cui, C., Hu, B., Zhao, L., & Liu, S. (2011). Titanium alloy production technology, market prospects and industry development. *Materials & Design*, *32*(3), 1684–1691. doi:10.1016/j.matdes.2010.09.011

de Formanoir, C., Brulard, A., Vivès, S., Martin, G., Prima, F., Michotte, S., Rivière, E., Dolimont, A., & Godet, S. (2017). A strategy to improve the work-hardening behavior of Ti–6Al–4V parts produced by additive manufacturing. *Materials Research Letters*, *5*(3), 201–208. doi:10.1080/21663831.2016.1245681

Dhanopia, A., & Bhargava, M. (2017). Finite element analysis of human fractured femur bone implantation with PMMA thermoplastic prosthetic plate. *Procedia Engineering*, *173*, 1658–1665. doi:10.1016/j.proeng.2016.12.190

Dodd, C. A., Hungerford, D. S., & Krackow, K. A. (1990). Total knee arthroplasty fixation. Comparison of the early results of paired cemented versus uncemented porous coated anatomic knee prostheses. *Clinical Orthopaedics and Related Research*, (260), 66–70. PMID:2225646

Donachie, M. J. (2000). *Titanium: a technical guide*. ASM international. doi:10.31399/asm.tb.ttg2.9781627082693

Elias, C. N., Lima, J. H. C., Valiev, R., & Meyers, M. (2008). Biomedical Applications of Titanium and its Alloys. *J. Miner. Met. Mater. Soc*, 46-9.

Emami, B., Lyman, J., Brown, A., Cola, L., Goitein, M., Munzenrider, J. E., & Wesson, M. (1991). Tolerance of normal tissue to therapeutic irradiation. *International Journal of Radiation Oncology* Biology* Physics, 21*(1), 109-122.

Emmelmann, C., Scheinemann, P., Munsch, M., & Seyda, V. (2011). Laser additive manufacturing of modified implant surfaces with osseointegrative characteristics. *Physics Procedia*, *12*, 364–368. doi:10.1016/j.phpro.2011.03.046

Gautam, R., Singh, R. D., Sharma, V. P., Siddhartha, R., Chand, P., & Kumar, R. (2012). Biocompatibility of polymethylmethacrylate resins used in dentistry. *Journal of Biomedical Materials Research. Part B, Applied Biomaterials*, *100*(5), 1444–1450. doi:10.1002/jbm.b.32673 PMID:22454327

Gaviria, L., Salcido, J. P., Guda, T., & Ong, J. L. (2014). Current trends in dental implants. *Journal of the Korean Association of Oral and Maxillofacial Surgeons*, *40*(2), 50. doi:10.5125/jkaoms.2014.40.2.50 PMID:24868501

Ghosh, S., Sanghavi, S., & Sancheti, P. (2018). Metallic biomaterial for bone support and replacement. In *Fundamental biomaterials: metals* (pp. 139–165). Woodhead Publishing. doi:10.1016/B978-0-08-102205-4.00006-4

Giannatsis, J., & Dedoussis, V. (2009). Additive fabrication technologies applied to medicine and health care: A review. *International Journal of Advanced Manufacturing Technology*, *40*(1), 116–127. doi:10.100700170-007-1308-1

Gu, X., Zheng, Y., Cheng, Y., Zhong, S., & Xi, T. (2009). In vitro corrosion and biocompatibility of binary magnesium alloys. *Biomaterials*, *30*(4), 484–498. doi:10.1016/j.biomaterials.2008.10.021 PMID:19000636

Gupta, P., Adhikary, M., Kumar, M., Bhardwaj, N., & Mandal, B. B. (2016). Biomimetic, osteoconductive non-mulberry silk fiber reinforced tricomposite scaffolds for bone tissue engineering. *ACS Applied Materials & Interfaces*, *8*(45), 30797–30810. doi:10.1021/acsami.6b11366 PMID:27783501

Gupta, R. K., Kumar, V. A., Mathew, C., & Rao, G. S. (2016). Strain hardening of titanium alloy Ti6Al4V sheets with prior heat treatment and cold working. *Materials Science and Engineering A*, *662*, 537–550. doi:10.1016/j.msea.2016.03.094

Hao, Y. L., Li, S. J., & Yang, R. (2016). Biomedical titanium alloys and their additive manufacturing. *Rare Metals*, *35*(9), 661–671. doi:10.100712598-016-0793-5

Harun, W. S. W., Asri, R. I. M., Alias, J., Zulkifli, F. H., Kadirgama, K., Ghani, S. A. C., & Shariffuddin, J. H. M. (2018). A comprehensive review of hydroxyapatite-based coatings adhesion on metallic biomaterials. *Ceramics International*, *44*(2), 1250–1268. doi:10.1016/j.ceramint.2017.10.162

Jagger, D. C., Harrison, A., & Al-Marzoug, K. (2000). Effect of the addition of poly (methyl methacrylate) beads on some properties of acrylic resin. *The International Journal of Prosthodontics*, *13*(5). PMID:11203657

Johnson, W. W. (1959). The history of prosthetic dentistry. *The Journal of Prosthetic Dentistry*, *9*(5), 841–846. doi:10.1016/0022-3913(59)90049-6

Jurczyk, M. (Ed.). (2012). *Bionanomaterials for dental applications*. CRC Press. doi:10.1201/b13255

Kedjarune, U., Charoenworaluk, N., & Koontongkaew, S. (1999). Release of methyl methacrylate from heat-curved and autopolymerized resins: Cytotoxicity testing related to residual monomer. *Australian Dental Journal*, *44*(1), 25–30. doi:10.1111/j.1834-7819.1999.tb00532.x PMID:10217017

Kereiakes, D. J., Cox, D. A., Hermiller, J. B., Midei, M. G., Bachinsky, W. B., Nukta, E. D., Leon, M. B., Fink, S., Marin, L., & Lansky, A. J.Guidant Multi-Link Vision Stent Registry Investigators. (2003). Usefulness of a cobalt chromium coronary stent alloy. *The American Journal of Cardiology*, *92*(4), 463–466. doi:10.1016/S0002-9149(03)00669-6 PMID:12914881

Khayat, S., Fanaei, H., & Ghanbarzehi, A. (2017). Minerals in pregnancy and lactation: A review article. *Journal of Clinical and Diagnostic Research : JCDR*, *11*(9), QE01. doi:10.7860/JCDR/2017/28485.10626 PMID:29207789

Khindria, S. K., Mittal, S., & Sukhija, U. (2009). Evolution of denture base materials. *Journal of Indian Prosthodontic Society*, *9*(2), 64. doi:10.4103/0972-4052.55246

Kumar, A., Kumar, P., Mittal, R. K., & Gambhir, V. (2023). Materials processed by additive manufacturing techniques. In Advances in Additive Manufacturing Artificial Intelligence, Nature-Inspired, and Biomanufacturing (pp. 217-233). Elsevier. doi:10.1016/B978-0-323-91834-3.00014-4

Kumar, A., Mittal, R. K., & Haleem, A. (Eds.). (2022). *Advances in Additive Manufacturing: Artificial Intelligence, Nature-Inspired, and Biomanufacturing*. Elsevier. doi:10.1016/C2020-0-03877-6

Kunchur, M. N., Dean, C., Liang, M., Moghaddam, N. S., Guarino, A., Nigro, A., Grimaldi, G., & Leo, A. (2013). Depairing current density of Nd2− xCexCuO4− δ superconducting films. *Physica. C, Superconductivity*, *495*, 66–68. doi:10.1016/j.physc.2013.08.005

Kunchur, M. N., Dean, C. L., Moghadam, N. S., Knight, J. M., He, Q. L., Liu, H., Wang, J., Lortz, R., Sou, I. K., & Gurevich, A. (2015). Current-induced depairing in the Bi 2 Te 3/FeTe interfacial superconductor. *Physical Review B: Condensed Matter and Materials Physics*, *92*(9), 094502. doi:10.1103/PhysRevB.92.094502

Li, N., & Zheng, Y. (2013). Novel magnesium alloys developed for biomedical application: A review. *Journal of Materials Science and Technology*, *29*(6), 489–502. doi:10.1016/j.jmst.2013.02.005

Li, Y., Yang, C., Zhao, H., Qu, S., Li, X., & Li, Y. (2014). New developments of Ti-based alloys for biomedical applications. *Materials (Basel)*, *7*(3), 1709–1800. doi:10.3390/ma7031709 PMID:28788539

Long, M., & Rack, H. J. (1998). Titanium alloys in total joint replacement—A materials science perspective. *Biomaterials*, *19*(18), 1621–1639. doi:10.1016/S0142-9612(97)00146-4 PMID:9839998

Luo, Y., Yang, L., & Tian, M. (2013). Application of biomedical-grade titanium alloys in trabecular bone and artificial joints. In *Biomaterials and Medical Tribology* (pp. 181–216). Woodhead Publishing. doi:10.1533/9780857092205.181

Mani, G., Feldman, M. D., Patel, D., & Agrawal, C. M. (2007). Coronary stents: A materials perspective. *Biomaterials*, *28*(9), 1689–1710. doi:10.1016/j.biomaterials.2006.11.042 PMID:17188349

Morsiya, C. (2022). A review on parameters affecting properties of biomaterial SS 316L. *Australian Journal of Mechanical Engineering*, *20*(3), 803–813. doi:10.1080/14484846.2020.1752975

Mulenga, C. (2019). A cross-section study to determine human height using femur length in Zambian population. *Acta Scientific Microbiology*, *2*(11), 57–61. doi:10.31080/ASMI.2019.02.0399

Nabiyouni, M., Brückner, T., Zhou, H., Gbureck, U., & Bhaduri, S. B. (2018). Magnesium-based bioceramics in orthopedic applications. *Acta Biomaterialia*, *66*, 23–43. doi:10.1016/j.actbio.2017.11.033 PMID:29197578

Nalwa, H. S. (2007). Handbook Of Nanostructured Biomaterials And Their Applications In Nanobiotechnology: : Biomaterials. American Scientific Publishers.

Narva, K. K., Lassila, L. V., & Vallittu, P. K. (2005). The static strength and modulus of fiber reinforced denture base polymer. *Dental Materials*, *21*(5), 421–428. doi:10.1016/j.dental.2004.07.007 PMID:15826698

Niinomi, M., Nakai, M., & Hieda, J. (2012). Development of new metallic alloys for biomedical applications. *Acta Biomaterialia*, *8*(11), 3888–3903. doi:10.1016/j.actbio.2012.06.037 PMID:22765961

Niu, J., Xiong, M., Guan, X., Zhang, J., Huang, H., Pei, J., & Yuan, G. (2016). The in vivo degradation and bone-implant interface of Mg-Nd-Zn-Zr alloy screws: 18 months post-operation results. *Corrosion Science*, *113*, 183–187. doi:10.1016/j.corsci.2016.10.009

Parry, L., Ashcroft, I. A., & Wildman, R. D. (2016). Understanding the effect of laser scan strategy on residual stress in selective laser melting through thermo-mechanical simulation. *Additive Manufacturing*, *12*, 1–15. doi:10.1016/j.addma.2016.05.014

Patel, N. R., & Gohil, P. P. (2012). A review on biomaterials: Scope, applications & human anatomy significance. *International Journal of Emerging Technology and Advanced Engineering*, *2*(4), 91–101.

Peter, I., & Rosso, M. (2015). Study of Ti-enriched CoCrMo alloy for dental application. *IEEE Access : Practical Innovations, Open Solutions*, *3*, 73–80. doi:10.1109/ACCESS.2015.2398312

Poinern, G. E. J., Brundavanam, S., & Fawcett, D. (2012). Biomedical magnesium alloys: A review of material properties, surface modifications and potential as a biodegradable orthopaedic implant. *American Journal of Biomedical Engineering*, *2*(6), 218–240. doi:10.5923/j.ajbe.20120206.02

Prasad, A. R., Ramji, K., & Datta, G. L. (2014). An experimental study of wire EDM on Ti-6Al-4V alloy. *Procedia Materials Science*, *5*, 2567–2576. doi:10.1016/j.mspro.2014.07.517

Rahmani, M., Ahmadi, M. T., Shayesteh, N., Amin, N. A., Rahmani, K., & Ismail, R. (2011, September). Current-voltage modeling of bilayer graphene nanoribbon schottky diode. In *2011 IEEE Regional Symposium on Micro and Nano Electronics* (pp. 256-258). IEEE. 10.1109/RSM.2011.6088337

Ratner, B. D., Hoffman, A. S., Schoen, F. J., & Lemons, J. E. (2004). *Biomaterials science: an introduction to materials in medicine.*

Sanchez, A. H. M., Luthringer, B. J., Feyerabend, F., & Willumeit, R. (2015). Mg and Mg alloys: How comparable are in vitro and in vivo corrosion rates? A review. *Acta Biomaterialia*, *13*, 16–31. doi:10.1016/j.actbio.2014.11.048 PMID:25484334

Santos, G. (2017). The importance of metallic materials as biomaterials. *Adv Tissue Eng Regen Med Open Access*, *3*(1), 300–302. doi:10.15406/atroa.2017.03.00054

Sasaki, H., Hamanaka, I., Takahashi, Y., & Kawaguchi, T. (2016). Effect of long-term water immersion or thermal shock on mechanical properties of high-impact acrylic denture base resins. *Dental Materials Journal*, *35*(2), 204–209. doi:10.4012/dmj.2015-291 PMID:27041009

Sehrawat, S., Kumar, A., Grover, S., Dogra, N., Nindra, J., Rathee, S., Dahiya, M., & Kumar, A. (2022a). Study of 3D scanning technologies and scanners in orthodontics. *Materials Today: Proceedings*, *56*, 186–193. doi:10.1016/j.matpr.2022.01.064

Sehrawat, S., Kumar, A., Prabhakar, M., & Nindra, J. (2022b). The expanding domains of 3D printing pertaining to the speciality of orthodontics. *Materials Today: Proceedings*, *50*, 1611–1618. doi:10.1016/j.matpr.2021.09.124

Singh, V., Gupta, M., Kumar, A., Luthra, S., & Kumar, A. (2023). Experimental investigations of thermo physical properties and convective heat transfer of Al2O3 and CuO nanofluids in a copper tube: Proposing New Correlations. *Biointerface Research in Applied Chemistry*, *13*(3), 229.

Staiger, M. P., Pietak, A. M., Huadmai, J., & Dias, G. (2006). Magnesium and its alloys as orthopedic biomaterials: A review. *Biomaterials*, *27*(9), 1728–1734. doi:10.1016/j.biomaterials.2005.10.003 PMID:16246414

Stevens, M. M. (2008). Biomaterials for bone tissue engineering. *Materials Today*, *11*(5), 18–25. doi:10.1016/S1369-7021(08)70086-5

Takahashi, Y., Hamanaka, I., & Shimizu, H. (2013). Flexural properties of denture base resins subjected to long-term water immersion. *Acta Odontologica Scandinavica*, *71*(3-4), 716–720. doi:10.3109/00016357.2012.715196 PMID:22900745

Tan, L., Dong, J., Chen, J., & Yang, K. (2018). Development of magnesium alloys for biomedical applications: Structure, process to property relationship. *Materials Technology*, *33*(3), 235–243. doi:10.1080/10667857.2017.1405890

Toh, W. Q., Tan, X., Bhowmik, A., Liu, E., & Tor, S. B. (2017). Tribochemical characterization and tribocorrosive behavior of CoCrMo alloys: A review. *Materials (Basel)*, *11*(1), 30. doi:10.3390/ma11010030 PMID:29278375

Uzun, G., Hersek, N., & Tincer, T. (1999). Effect of five woven fiber reinforcements on the impact and transverse strength of a denture base resin. *The Journal of prosthetic dentistry*, *81*(5), 616-620.

Wang, J. L., Xu, J. K., Hopkins, C., Chow, D. H. K., & Qin, L. (2020). Biodegradable magnesium-based implants in orthopedics—A general review and perspectives. *Advancement of Science*, *7*(8), 1902443. doi:10.1002/advs.201902443 PMID:32328412

Wanhill, R., & Barter, S. (2011). *Fatigue of beta processed and beta heat-treated titanium alloys*. Springer Science & Business Media.

Wik, T. S., Foss, O. A., Havik, S., Persen, L., Aamodt, A., & Witsø, E. (2010). Periprosthetic fracture caused by stress shielding after implantation of a femoral condyle endoprosthesis in a transfemoral amputee—A case report. *Acta Orthopaedica*, *81*(6), 765–767. doi:10.3109/17453674.2010.533937 PMID:21067436

Wong, J. Y., & Bronzino, J. D. (2007). *Biomaterials*. CRC press.

Yang, Y., He, C., Dianyu, E., Yang, W., Qi, F., Xie, D., & Shuai, C. (2020). Mg bone implant: Features, developments and perspectives. *Materials & Design*, *185*, 108259. doi:10.1016/j.matdes.2019.108259

Yazdimamaghani, M., Razavi, M., Vashaee, D., & Tayebi, L. (2014). Development and degradation behavior of magnesium scaffolds coated with polycaprolactone for bone tissue engineering. *Materials Letters*, *132*, 106–110. doi:10.1016/j.matlet.2014.06.036

Zhang, S., Zhang, X., Zhao, C., Li, J., Song, Y., Xie, C., Tao, H., Zhang, Y., He, Y., Jiang, Y., & Bian, Y. (2010). Research on an Mg–Zn alloy as a degradable biomaterial. *Acta Biomaterialia*, *6*(2), 626–640. doi:10.1016/j.actbio.2009.06.028 PMID:19545650

Zheng, Y. F., Gu, X. N., & Witte, F. (2014). Biodegradable metals. *Materials Science and Engineering R Reports*, *77*, 1–34. doi:10.1016/j.mser.2014.01.001

Chapter 5
Review on the Development of Smart Materials, Including Shape Memory Alloys, Characterization, and Recent Applications

Preeti Singh Bahadur
Amity University, Greater Noida, India

Ruchira Srivastava
Amity University, Greater Noida, India

Vijay Srivastava
Amity University, Greater Noida, India

ABSTRACT

Smart materials have made a substantial contribution to their widespread use across a number of industries. Numerous intelligent materials have the ability to alter their characteristics and behavior in response to an outside catalyst or stimulus. These materials are altering how typical engineering materials are used. To improve the desirable features of existing materials, smart materials can either be used alone or combined with them. Due to its transdisciplinary uses, smart materials are found in all human innovations. While some of these materials are in use now, some have been for a very long time. Therefore, suffice it to remark that further study is still needed on the recently produced smart materials. The development of composite materials to shape memory alloys is the subject of this review study. This literature describes the development of smart materials from traditional composites to cutting-edge shape memory alloys as well as how they are utilized.

DOI: 10.4018/978-1-6684-9224-6.ch005

1. INTRODUCTION

Materials have been utilized by humans since very long time for a variety of purposes to make daily life more comfortable. From stone age to the current world, materials have undergone technological improvements. One of the major discoveries in the science of metallurgy that enabled people since ancient times discover more advanced materials was bronze. Technology was essential in the synthesis and characterization of more sophisticated materials, known as smart materials, which are now becoming very commonplace. Smart materials are also referred to as intelligent materials, because their capacity to sense stimuli and respond to them improves the performance of materials. Materials like these can change the essential structure, performance, sturdiness, and weight. Metals, ceramics, polymers, and smart materials are the four main classifications of materials. When an external force is applied, these materials have a propensity to change size and shape. Some of them exhibit self-healing and self-destructive qualities that are extraordinary compared to those of ordinary materials and influence society. These materials have a propensity to adapt to changing environmental circumstances.

Based on their qualities, smart materials are divided into both passive and active materials, as shown in Figure 1. Active smart materials are those that can change one type of energy into another while remaining impermeable to the outside environment, as opposed to passive smart materials, which can merely change or transform energy. Active materials are self-diagnostics because of their quick form and size changes and reduced time to response, self-actuation, and increased exposure to external stimuli. Self-diagnostics, often known as self-healing properties, enable them to detect surface cracks and autonomously repair them. Multifunctional materials are being discovered recently, and their properties have been examined using a variety of characterization approaches. Materials have evolved, through composites to smart memory alloys. These advanced materials have transcended normal materials, and current procedures have been used to create them instead of older, more traditional methods (Kim, 2017), (Hu, 2016), (Kamila, 2013), (Swait, et al., 2012), (Roytburd et al., 2000), (Dolez & Mlynarek, 2016) and (Ajay et al., 2023).

These contemporary materials are self-adaptive and have a memory function that mimics the human brain and alters the characteristics of the material to the past after it has been changed in accordance with the surroundings. These materials are widely used in the creation of products related to the environment, civil infrastructure, and the health sector. Since the environment outside controls these materials, they have an inherent sensor, an actuator, plus a control mechanism built right in to help them react to their surroundings. In other words, going back to how it was. One class of intelligent material is piezoelectric, which generates voltages in response to stress. Due of their mechanical and physical characteristics, composites have recently played an unprecedented role. Amongst, its many qualities are excellent wear resistance, resistant to corrosion and exposure to elevated temperatures. Due to their fabrication techniques, the manufacturing cost was decreased even though they had a high scattering strength and low thermal shock resistance. In the production of various components like breaks, tanks, shafts, and vessels, as well as in processes requiring high wear resistance in the onshore oil production, marine structures, shipping, and health sectors, composites have played a significant role in the aerospace, automotive, machine, and thermal industries. Wood, bones, shells, and other naturally occurring composites are a few examples. These typical composites have a number of flaws, including the inability to react to external stimuli, which prevents them from having self-healing, crack detection, and shape memory capabilities. Smart materials were developed to address these shortcomings. Smart materials include smart composites, shape memory alloys, piezoelectric materials, magnetorheological fluids, magnetostrictive materials, and optical fibres, but shape memory alloys were a major engineering breakthrough that had a positive impact on many societal sector(Bearat & Vernon, 2011), (Du et al., 2011), (Drossel et al., 2015) and (Bahl et al., 2015).

Figure 1. Smart materials

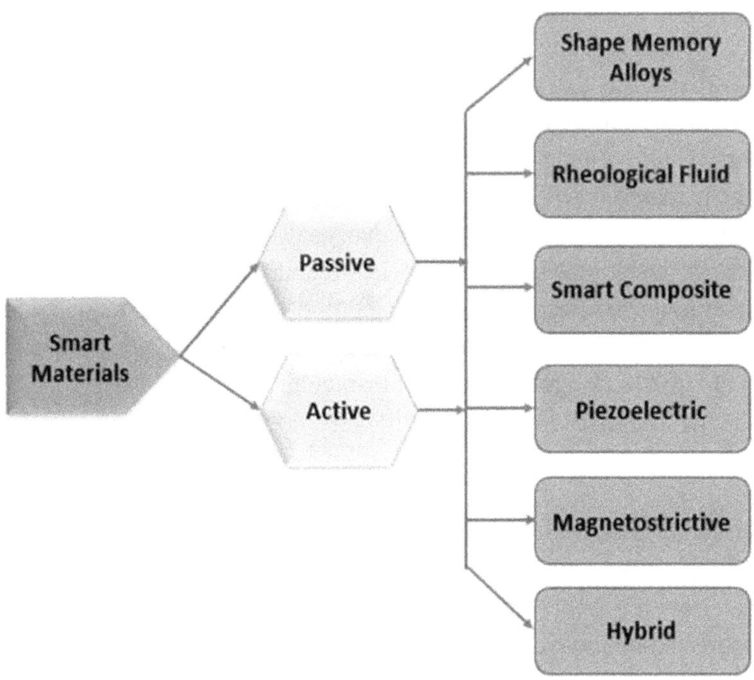

Shape memory alloys are distinct from other materials because of their ability to change their shape and size in response to temperature. In other words, they possess a brain that holds memories and can recall the original shape. When exposed to the outside environment, this characteristic aids in regaining its original structure's shape and size. Due of their thermo-mechanical characteristics and the cyclic solid-solid phase transformation, SMAs exhibit shape change as a result of phase transformation. Large elastic strains can be recovered with the aid of superelasticity, which also enables the material to withstand significant deformations. One of the most widely used nickel-titanium alloys is known by the brand name Nitinol and contains 45–50% titanium. Copper-zinc-aluminum alloys and copper-aluminum-nickel alloys, which are also referred to as non-titanium alloys, are two further shape-memory alloys. However, compared to nickel-titanium alloys, these non-titanium shape-memory alloys perform less well in terms of characteristics and shape memory. Shape memory effect and pseudoelasticity are the two main effects of the group of cutting-edge materials known as SMAs. Shape memory polymers, a type of stimuli-responsive material that may alter its shape via a temporary shape to a permanent one, are one of the other classes. The material can now have three shapes, or more accurately, many shapes, and is distinguished from shape-memory alloys (SMAs) via the melting transaction. Compared to SMAs, they have a lower density, a larger elastic strain, and an easy process. These materials can be used for a variety of industrial applications because of their inexpensive manufacturing and processing costs. Active materials are another name for shape-memory alloys. Active materials are currently utilized extensively in a wide range of applications, one of which is sensing. In this use, the material has the ability to translate mechanical deviation into non-mechanical output. Direct coupling behavior can be seen in a family of active materials that includes shape-memory alloys and piezo-ceramics. Shape memory alloys are a specific class of intelligent material where heating recovers strain. These characteristics make shape-memory alloys suitable for use in everyday situations. This article discusses contemporary developments in composites technology, including their evolution from conventional composites to smart composites to shape-memory alloys, as well as their characterization and applications that benefit many societal sectors (Aktas et al., 2020), (Bogue, 2014), (Naresh et al., 2016), (Hamdi et al.,2018) and (Kulkarni et al., 2018).

2. GENERAL SMART MATERIALS

Smart materials are either man-made or natural and develop useful qualities in response to environmental change. According to studies, "smart materials" are those that can modify their properties in a predetermined way in response to any outside stimulation. The stimuli can include chemicals, pressure, UV radiation, friction, electromagnetic waves, electric currents, and more. Properties like shape, size, ductility, colour, etc. may vary as a result (Sharp & Clemen, 2004) and (Rogers, 1989). Types of smart materials, its advantages, disadvantages and applications are discussed below:

2.1. Piezoelectric Materials

Whenever a force is put on it, materials that are piezoelectric change their characteristics. Whenever mechanical force is applied to these materials, electric pulses are visible, and vice versa. The afore mentioned occurrence is known as a converse and direct consequence is shown in Figure 2. When the size of the material changes, electric current is produced. These materials can be used in a variety of fields, including civil engineering (Garner et al., 2000) and (Sodano et al., 2004).

Figure 2. Piezoelectric materials
(Garner et al., 2000) and (Sodano et al., 2004)

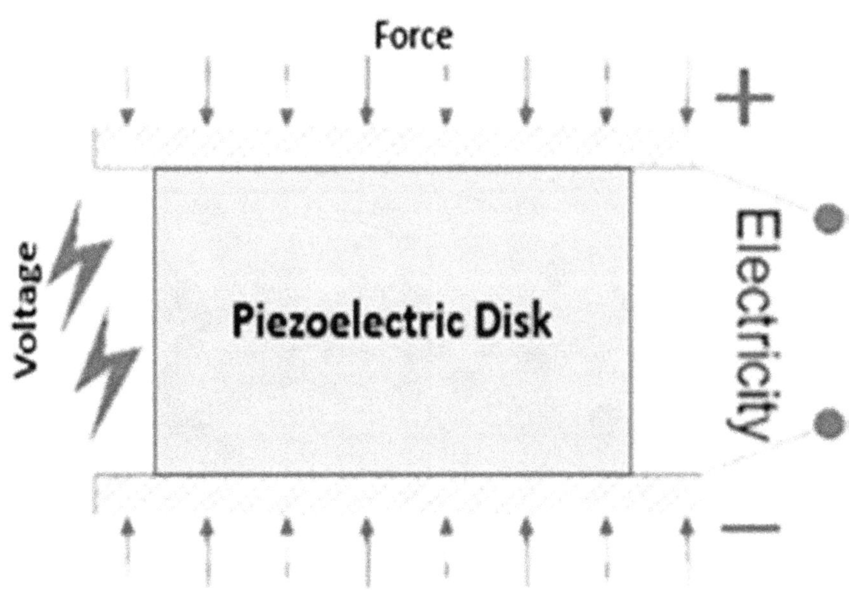

Some of the applications are:

a. When harvesting energy on a small scale, piezoelectric transformers are employed.
b. Use in autos and other sensors and actuators, such as the accelerometer.
c. Standard frequency for clocks,
d. Ultrasonography machines are submersible mine detectors.
e. Camera lenses with piezo motors are employed in medical applications.

Some of the advantages are high stability, high sensitivity (Chen et al., 2019) and disadvantages are not used for static measurement (Chen et al., 2019). Its major applications are applicable in construction of Roads and buildings etc. (Xuet al., 2018).

2.2. Electrostrictive Materials

Regarding their manner of operation, electrostrictive materials are comparable to piezoelectric materials. The distinction is a proportionate change of about 2 degrees is shown in Figure 3.

Figure 3. Electrostrictive materials
(Damjanovic & Newnham, 1992) and (Uchino, 1986)

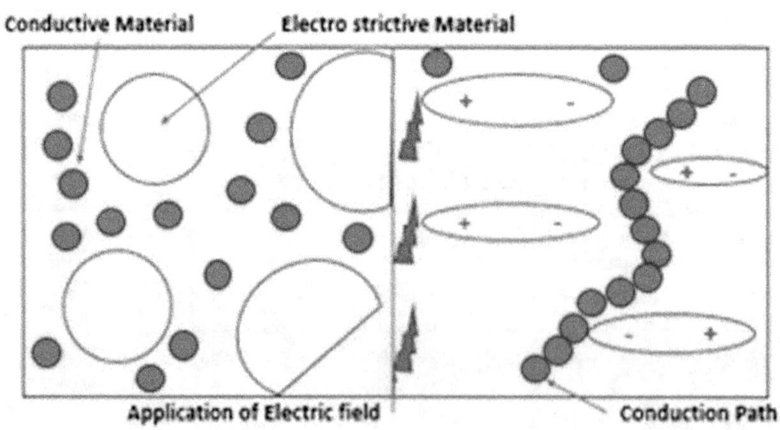

All of the particles in such materials move and rearrange in a specific direction when an electric charge is applied. To improve the characteristics, all the conductivity-containing particles align. There are numerous uses for this characteristic in smart structures (Damjanovic & Newnham, 1992) and (Uchino, 1986).

Some of the advantages are High energy density, Intrinsic robustness and disadvantages are Intricacy of material gets increased. Enough experimental evidence were not found. Some of its applications are in Damper, sensor, energy storage etc. (Olabi & Grunwald, 2008) and (Nia et al., 2017).

2.3. Magnetostrictive Materials

When a magnetic field is applied, the materials exhibit a change in strain (deformation), and the opposite is also true is shown in Figure 4. These materials are used as sensors and actuators. These act similarly to ferromagnetic materials in that all the particles align with the applied magnetic field (Olabi & Grunwald, 2008).

Figure 4. Magnetostrictive materials
(Olabi & Grunwald, 2008)

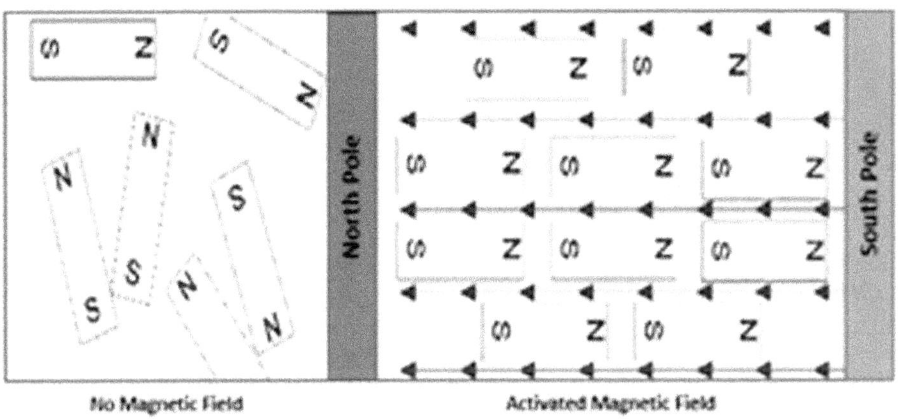

Because the state of MR materials can be controlled by the strength of an applied magnetic field, it is useful in applications where variable performance is desired. Many smart systems and structures would benefit from the change in viscosity or other material properties of MR materials. Beginning with the commercialization of MR fluid rotary brakes for use in aerobic exercise equipment in 1995. The past few years have witnessed a blossoming commercialization of MR fluid technology. MR fluid technology has been embraced by several manufacturers for inclusion in adverse spectrum of products that are now commercially available.

2.4. Rheological Materials

These substances are often found in a liquid condition, but they instantly change their physical state when a magnetic or electrical field is introduced. Such materials' fluidity can shift from stiff to more liquid form is shown in Figure 5. Rheological materials have primarily been used in the automotive industry. This substance is typically used in locations where the viscosity of construction materials needs to be changed (Parihar et al.,2007).

Figure 5. Rheological materials
(Parihar et al., 2007)

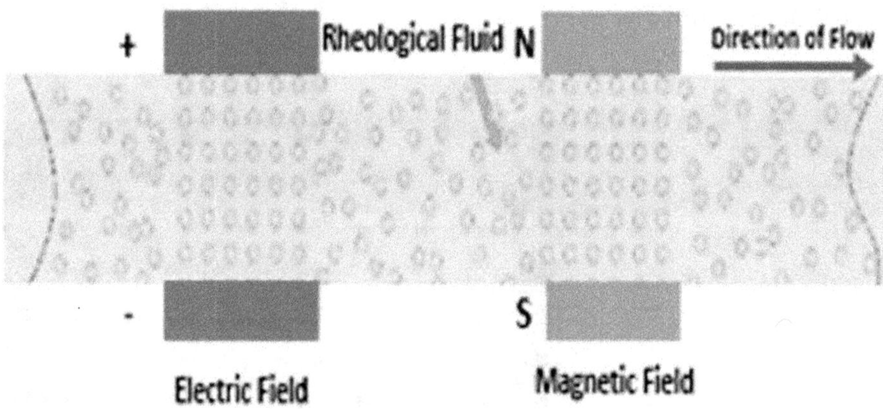

Some of its advantages are application of electric or magnetic field changes its physical state (Ahamed et al., 2018) and (Choi et al., 2016). And disadvantages are not easily available to work with. Some of its applications are in automobiles, buildings etc. (Parihar et al.,2007) and Ahamed et al., 2018).

2.5. Thermo-Responsive Materials

These substances, which come in the form of polymers, substantially alter their physical characteristics when exposed to temperature changes is shown in Figure 6. This material is employed in vehicles and aircraft, as well as in thermostats. It fluctuates continuously, ranging in size from extremely little to very huge. Additionally, they are known as stimulus responsive materials (Roy et al., 2013) and (Lyon et al., 2009).

Figure 6. Thermo-responsive materials
(Roy et al., 2013) and (Lyon et al., 2009)

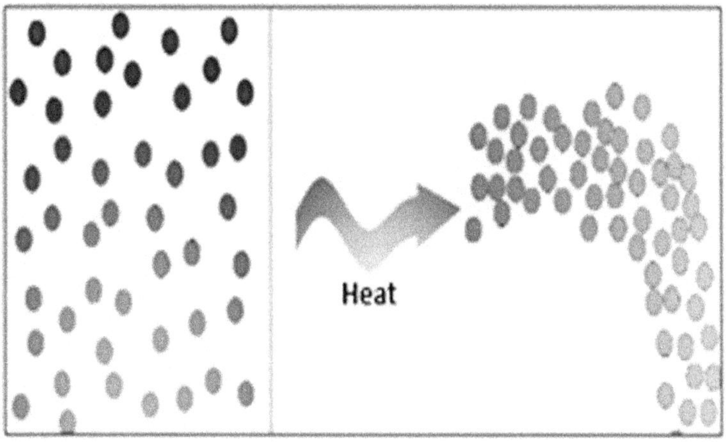

Some of its advantages are that they change physical property when exposed to temperature variation and disadvantages is that this material is very costly. These materials are applied in Vehicles, aircraft, thermostat etc. (Chassenieux & Tsitsilianis, 2016).

2.6. Electrochromic Materials

These are a crucial component of liquid crystal display devices today. When such materials are exposed to an electric current, their optical characteristics alter is shown in Figure 7. These components are also found in lithium-ion batteries. Additionally, if necessary, redox reaction alters its colour. Ions pass through a separator and reach the outer layer when voltage is applied. The layer becomes saturated and begins to reflect light, which causes it to become opaque or acquire any other necessary quality (Mortimer, 1997) and (Mortimer, 2011).

Some of its advantages are change of optical properties when an electric current is passed, and disadvantages are low coloration efficiency and durability. Its major application is in liquid crystal display units, lithium ion batteries etc. (Mortimer, 1997).

Figure 7. Electrochromic materials
(Mortimer, 1997) and (Mortimer, 2011)

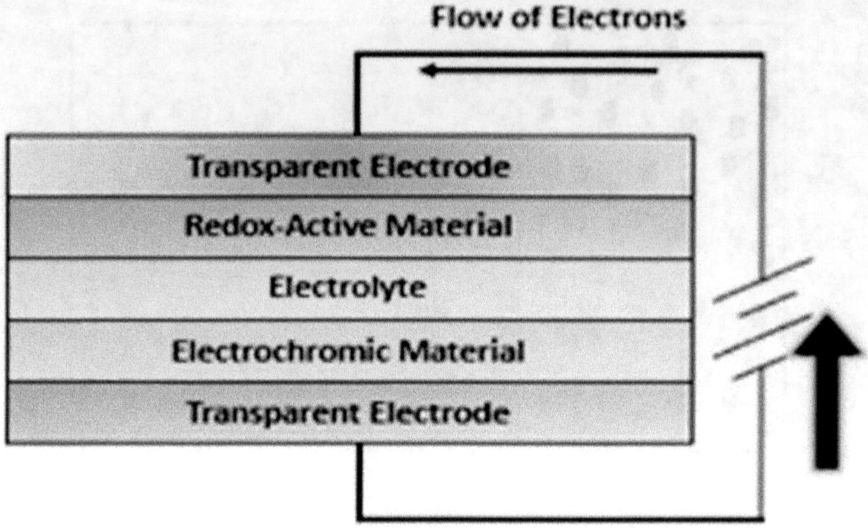

2.7 Fullerenes

These are composed of pentagons and hexagons. The caged sphere-shaped geometric structure has nodes made up entirely of carbon atoms is shown in Figure 8. These components are used in electrical equipment (Harvey, 2000). These can be found in a variety of shapes, including tubes, ellipses, hollow spheres, and more. Carbon atoms are bound together by single or double bonds to form fullerene. Its form makes it very stable and adaptable by nature. The use of these materials ranges from mechanical durability to crack prevention and corrosion resistance (Lee et al., 2010).

Some of its advantages are that they are highly stable, versatile in nature and disadvantages are health and environmental impacts. Their applications are in corrosion resistance, crack prevention to mechanical durability, electronic appliances etc. (Nierengarten, 2017) and (Chen et al., 2018).

2.8. Biomimetic Materials

These are the materials that drew inspiration from the geometric patterns seen in nature, as depicted in Figure 9. Nature offers several extremely straightforward yet useful shapes that provide functions like strength, concealment, water resistance, movement, etc. Such a material produces favourable qualities by simulating the geometry of a honeycomb structure. Examples include the leaves of a coconut tree, a bird's flight, a beehive, a spider web, a fish's skin that repels water, and more. Applications include everything from self-sensing to self-repairing in structures (Aizenberg & Fratzl, 2009) and (Al-Tabbaa et al., 2018).

Figure 8. Fullerenes
(Lee et al., 2010)

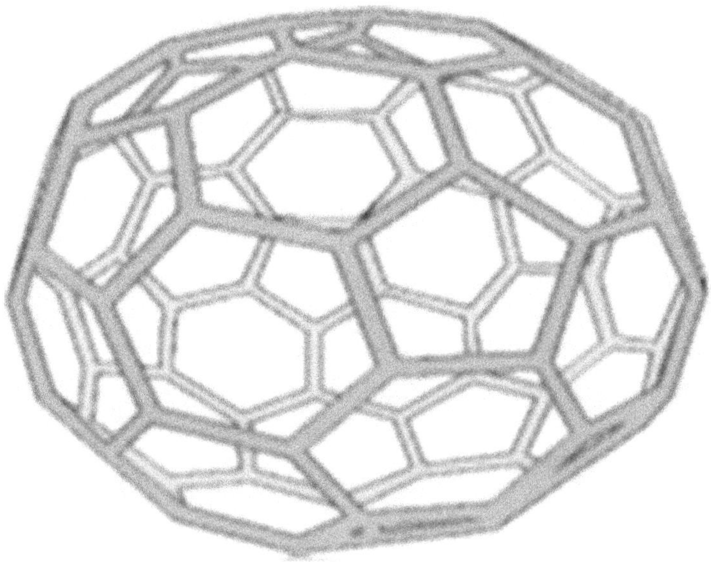

Figure 9. Biomimetic materials
(Aizenberg et al., 2009) and (Al-Tabbaa et al., 2018)

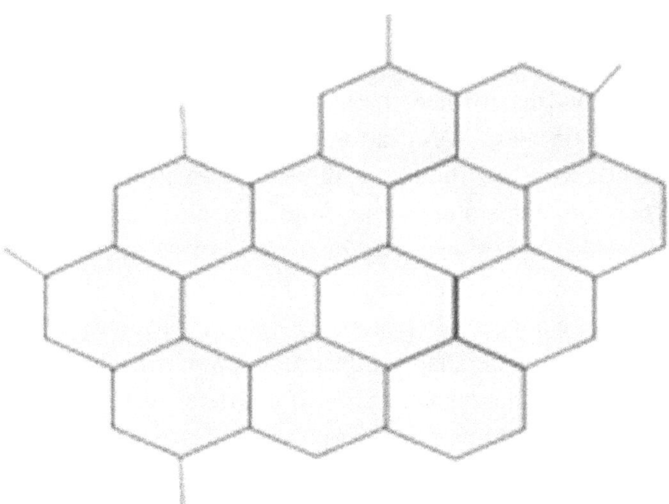

Some of its advantages are strength, camouflage, water proofing (Yaraghi et al., 2018) and disadvantages are poor abrasion resistance, Sensitivity to moisture. Biomimetic materials find its applications in various fields like architecture, robotics, superhydrophobic, self-cleaning materials, adhesives, nanotechnology, and biomedical.

3. DESCRIPTION OF SHAPE-MEMORY ALLOYS(SMAS)

When subjected to an appropriate thermal procedure, these alloys have the ability to regain their previous size and shape. SMAs also produce thermoelastic martensite, which implies that it may change from austenite into martensite and back again by applying the correct temperature. The hysteresis of the SMAs' heating and cooling performances is one of their key properties. Copper-containing Ni-Ti alloys go through a premartensitic phase change. Motion and strength are produced via the shape memory effect, while energy for deformation is retained in the form of extra elasticity. reme elasticity, hysteresis, significant damping effect, etc. are among the characteristics of SMAs (Banerjee et al., 1996) and (Otsuka et al., 1998).

3.1. Shape-Memory Effect

The shape-memory effect is essentially the ability of thermal procedures to recover the original size and shape of a distorted product. This process, also known as thermo elastic martensitic transition, results in a change in crystalline phase. The alloy returns to its previous shape and size and produces high strength materials when heated above the transformation temperature. When stress is created in alloys, martensite can be deformed with less energy than the austenitic phase using conventional methods. Repeated mechanical stress of the martensitic phase under high strains causes the formation of faults and disturbances that stabilise the corresponding plate alignments during the martensitic phase. We are aware that martensite has lesser symmetry than austenite, which accounts for the ease with which martensite can be generated from austenite. When the temperature drops from austenite, self-adjusting martensite varieties are formed. When stressed, identical borders will form, which causes martensite variants to subjectively disperse. Therefore, after reverting to austenite, martensite varieties can be easily restored to their original shape. Considering the properties of temperature the shape of the product won't change if it is cooled from above Af to below Mf, and it won't change if it is deformed below Mf temperature till an external source of thermal energy is provided. The two types of shape memory effects are one-way shape memory effect and two-way shape memory effect as shown in figure 10.(Lee et al., 2019) and (Facolt, 2008).

3.1. 1. One Way Shape Memory Effect (OWSME)

By simply raising the temperature after deformation, the material will, in effect, return to its original size and shape. Due to this action, material that is subjected to some force can produce strain up to 9%. The sample should start out in an austenitic state, and then, by lowering the temperature and applying no external force, martensite production will take place in the material without having any evident macroscopic effects. Finally, reverse strain occurs in the specimen by transforming it to martensite and austenite can be observed by this sample maintaining its original shape and size. This is caused by the orientation once again of different kinds of present martensites, which can be seen as an increase in sample strain and the external force that has been applied as well as temperature should stay constant, all of this is possible by just raising the temperature of the material without any external load(Mantovani, 2000).

Therefore, all different forms of martensite variants combine into a single version that is under load if the externally applied load is high. Twin orientation will arise by the conversion of a single variety of the martensite as a result of the significant mobilisation of the twin boundary. The inverse transformation starts when the sample's temperature exceeds Af. Heating, which recovers the sample's inelastic strain, will have been a temptation during this procedure.

3.1.2. Two Way Shape Memory Effect (TWSME)

The size plus shape of the austenitic state, under certain crucial conditions and the shape of the martensitic state, are considered in this process by the materials. However, OWSME is the intrinsic property of the SMAs, not this effect, which is a frequent behaviour of SMAs. However, after using the appropriate thermomechanical procedures, this is doable. Due to this effect, there will be a reversible & spontaneous change in the shape and size of the material along with the temperature cycle. We can also state that this effect will cause a spontaneous change in shape during cooling and heating If the material recalled this effect, we could update the material's reversible shape and size between two dissimilar ones by changing the temperature simply between Af and Mf. Because disturbances are brought about by significant deformation in the martensitic phase, the macroscopic sample recalls its shape and size to stabilise the measurements of this phase. These alterations are seen both in the initial stages and following the reversible creation of martensite through heating (Lexcellent et al., 2000). In this case, the material can reach its initial condition by simply raising the temperature, and it also permits the sample to take on a secondary shape by significantly lowering the temperature. The sample material's ability to remember this effect is a result of the learning processes' use of energy-storing abilities.

Figure 10. One way and Two-way shape memory effect
(Otsuka et al., 1998)

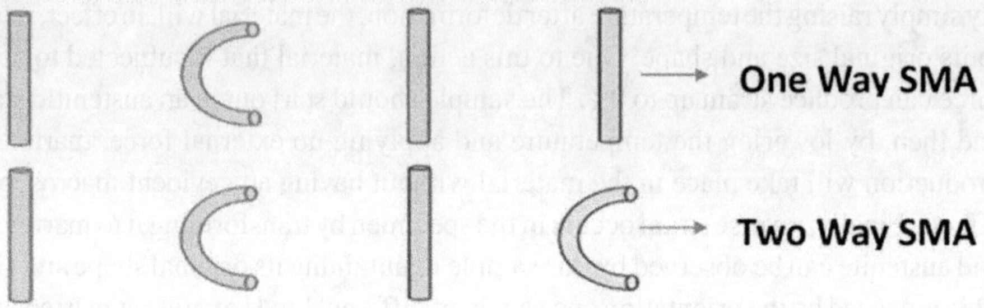

The sample will initially develop in the austenitic phase, and after that, martensite will form in the material with no evident macroscopic effects by lowering the temperature and applying no external force. These are some suitable thermodynamic mechanisms to achieve this outcome, at this time, the sample remembered its secondary shape in the martensitic phase because the material's memory had been improved by the mechanical cyclic processes (repeatedly increasing then lowering stress), as a result, the sample is allowed to repeatedly transition from its initial form to the recalled shape as a result of the learning process with only a small temperature increase and no external strain (Chen et al., 2008) and (Singh et al., 2022).

3.2. Superelasticity

Shape memory alloys were separated from other materials by the superelastic effect, which is a characteristic of SMAs. This effect aids in restoring the SMAs to their initial state under no strain after they have been distorted. When loading and unloading take place over the Af temperature, this effect is felt. Due to the application of a large load over the temperature Af, twin variations of martensite will form, which are unstable at high temperatures. Both the loaded and unloaded pathways do not overlap during the loading cycle, and the loaded path has a larger stress than the unloaded plateau. In certain circumstances, pseudo elasticity, an evolved version of transformational pseudo elasticity, is also known as superelasticity. At a certain temperature, nearly all shape memory alloys have the superelastic feature. Simply decreasing the rate of stress, raising austenitic strength, and decreasing hysteresis will boost the material's superelasticity. Because nickel-titanium hybrid materials exhibit extremely high superelasticity, Ni-Ti alloys make superior SMAs. The sample will initially form in the austenitic phase, and then deform elastically when under external load, then martensite will form internally in the austenitic phase as a result of an increase in external load, then final recovery happens to all of the combined strains whenever external loads are removed, and the specimen is restored to its initial state. If the material has a high superelasticity, the entire procedure will be swift. Although this characteristic resembles elasticity on the surface, it is incredibly different and sophisticated (Dobrescu & Vasilescu, 2015) and (Parvizi et al., 2022).

3.3. Damping Effect

Due to their strong damping properties, SMAs are able to efficiently transform mechanical energy to thermal energy (which dissipates as heat energy). Since this transfer cannot be undone, the material may now absorb shocks and vibrations. Internal friction, which looks at how martensite and austenite surfaces move when the two phases (austenite and martensite) are present, resulting in an increased high-damping effect since it allows for the distribution of a lot of energy. Particularly in the martensite state, friction studies the reversible shifting of martensite variant interfaces that permit better energy distribution than in the combination of both phases because friction in the austenitic state is minimal in comparison to other alloys and metal compositions. Strong modulus minima and an internal friction peak are both indicators of martensitic transition. The performance of tools, noise control, and longevity can all be increased if the materials have strong damping properties. Although damping is frequency independent, increasing the amplitude of an active vibration improves damping capabilities. In hybrid material composites, shape memory alloy stresses and strains can be regulated to provide active damping. We note that the damping coefficient values of every material vary. Some of the key factors that greatly influence the damping values include temperature, granule size, and martensitic interface density. Some fundamental SMA mechanisms, such as taking into account internal friction, generated martensite by stress, and temperature (thermally), can effectively apply the damping effect. By regulating the internal energy of the structure, the damping effect regulates the overlapping of the dynamical properties of the material elements. It also regulates impact damage, dynamic form control, and sonic radiation (Liu, 2000) and (Cho, 2017).

3.4. Hysteresis in SMAs

The difference between the temperatures at which SMA is heated to produce half of its austenite and cooled to produce the other half of its martensite is known as the hysteresis temperature. When the alloy reaches its original shape, which could be in the total martensite or complete austenite phase, if the cycle continues within the domain of Ms to As, there will be a change in hysteresis over a number of cycles (Ortín and Delaey, 2002).

3.4.1. Temperature Dependent Transformations

The hysteresis loop is created when the martensitic state changes cyclically as a result of temperature variation. The waste of the energy of a single crystal is indicated by the difference between two uniform temperature at which transformations of a single interface take place. In contrast to single interface transformations, multiple interface transformations widen the temperature range across which elastic strain energy can be partially stored and mitigate energy dissipation. The region of half cycles that they contain can be used as a model for how energy is lost during transformation. Compared to the rombohedral phase, a larger amount of energy is lost during the transition to the hexagonal phase. All of this occurs because there are no tensions in the thermally induced changes. Additionally, as more stress is applied in the low-stress region, more strain is created throughout this process. The hysteresis effect caused by temperature is diminished for high nickel content in SMA by an increase in external stress. With the implementation of stress, transition temperature also varies. Hysteresis loops represent the complete change of austenite to martensite, and minor loops, which are located beneath the big loops, represent the partial transformation. While both major and minor loops have a similar geometry and transition temperature (Dutta & Ghorbel, 2005).

3.4.2. Stress Dependent Transformations

There is a random upsurge in the superelastic effect for the single as well as poly crystals in this transformation, which has uniaxial stresses and is more temperature uniform than the austenitic phase temperature. Both crystal types will experience a hysteresis effect cycle, with single crystals experiencing a horizontal slope and polycrystalline systems experiencing an ascending slope. Every internal loop is independently specified by the initiation point, or as we might say, return point, which is a very important property of hysteresis can can be seen in SMAs. In both transformations, poly crystal SMAs have a significantly higher RPM effect than single crystals. In both diagrams (stress-strain hysteresis curve and strain-temperature hysteresis curve), the transformations to diverse martensitic crystalline composition by tension or compression uniaxially at varying values of temperatures create a variety of hysteresis cycles. The stress hysteresis in hysteresis loops is the middle of the loading stress curve. Hysteresis of solutionized low & old compositions of nickel will grow as external stress increases, and the same modification of the hysteresis curve has been seen for both single and multiple variations of martensite. Thermal stress hysteresis is the diameter of the strain temperature loops calculated at half of the forward strain transformations curve. Low and high concentrations of nickel alloys deform at different temperatures, and the importance of stress is shown by the pseudoelastic stress hysteresis. In strain-temperature plots, a constant slope may be noticed during cooling and heating (Hamilton et al., 2004).

4. APPLICATION OF SMART MATERIAL

Smart materials are widely used in most technological fields today, including civil engineering and building, healthcare, the military, robotic machinery, aerospace technology, business, nuclear power plants, and several active research organizations. Smart material in various fields are shown in figure 11.

Figure 11. Smart materials in various fields

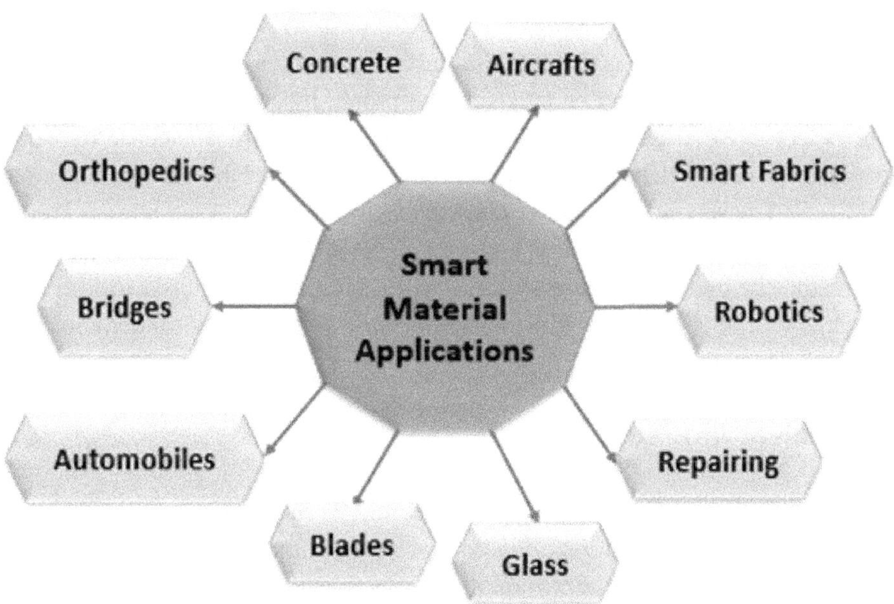

4.1. Walking Piezo Actuator

Walking Piezo Actuators, which comes in a variety of piezoelectric materials, are motors that can move both transversely and longitudinally. Each motor has four legs, each of which is made up of two pieces. While the foot is made to allow for forward and backward bending, the top part can be used to lengthen the leg. The longitudinal piezo motors are initially axially stretched out. Then, to create a bend, an orthogonal voltage is delivered to the foot of all legs. Two of the bended feet, which are on a plane at that point, then change their position and bend to the other side. Then, while the additional two legs continue the procedure, these two legs are contracted. Walking Piezo Actuators are used in most precise microscopes and exceedingly tiny robots because they may be controlled in the nanoscale range. Additionally, encoder precision is lower than 5 nm for constant speeds between 100 nm/s and 1 m/s, and is below 400 nm for speeds greater than 2 mm/s (Hawkins et al., 2008), (Dietz & Jaeger, 2013) and (Spanner & Koc, 2016).

4.2. Aeronautical Technology

Numerous cutting-edge materials have been used in various aeronautical technology fields. Since these materials are used in their manufacturing, detailing or even listing all of them would go beyond the scope of this review study. A lot of fringes have been discussed in this area.

The first illustration is the precise indicator known as the ultrasound fuel probe. The level of fuel can be precisely monitored by creating an ultrasonic with a piezoelectric electrical stimulus and detecting it with a piezoelectric transducer piezoelectric sensor when the incident wave is reflected by the surface of the liquid fuel. The second illustration is smart wings or SMA-based wings. Through aeroplane ascent and descent, this system uses SME to alter the form of the wings. As a result, the hydraulic motor systems will no longer be required. Additionally, the wings will be more compatible and produce less noise (Toozandehjani, 2018) and (Barbarino et al., 2008).

4.3. Nuclear Industries

In this expensive and risky industry, smart materials may offer several advantages. They can be utilised for automation systems to prevent people from being exposed to radioactive materials, and they are more effective and less expensive. However, one of the difficulties in this area is effectiveness to radioactive radiation and the effective rate of exposure (Giurgiutiu & Zagrai, 2003).

4.4. Intelligent Textile

One of the most significant human inventions is textile, which helps keep our bodies warm on chilly days and shields them from the sun's rays. However, in today's world, combining textiles with smart materials can give them additional features like changing colour in reaction to agitations in the surrounding environment. Shape memory polymers (SMPs) can improve the performance of various kinds of intelligent clothing. Intelligent fabrics can have one or more of the following characteristics: self-moving, emotional sensitivity, intrinsic cleaning, colour or shape alteration. Since polymers lose their softness after going through the glass transition at a certain temperature, they can be used to create textiles for body temperature regulation and ventilation. As a result of the polymer's enlarged molecular volume above the transition temperature of glass (Tg), there is more room for swapping evaporated water, which serves as a key factor in regulating body temperature on hot days. On the other hand, these waterproof polymer fabrics are breathable. In order to minimise evaporation from the body or infiltration of moisture from the environment, textiles can be created with lowest volume molecules at room temperature. However, at a greater temperature (Tg), the molecules expanded, creating a few tiny ventilation holes. By using multi-walled carbon nanotubes, the water vapour permeability has been enhanced (Thakur, 2017), (Rock, 2012) and (Mondal & Hu, 2006).

Additionally, because SMPs' vertical stress is lower than that of conventional textile, they are more flexible and may be made to fit the body.

4.5. Micro Actuator of Ni-Ti Film

Making Ni-Ti film allows for the combination of superelasticity and shape memory effects in a single object. The initial layers on the substrate must be sputtered using nickel-rich material before continue the process by lowering the Ni rate in the subsequent layers to create the film. As a result, the film acquires two different characteristics that could function as a bilayer system, or a two-way shape memory effect (Fu et al., 2009) and (Crăciunescu & Ercuta, 2015).

4.6. Linear Magneto Strictive Actuator

A linear rod built of MS materials is able to provide a force suitable for large transducer actuators. With a rod at the centre and a coil encircling it, it resembles a solenoid. By using a DC current to create a magnetic field, which causes the rod to expand, various mechanical uses can be made of the movement of the rod. Additionally, MS actuators have been employed in a wide variety of other applications, including response mass actuators, sonar transducers, wireless rotational motors, wireless linear micro-motors, thin-film applications in valves, and contactless torque sensors(Qader & Omar, 2019), (Shang et al., 2008) and (Olabi & Grunwald, 2008).

4.7. Nano Systems

Nanomaterials have a larger surface-to-volume ratio than their bulk counterparts, making them more active. Additionally, researchers are examining the impact of size on other physical parameters including thermal conductivity, lattice constant, volume per atom, melting temperature, Debye temperature and more. Quantum dots, nanowires, thin films, and even three-dimensional carbon nanotubes (CNTs) are all examples of nano-engineered materials that can be produced. In a similar size, CNT has electromechanical property that might be employed as an actuator. CNT is utilised to make nano tweezers, which can be used in very accurate contemporary microscopes to grasp microscopic things. By applying a voltage, tweezers' arms are elastically bent, and when the biassing voltage is zero, they return to their relaxed form. Additionally, the CNT elasticity value and nanotube diameter may be closely connected. By grabbing the sample with such arms, users of scanning probe microscopes may perform electrical measurements and other manipulations on it.

In a modern technology TiNi/DLC microcage with TWSME, we can use thermal or electrical energy to initiate the opening or closing procedure. It attracts more attention than traditional tweezers because it doesn't use extra force to grasp tiny objects (Qader & Omar, 2017), (Purushothaman & Muralidharan, 2009), (Mamand et al., 2012), (Yang et al., 2015), (Li et al., 2003), (Omar, 2016), (Abdullah et al., 2017), (Abdullah, 2016), (Li et al., 2008), (Kim et al., 1999) and (Li et al., 2003).

4.8. Power Generating Sidewalk

By using particular piezoelectric tiles, it is possible to harvest clean energy from people's footfalls. Using piezoelectric materials in pedestrian applications to supply energy to some pedestrian facilities is advantageous since they require a dynamic force to generate electricity(Cramm et al., 2011).

4.9. Medical Field

4.9.1.Orthodontic Arch Wire

Arch wire is a fantastic idea to address the issue of teeth that are unorganized. Although this method is not new, more effective materials have been employed to make the arch-wire. For a while, materials including gold alloy, stainless steel, beta titanium, and Cr-Co-Ni alloys were widely used. In contrast to the earlier conventional materials, Ni-Ti alloys have several unique characteristics, such as pseudoelasticity, which provides a constant force over a considerable distance with enough stiffness (Miyazaki and Sachdeva, 2009), (Ölander et al., 1932) and (Greninger et al., 1938).

The ideal range, or acceptable stress range, is identified. More forces have the potential to harm tissue, while fewer forces are more ineffective, which are referred to as excessive and suboptimal, respectively. The displacement of a wire that has been deformed, or the effective strain recovery, might thus be determined from this range. It is more effective to use them if the effective strain is longer.

Greater effective strain exists in Ni-Ti than in stainless steel. Additionally, because the martensite transformation of superelastic nitinol occurs in the ideal region, it has a higher value than other standard materials. Additionally, the archwire may continuously pull the misplaced teeth back into their proper position since a steady force is generated in the range of phase transformation.

4.9.2. Plates for Fractured Bones

People have been physically fixing broken bones since prehistoric times using materials like wood. Today, however, surgeons attempt to realign all broken pieces using surgery, and they can be joined together using some conformable plates made of SMAs. These types of alloys can be utilised to apply a similar stress to broken bones and aid in welding them quickly because they can store potential energy when they are deformed (Machado et al., 2003).

4.9.3. Spinal Vertebral Spacer

33 vertebrate bones that are joined to one another make up the spine. Since they can adjust their orientation for various body types, a washer-like SMA known as the vertebral spacer disc is required to shield them from friction. The most common issue is brought on by negative attitudes, which harm these discs. They can endure significant deformations without restricting movement of the body. Similarly, the SMA-based disk's superelasticity is used to conserve energy during deformation in terms of elasticity and martensite transformation, and by releasing this energy, the body can be forced to return to its original form (Kaplan et al., 2005).

4.9.4. Biomedical Application

To determine the precise site of the sickness, doctors must diagnose certain human organs. Today, a colonoscopy can be performed using an inchworm robot that uses SMA actuators. A prototype with three possessive degrees of freedom based on SMA-actuators (Hartl et al., 2009). When it passes through the colon, it is plainly observable. SMA-actuators on the robot expanded, lengthening it. Some of these actuators are additionally utilised to attach the robot to the wall. The robot has a one-step forward and backward movement range (Petrini & Migliavacca, 2011) more success when I suggested an earthworm-shaped robot with several actuator modules and a silicon casing to make it pliable and supple.

4.9.5. Bioengineered Robotic Hand

With the help of intelligent materials, it may be possible to recreate lost human bodies that could carry out their responsibilities with ease and accuracy. An example of a product made with SMAs is a bioengineered robotic hand. It is a small, light, and potent piece of technology that can operate just like artificial muscle. Aluminium and SMAs are used to create its fingers (Rush, 1997) and (Schetky, 1993).

4.10. Reducing Waste

Numerous artificial contaminations, including poisonous compounds and nuclear waste, pose serious health risks and have even altered some crucial environmental conditions. Therefore, certain smart materials are excellent candidates and have a significant impact for this goal in eliminating or lowering these hazardous wastes. Smart catalysts can recover themselves without generating additional impurities while eliminating wastes in a programmable manner. In addition, graphene-based nanoparticles could be used for water treatment. Similarly, carbon nanotube (CNT) can absorb medicines, endocrine-disrupting substances, polycyclic aromatic hydrocarbons, heavy metals, metalloids, and waste water (Gray & Bergbreiter, 1997), (Bergbreiter et al., 1993), (Yin et al., 2013), (Bao et al., 2011), (Xie et al., 2013), (Mishra, 2016), (Tofighy & Mohammadi, 2011), (Bagheri et al., 2011), (Yang & Xing, 2007), (Jung et al., 2015) and (Lu & Tsai, 2011).

4.11. Electrochromic Mirrors and Smart Windows

One of the challenges of nighttime driving is the backlighting of approaching vehicles from mirrors. In order to address this issue, smart mirrors that can substantially lessen the reflected light were developed. These mirrors are typically made of seven thin sputtered layers, including a clear glass, two transparent electrodes (anode and cathode), an ionic storage layer, an electrolyte, an electrochromic layer (such as WO3), and a standard mirror. When the electrical connection is open, the mirror will typically reflect almost all incident light. However, when the circuit is shut off, an electric field forms between the electrodes, which causes electrons from the ion storage layer to begin travelling through the electrolyte layer and towards the electrochromic layer. As a result, the mirror becomes black and the reflected light diminishes as the electric potential difference increases.

Similar to this, all that needs to be done to build a smart window is to switch out the backward mirror for a different piece of clear glass. In order to manage the transparent light that is irradiated. Utilising smart windows in intelligent homes has various benefits, including reducing energy use and enhancing privacy (Varaprasad 1998), (Niklasson & Granqvist, 2007), (Kumar et al., 2022) and (Kumar et al., 2023).

4.12. Magnetorheological Shock Absorber

A spring and a shock absorber are both parts of the damper system known as a magnetorheological shock absorber. They are an important component in practically all cars that reduces incoming shocks. Additionally, they can give the driver and passengers a comfortable feeling while protecting all of the car's components. The shock absorbers typically consist of a moving piston and an oily fluid in a closed cylinder. A piston divides the fluid inside the cylinder into up and down chambers. Therefore, a fluid's viscosity can aid in reducing the piston's up-and-down velocity. Additionally, a coil that may be regulated by electrical current is integrated in the MRF shock absorber in place of the traditional fluid. When magnetic flux affects the MRF flow through the channels and lessens the shock's impact (Saito, 2017), (Zuo et al., 2015) and (Li et al., 2000)

5. CONCLUSION

This review paper discusses the characteristics, contemporary uses, and the growth of smart materials, such as shape memory alloys. Smart material technology is a very interdisciplinary field by nature. Starting with the fundamental sciences like physics, chemistry, mechanics, computing, and electronics, it also includes the applied sciences and engineering, such mechanical and aeronautical engineering. This may help to explain why the development of smart structures for engineering systems is proceeding more slowly than expected, despite the rapid advancement of the science behind smart materials. The use of intelligent materials and structures is one of the most promising technologies for lifetime effectiveness and increased reliability at the moment. The ultimate goals of study in this area and are essential to the creation of high-quality smart materials are understanding and controlling the composition as well as microstructure of new materials. While all smart materials are significant, several of them are attracting more attention from academics because of their use in delicate industries. Robots, medical gadgets, automobile engines, and highly compact electronics are some of the key areas where smart materials play a significant role. In general, piezoelectric and shape memory materials are used more than other varieties, particularly for actuation reasons. Modern advanced shape memory alloys are compared to different manufacturing and development methods to construct intelligent materials. Shape memory effects include effects that are one-way and two-way. However, the healthcare industry's broad use of smart materials for implants and various other applications is the most significant.

6. FUTURE OF SMART MATERIALS

Smart materials and structures have a bright future. The ultimate goal of study in this area is to comprehend and control the composition and microstructure of any new materials, which is essential for the creation of high-quality smart materials. The creation of novel materials for use in various applications will be sped up by the insights acquired by accumulating data on how a material's crystal core structure behaves as it heats up, cools down, deforms, and changes. Superconducting wires, nanostructured materials, and structural ceramics are a few examples of the sophisticated materials that will shape nanotechnology.

REFERENCES

Abdullah, B. J., Jiang, Q., & Omar, M. S. (2016). Effects of size on mass density and its influence on mechanical and thermal properties of ZrO2 nanoparticles in different structures. *Bulletin of Materials Science*, *39*(5), 1295–1302. doi:10.100712034-016-1244-5

Abdullah, B. J., Omar, M. S., & Jiang, Q. (2017). Size Effects on Cohesive Energy, Debye Temperature and Lattice Heat Capacity from First-Principles Calculations of Sn Nanoparticles. *Proceedings of the National Academy of Sciences. India. Section A, Physical Sciences*, *88*(4), 629–632. doi:10.100740010-017-0417-y

Ahamed, R., Choi, S.-B., & Ferdaus, M. M. (2018). A state of art on magneto-rheological materials and their potential applications. *Journal of Intelligent Material Systems and Structures*, *29*(10), 2051–2095. doi:10.1177/1045389X18754350

Aizenberg, J., & Fratzl, P. (2009). Biological and Biomimetic Materials. *Advanced Materials*, *21*(4), 387–388. doi:10.1002/adma.200803699

Ajay, P. Ahmad, S., Sharma, J., & Gambhir, V. (Eds.). (2023). Handbook of Sustainable Materials: Modelling, Characterization, and Optimization (1st ed.). CRC Press.] doi:10.1201/9781003297772

Aktas Eken, G., & Hayri Acar, M. (2020). *PVDF-Based Shape Memory Materials, Oppor*. Fluoropolymers. doi:10.1016/B978-0-12-821966-9.00009-2

Al-Tabbaa, A., Lark, B., Paine, K., Jefferson, T., Litina, C., Gardner, D., & Embley, T. (2018). Biomimetic cementitious construction materials for next-generation infrastructure. *Proceedings of the Institution of Civil Engineers, Smart Infrastructure and Construction, 171*(2), 67–76, 10.1680/jsmic.18.00005

Bagheri, H., Ayazi, Z., & Aghakhani, A. (2011). A novel needle trap sorbent based on carbon nanotube-sol–gel for microextraction of polycyclic aromatic hydrocarbons from aquatic media. *Analytica Chimica Acta*, *683*(2), 212–220. doi:10.1016/j.aca.2010.10.026 PMID:21167973

Bahl, S., Nagar, H., Singh, I., & Sehgal, S. (2020). Smart materials types, properties and applications: a review. Elsevier. . doi:10.1016/j.matpr.2020.04.505

Bao, Q., Zhang, D., & Qi, P. (2011). Synthesis and characterization of silver nanoparticle and graphene oxide nanosheet composites as a bactericidal agent for water disinfection. *Journal of Colloid and Interface Science*, *360*(2), 463–470. doi:10.1016/j.jcis.2011.05.009 PMID:21628064

Barbarino, S., Ameduri, S., Lecce, L., & Concilio, A. (2008). Wing Shape Control through an SMA-Based Device. *Journal of Intelligent Material Systems and Structures*, *20*(3), 283–296. doi:10.1177/1045389X08093825

Bearat, H.H. & Vernon, B. L. (2011). Environmentally responsive injectable materials, Inject. *Biomater*, 263–297. . doi:10.1533/9780857091376.3.263

Bergbreiter, D. E., Zhang, L., & Mariagnanam, V. M. (1993). Smart ligands that regulate homogeneously catalyzed reactions. *Journal of the American Chemical Society*, *115*(20), 9295–9296. doi:10.1021/ja00073a058

Bogue, R. (2014). Smart materials: A review of capabilities and applications. *Assembly Automation*, *34*(1), 16–22. doi:10.1108/AA-10-2013-094

Chassenieux, C., & Tsitsilianis, C. (2016). Recent trends in pH/thermo-responsive self-assembling hydrogels: From polyions to peptide-based polymeric gelators. *Soft Matter*, *12*(5), 1344–1359. doi:10.1039/C5SM02710A PMID:26781351

Chen, J., Qiu, Q., Han, Y., & Lau, D. (2019). Piezoelectric materials for sustainable building structures: Fundamentals and applications. *Renewable & Sustainable Energy Reviews*, *101*, 14–25. doi:10.1016/j.rser.2018.09.038

Chen, M., Guan, R., & Yang, S. (2018). Hybrids of Fullerenes and 2D Nanomaterials. *Advanced Science, 6*(1), 1800941. Portico. . doi:10.1002/advs.201800941

Chen, S., Hu, J., Zhuo, H., & Zhu, Y. (2008). Two-way shape memory effect in polymer laminates. *Materials Letters*, *62*(25), 4088–4090. doi:10.1016/j.matlet.2008.05.073

Choi, S.-B., Li, W., Yu, M., Du, H., Fu, J., & Do, P. X. (2016). State of the art of control schemes for smart systems featuring magneto-rheological materials. *Smart Materials and Structures*, *25*(4), 043001. doi:10.1088/0964-1726/25/4/043001

Crăciunescu, C., & Ercuta, A. (2015). Modulated interaction in double-layer shape memory-based micro-designed actuators. *Science and Technology of Advanced Materials*, *16*(6), 065003. doi:10.1088/1468-6996/16/6/065003 PMID:27877845

Dagdelen, F., Kok, M., & Qader, I. N. (2019). Effects of Ta Content on Thermodynamic Properties and Transformation Temperatures of Shape Memory NiTi Alloy. *Metals and Materials International*, *25*(6), 1420–1427. doi:10.100712540-019-00298-z

Damjanovic, D., & Newnham, R. E. (1992, April). Electrostrictive and Piezoelectric Materials for Actuator Applications. *Journal of Intelligent Material Systems and Structures*, *3*(2), 190–208. doi:10.1177/1045389X9200300201

Dietz, T. G., & Jaeger, H. (2013). *Magnetostrictive actuator of a medical ultrasound transducer assembly, and a medical ultrasound handpiece and a medical ultrasound system having such actuator*. Google Patents.

Dobrescu, M., & Vasilescu, M. (2015). Processing and superelasticity of Ni-Ti shape memory alloys. *Key Engineering Materials*, *660*, 103–107. . doi:10.4028/www.scientific.net/KEM.660.103

Dolez, P. I., & Mlynarek, J. (2016). *Smart materials for personal protective equipment:tendencies and recent developments, Smart Text.* Their Appl. doi:10.1016/B978-0-08-100574-3.00022-9

Drossel, W. G., Kunze, H., Bucht, A., Weisheit, L., & Pagel, K. (2015). Smart3 - smart materials for smart applications. *Procedia CIRP*, *36*, 211–216. doi:10.1016/j.procir.2015.01.055

Dutta, S. M., & Ghorbel, F. H. (2005). Differential hysteresis modeling of a shape memory alloy wire actuator. *IEEE/ASME Transactions on Mechatronics*, *10*(2), 189–197. doi:10.1109/TMECH.2005.844709

Fu, Y. Q., Luo, J. K., Flewitt, A. J., Huang, W. M., Zhang, S., Du, H. J., & Milne, W. I. (2009). Thin film shape memory alloys and microactuators. *International Journal of Computational Materials Science and Surface Engineering*, *2*(3/4), 208. doi:10.1504/IJCMSSE.2009.027483

Garner, L. J., Wilson, L. N., Lagoudas, D. C., & Rediniotis, O. K. (2000, September). Development of a shape memory alloy actuated biomimetic vehicle. *Smart Materials and Structures*, *9*(5), 673–683. doi:10.1088/0964-1726/9/5/312

Giurgiutiu, V., & Zagrai, A. N. (2000). Use of smart materials technologies in radiation environments and nuclear industry. *Smart Structures and Materials 2000: Smart Structures and Integrated Systems.* SPIE. . doi:10.1117/12.388812

Gray, H. N., & Bergbreiter, D. E. (1997). Applications of Polymeric Smart Materials to Environmental Problems. *Environmental Health Perspectives*, *105*, 55. doi:10.2307/3433397 PMID:9114277

Hamdi, O., Mighri, F., & Rodrigue, D. (2018). Piezoelectric cellular polymer films: Fabrication, properties and applications. *AIMS Materials Science*, *5*(5), 845–869. doi:10.3934/matersci.2018.5.845

Hamilton, R. F., Sehitoglu, H., Chumlyakov, Y., & Maier, H. J. (2004). Stress dependence of the hysteresis in single crystal NiTi alloys. *Acta Materialia*, *52*(11), 3383–3402. doi:10.1016/j.actamat.2004.03.038

Han, C. D. (2007). *Experimental Methods for Measurement of the Rheological Properties of Polymeric Fluids. Rheology and Processing of Polymeric Materials* (Vol. 1). Polymer Rheology. doi:10.1093/oso/9780195187823.001.0001

Hartl, D. J., Mooney, J. T., Lagoudas, D. C., Calkins, F. T., & Mabe, J. H. (2009). Use of a Ni60Ti shape memory alloy for active jet engine chevron application: II. Experimentally validated numerical analysis. *Smart Materials and Structures*, *19*(1), 015021. doi:10.1088/0964-1726/19/1/015021

Harvey, J. A. (2000, December). Smart materials, Kirk-Othmer Encycl. *Chem. Technol.*, *1*. doi:10.1002/0471238961.1913011808011822.a01

Hawkins, M. J., Soon-Shiong, P., & Desai, N. (2008). Protein nanoparticles as drug carriers in clinical medicine. *Advanced Drug Delivery Reviews*, *60*(8), 876–885. doi:10.1016/j.addr.2007.08.044 PMID:18423779

Hu, J. L. (2016). *Introduction to Active Coatings for Smart Textiles, Act.* Coatings Smart Text. doi:10.1016/B978-0-08-100263-6.00001-0

Joshi, G. P., Tyagi, A., Meena, S. L., Chourasia, S., Saxena, N. S., & Kumar, A. (2021, August). Optical Band Gap Determination of Ni-Al Doped Polyaniline at Room Temperature and Different Annealing Temperatures. [). IOP Publishing.]. *Journal of Physics: Conference Series*, *1950*(1), 012091. doi:10.1088/1742-6596/1950/1/012091

Jung, C., Son, A., Her, N., Zoh, K.-D., Cho, J., & Yoon, Y. (2015). Removal of endocrine disrupting compounds, pharmaceuticals, and personal care products in water using carbon nanotubes: A review. *Journal of Industrial and Engineering Chemistry*, *27*, 1–11. doi:10.1016/j.jiec.2014.12.035

Kamila, S. (2013). Introduction, classification and applications of smart materials: Anoverview. *American Journal of Applied Sciences*, *10*(8), 876–880. doi:10.3844/ajassp.2013.876.880

Kaplan, K. M., Spivak, J. M., & Bendo, J. A. (2005). Embryology of the spine and associated congenital abnormalities. *The Spine Journal*, *5*(5), 564–576. doi:10.1016/j.spinee.2004.10.044 PMID:16153587

Kim, J. (2017). Multifunctional smart biopolymer composites as actuators. *Biopolym. Compos. Electron*, 311–331. . doi:10.1016/B978-0-12-809261-3.00011-5

Kim, P., & Lieber, C. M. (1999). Nanotube Nanotweezers. *Science*, *286*(5447), 2148–2150. doi:10.1126cience.286.5447.2148 PMID:10591644

Kulkarni, H., Zohaib, K., Khusru, A., & Shravan Aiyappa, K. (2018). Application of piezoelectric technology in automotive systems. *Materials Today: Proceedings*, *5*(10), 21299–21304. doi:10.1016/j.matpr.2018.06.532

Kumar, A., Kumar, P., Mittal, R. K., & Gambhir, V. (2023). Materials processed by additive manufacturing techniques. In Advances in Additive Manufacturing Artificial Intelligence, Nature-Inspired, and Biomanufacturing (pp. 217-233). Elsevier.] doi:10.1016/B978-0-323-91834-3.00014-4

Kumar, A., Mittal, R. K., & Haleem, A. (Eds.). (2022). *Advances in Additive Manufacturing: Artificial Intelligence, Nature-Inspired, and Biomanufacturing*. Elsevier. doi:10.1016/C2020-0-03877-6

Lee, A. Y., An, J., Chua, C. K., & Zhang, Y. (2019). Preliminary investigation of the reversible 4D printing of a dual-layer component. *Engineering (Beijing)*, *5*(6), 1159–1170. doi:10.1016/j.eng.2019.09.007

Lee, H.-S., & Cho, C. (2008). Magnetoelastic Characteristics of the Multilayered Magnetostrictive Thin Film With Polyimide Substrate for Micro Actuator. *IEEE Transactions on Magnetics*, *44*(11), 4049–4052. doi:10.1109/TMAG.2008.2003070

Lee, J., Mahendra, S., & Alvarez, P. J. J. (2010, July). Nanomaterials in the Construction Industry: A Review of Their Applications and Environmental Health and Safety Considerations. *ACS Nano*, *4*(7), 3580–3590. doi:10.1021/nn100866w PMID:20695513

Lexcellent, C., Leclercq, S., Gabry, B., & Bourbon, G. (2000). Two way shape memory effect of shape memory alloys: an experimental study and a phenomenological model. *Int.J. Plast. 16*(10), 1155–1168. https://doi.org/ 00005-X. doi:10.1016/S0749-6419(00)

Li, C. (2003). Elastic moduli of multi-walled carbon nanotubes and the effect of van der Waals forces. *Composites Science and Technology*, *63*(11), 1517–1524. doi:10.1016/S0266-3538(03)00072-1

Li, C., Thostenson, E. T., & Chou, T.-W. (2008). Sensors and actuators based on carbon nanotubes and their composites: A review. *Composites Science and Technology*, *68*(6), 1227–1249. doi:10.1016/j.compscitech.2008.01.006

Li, D., Wu, Y., Kim, P., Shi, L., Yang, P., & Majumdar, A. (2003). Thermal conductivity of individual silicon nanowires. *Applied Physics Letters*, *83*(14), 2934–2936. doi:10.1063/1.1616981

Li, W. H., Yao, G. Z., Chen, G., Yeo, S. H., & Yap, F. F. (2000). Testing and steady state modeling of a linear MR damper under sinusoidal loading. *Smart Materials and Structures*, *9*(1), 95–102. doi:10.1088/0964-1726/9/1/310

Liu, Y. (2000). *Shape memory implants, Shape Mem*. Implant. . doi:10.1007/978-3-642-59768-8

Lu, T.-L., & Tsai, Y.-C. (2011). Sensitive electrochemical determination of acetaminophen in pharmaceutical formulations at multiwalled carbon nanotube-alumina-coated silica nanocomposite modified electrode. *Sensors and Actuators. B, Chemical*, *153*(2), 439–444. doi:10.1016/j.snb.2010.11.013

Lyon, L. A., Meng, Z., Singh, N., Sorrell, C. D., & St. John, A. (2009). Thermoresponsive microgel-based materials. *Chemical Society Reviews*, *38*(4), 865. doi:10.1039/b715522k PMID:19421566

Machado, L. G., & Savi, M. A. (2003). Medical applications of shape memory alloys. *Brazilian Journal of Medical and Biological Research*, *36*(6), 683–691. doi:10.1590/S0100-879X2003000600001 PMID:12792695

Madangopal, K., Banerjee, S., & Lele, S. (1994). Thermal arrest memory effect. *Acta Metallurgica et Materialia*, *42*(6), 1875–1885. doi:10.1016/0956-7151(94)90012-4

Mamand, S., Omar, M., & Muhammed, A. (2012). Calculation Of Lattice Thermal Conductivity Of Suspended GaAs nanobeams: Effect Of Size Dependent Parameters. *Advanced Materials Letters*, *3*(6), 449–458. doi:10.5185/amlett.2012.icnano.102

Mantovani, D. (2000). Shape memory alloys: Properties and biomedical applications, JOM. *Journal of the Minerals Metals & Materials Society*, *52*(10), 36–44. doi:10.100711837-000-0082-4

Mishra, A. K. (Ed.). (2016). *Smart Materials for Waste Water Applications*., doi:10.1002/9781119041214

Miyazaki, S., & Sachdeva, R. L. (2009). Shape memory effect and superelasticity in Ti—Ni alloys. *Shape Memory Alloys for Biomedical Applications*, 3–19. Elsevier. . doi:10.1533/9781845695248.1.3

Mondal, S., & Hu, J. L. (2006). A novel approach to excellent UV protecting cotton fabric with functionalized MWNT containing water vapor permeable PU coating. *Journal of Applied Polymer Science*, *103*(5), 3370–3376. doi:10.1002/app.25437

Mortimer, R. J. (1997). Electrochromic materials. *Chemical Society Reviews*, *26*(3), 147. doi:10.1039/cs9972600147

Mortimer, R. J. (2011). Electrochromic Materials. *Annual Review of Materials Research*, *41*(1), 241–268. doi:10.1146/annurev-matsci-062910-100344

Naresh, C., Bose, P. S. C., & Rao, C. S. P. (2016). Shape memory alloys: a state of art review. *IOP Conf. Ser. Mater. Sci. Eng. 149* (1). 10.1088/1757-899X/149/1/012054

Nia, E. M., Zawawi, N. A. W. A., & Singh, B. S. M. (2017). A review of walking energy harvesting using piezoelectric materials. *IOP Conference Series. Materials Science and Engineering*, *291*, 012026. doi:10.1088/1757-899X/291/1/012026

Nierengarten, J.-F. (2017). Fullerene hexa-adduct scaffolding for the construction of giant molecules. *Chemical Communications (Cambridge)*, *53*(87), 11855–11868. doi:10.1039/C7CC07479D PMID:29051931

Niklasson, G. A., & Granqvist, C. G. (2007). Electrochromics for smart windows: Thin films of tungsten oxide and nickel oxide, and devices based on these. *Journal of Materials Chemistry*, *17*(2), 127–156. doi:10.1039/B612174H

Olabi, A. G., & Grunwald, A. (2008). Design and application of magnetostrictive materials. Materials &amp. *Materials & Design*, *29*(2), 469–483. doi:10.1016/j.matdes.2006.12.016

Ölander, A. (1932). An Electrochemical Investigation Of Solid Cadmium-Gold Alloys. *Journal of the American Chemical Society*, *54*(10), 3819–3833. doi:10.1021/ja01349a004

Omar, M. S. (2016). Structural and Thermal Properties of Elementary and Binary Tetrahedral Semiconductor Nanoparticles. *International Journal of Thermophysics*, *37*(1), 11. doi:10.100710765-015-2026-9

Ortín, J., & Delaey, L. (2002). Hysteresis in shape-memory alloys. *International Journal of Non-linear Mechanics*, *37*(8), 1275–1281. doi:10.1016/S0020-7462(02)00027-6

Otsuka, H., Tanahashi, H., Maruyama, T., Murakami, M., & Yamada, H. (1998). Development of Fe-Mn-Si shape memory alloys for pipe joints in steel construction. *Materia (Sendai)*, *37*(4), 283–285. doi:10.2320/materia.37.283

Panthongsy, P., Isarakorn, D., Hamamoto, K., & Janphuang, P. (2019). Performance and Behavior Analysis of Piezoelectric Energy Harvesting Floor Tiles. *2019 5th International Conference on Engineering, Applied Sciences and Technology (ICEAST)*. IEEE. 10.1109/ICEAST.2019.8802555

Parvizi, S., Hashemi, S., & Moein, S. (2022). NiTi shape memory alloys: Properties Nickel- Titanium Smart Hybrid Mater. From *Micro- to Nano-structured Alloy, Emerg. Appl.* 399–426. Elsevier. . doi:10.1016/B978-0-323-91173-3.00021-3

Petrini, L., & Migliavacca, F. (2011). Biomedical Applications of Shape Memory Alloys. *Journal of Metallurgy*, *2011*, 1–15. doi:10.1155/2011/501483

Phillips, A., & Thelin, L. W. (1927). An x-ray study of the beta transformation in copper-zinc alloys. *Journal of the Franklin Institute*, *204*(3), 359–368. doi:10.1016/S0016-0032(27)92102-4

Purushothaman, K. K., & Muralidharan, G. (2009). The effect of annealing temperature on the electrochromic properties of nanostructured NiO films. *Solar Energy Materials and Solar Cells*, *93*(8), 1195–1201. doi:10.1016/j.solmat.2008.12.029

Qader, I. N., & Omar, M. S. (2017). Carrier concentration effect and other structure-related parameters on lattice thermal conductivity of Si nanowires. *Bulletin of Materials Science*, *40*(3), 599–607. doi:10.100712034-017-1393-1

Rock, M. (2012). *Temperature responsive smart textile*. Google Patents.

Rogers, C. A. (1989). *U.S. Army Research Office Workshop on Smart Materials*. Structures and Mathematical Issues Held in Blacksburg. 10.21236/ADA218528

Roy, D., Brooks, W. L. A., & Sumerlin, B. S. (2013). New directions in thermoresponsive polymers. *Chemical Society Reviews*, *42*(17), 7214. doi:10.1039/c3cs35499g PMID:23450220

Roytburd, A., Slutsker, J., & Wuttig, M. (2000). Smart composites with shape memory alloys. *Compr. Compos. Mater.* . doi:10.1016/B0-08-042993-9/00212-6

Rush, J. A. (1997). *Memory wire robotic hand*. Google Patents.

Saito, K. (2017). *Magnetic rheological fluid shock absorber*. Google Patents.

Schetky, L. (1993). *A review of: "Engineering Aspects Of Shape Memory Alloys" by T.W. Duerig, K.N. Melton, D. Stockel.* Taylor & Francis. doi:10.1080/10426919308934859

Shang, X., Pan, E., & Qin, L. (2008). Mathematical modeling and numerical computation for the vibration of a magnetostrictive actuator. *Smart Materials and Structures*, *17*(4), 045026. doi:10.1088/0964-1726/17/4/045026

Singh, V., Kumar, A., Alam, M., Kumar, A., Kumar, P., & Goyat, V. (2022). A study of morphology, UV measurements and zeta potential of zinc ferrite and Al2O3 nanofluids. *Materials Today: Proceedings*, *59*, 1034–1039. doi:10.1016/j.matpr.2022.02.371

Sodano, A., Inman, D., & Park, G. (2004). A Review of Power Harvesting from Vibration Using Piezoelectric Materials. *Shock Vib Digest, 36*(3), 197–205. . doi:10.1177/0583102404043275

Soutis, C. (2010). ACM Special Issue: ICCM-17: Deformation and Fracture of Composites. *Applied Composite Materials*, *18*(1), 1–1. doi:10.100710443-010-9183-6 PMID:35095206

Spanner, K., & Koc, B. (2016). Piezoelectric Motors, an Overview. *Actuators*, *5*(1), 6. doi:10.3390/act5010006

Swait, T. J., Rauf, A., Grainger, R., Bailey, P. B. S., Lafferty, A. D., Fleet, E. J., Hand, R. J., & Hayes, S. A. (2012). Smart composite materials for self-sensing and self-healing. *Plastics, Rubber and Composites*, *41*(4–5), 215–224. doi:10.1179/1743289811Y.0000000039

Thakur, S. (2017). *Shape Memory Polymers for Smart Textile Applications*. Textiles for Advanced Applications., doi:10.5772/intechopen.69742

Tofighy, M. A., & Mohammadi, T. (2011). Adsorption of divalent heavy metal ions from water using carbon nanotube sheets. *Journal of Hazardous Materials*, *185*(1), 140–147. doi:10.1016/j.jhazmat.2010.09.008 PMID:20926186

Toozandehjani, M. (2018). Conventional and Advanced Composites in Aerospace Industry: Technologies Revisited. *American Journal of Aerospace Engineering*, *5*(1), 9. doi:10.11648/j.ajae.20180501.12

Uchino, K. (1986). Piezoelectric and Electrostrictive Actuators. *Sixth IEEE International Symposium on Applications of Ferroelectrics.* IEEE. 10.1109/ISAF.1986.201218

Varaprasad, D. V. (1998). *Electrochromic mirrors and devices*. Google Patents.

Wada, K., & Liu, Y. (2008). On the mechanisms of two-way memory effect and stress-assisted two-way memory effect in NiTi shape memory alloy. *Journal of Alloys and Compounds*, *449*(1–2), 125–128. doi:10.1016/j.jallcom.2006.01.128

Xie, G., Zhang, K., Guo, B., Liu, Q., Fang, L., & Gong, J. R. (2013). Graphene-Based Materials for Hydrogen Generation from Light-Driven Water Splitting. *Advanced Materials*, *25*(28), 3820–3839. doi:10.1002/adma.201301207 PMID:23813606

Xu, X., Cao, D., Yang, H., & He, M. (2018). Application of piezoelectric transducer in energy harvesting in pavement. *International Journal of Pavement Research and Technology*, *11*(4), 388–395. doi:10.1016/j.ijprt.2017.09.011

Yang, J., Tang, H., Zhao, Y., Zhang, Y., Li, J., Ni, Z., Chen, Y., & Xu, D. (2015). Thermal conductivity of zinc blende and wurtzite CdSe nanostructures. *Nanoscale*, *7*(38), 16071–16078. doi:10.1039/C5NR04117A PMID:26372172

Yang, K., & Xing, B. (2007). Desorption of polycyclic aromatic hydrocarbons from carbon nanomaterials in water. *Environmental Pollution*, *145*(2), 529–537. doi:10.1016/j.envpol.2006.04.020 PMID:16777283

Yaraghi, N. A., & Kisailus, D. (2018). Biomimetic Structural Materials: Inspiration from Design and Assembly. *Annual Review of Physical Chemistry*, *69*(1), 23–57. doi:10.1146/annurev-physchem-040215-112621 PMID:29237136

Yin, H., Zhao, S., Wan, J., Tang, H., Chang, L., He, L., Zhao, H., Gao, Y., & Tang, Z. (2013). Three-Dimensional Graphene/Metal Oxide Nanoparticle Hybrids for High-Performance Capacitive Deionization of Saline Water. *Advanced Materials, 25*(43), 6270–6276. Portico. . doi:10.1002/adma.201302223

Zuo, L., Tang, X., & Zhang, P. S. (2015). *Electricity generating shock absorbers*. Google Patents.

Chapter 6
Experimental Study of Impact Parameters on the Deforming Loads in Incremental Forming

Lokesh Rathee
University Institute of Engineering and Technology, Maharshi Dayanand University, India

Rakesh Rathee
University Institute of Engineering and Technology, Maharshi Dayanand University, India

Ajay Kumar
School of Engineering and Technology, JECRC University, India

ABSTRACT

Single point incremental forming (SPIF) is an emerging sheet forming method and has been widely accepted by the researchers for fabricating the customized items for the industrial uses and the end users. The measurement of the deforming loads during this viable method would assess the need of specific capacity of the SPIF machines for the given process conditions and materials. Therefore, this work aims at investing the influence of the punch feed rate and the forming angle on the peak values of the deforming loads. The conical frustums of the varied forming angles are produced from the AA2024 aluminum alloy sheets. The CNC milling machine was utilized for this die-less sheet forming. The table type dynamometer was utilized for recording the peak values of the deforming loads during the fabrication of the frustums. Experimental results confirmed that both the input factors were significant to impact the deforming loads. It was also confirmed that the small increment in the forming angle led to the significant rise in the deforming loads for all punch feed rates.

DOI: 10.4018/978-1-6684-9224-6.ch006

1. INTRODUCTION

A dramatic development in the obsolescence of manufacturing techniques has been well accepted by the global market for producing user-ready and user-oriented components. The production sector has been attempting several approaches to upgrade its business by hatching the customized products economically (Mittal, 2020). Customized and batch-type production can be executed by employing the emerging techniques of manufacturing of less cycle-time and economical set-up. Incremental Sheet Forming (ISF) can be refrred as an emerging method for manufacturing the customized sheet material parts for various applications (Kumar and Gulati, 2019, Kumar et al., 2018a). The end-user components can be formed by the ISF technique with negligible wastage. Also, the force needed to produce the deformation during the ISF process is quite less in comparison of other sheet forming methods. The requirement of a smaller forming force leads to the employment of the light-weight and small size forming machinery to execute the process (Kumar et al., 2018b; Jeswiet et al., 2005; Bagudanch et al., 2019). Moreover, the cost of the forming machinery and tools involved in the ISF technique becomes lower that leads to producing a smaller number of components to reach the breakeven point of the process. Furthermore, the absence of dedicated and costly experimental set-up, during the ISF process, opens the gates for producing customized parts economically according to the cutting-edge requirements of product users (Gulati and Kumar, 2018; Manju and Kumar, 2020; Centeno et al., 2014).

Moreover, various economic aspects, such as low emissions and energy requirements, can be controlled up to the large extent by ensuring the utilization of the potential techniques and processes that can fabricate the parts of lightweight materials with reduced power and machinery to ensure the saving of resources (Suresh et al., 2015; Kumar et al., 2019a). The principle of ISF includes deforming the sheet material plastically, contour by contour, by governing the punch on the specimen using the CNC machine (Kumar et al., 2020). SPIF is a purely “die-less process” that refuses the requirement of specific dies and is capable of fabricating the intricate and complex shapes successfully (Kumar et al., 2019b; Kumar etal., 2018c). SPIF, also called “negative incremental forming”, is based on deforming the sheet material, contour by contour, by a hemispherical-ended punch that is normally controlled by a predetermined toolpath as illustrated in Figure 1. Furthermore, the lower deforming loads are needed for deforming the material into the designed shape. The need for a smaller amount of required forces, for producing deformation, leads to reducing the power and energy requirements up to a greater extent for accomplishing the process that further reduces the cost of process and machinery. Hence, the peak deforming loads to produce the desired shape, for specific operating conditions and materials, can determine the size and capacity of forming hardware. Moreover, estimation and investigation of the maximum value of deforming loads may assure the secure and effective employment of forming machinery during this die-less process (Kumar et al., 2019c). The measurement of forming force can be accomplished by mounting the force dynamometers either on the forming tool or between the ISF fixture and machine table. Most of the researchers (Kumar et al., 2018c; 2019c, Alsamhan et al., 2019; Honarpisheh et al., 2018; Zhai et al., 2020; Kumar and Gulati, 2018, 2021; Oleksik et al., 2010; Fiorentino et al., 2009) have used a later method (in which a load cell or force dynamometer is fixed between the clamping device and machine table) for recording the deforming loads as represented by Figure 2.

Kumar et al. (2018c) explored the axial peak forces to check the influence of input variables and found that the deforming loads increased as the wall angle and punch radius was increased. Kumar et al. (2019c) also explored the impact of punch shape and formimg angle on the maximum axial loads and they found that the deforming loads were raised on the increase of punch radius and wall angle. Alsamhan et al. (2019) designed a model for estimating the forming force using artificial intelligence techniques for SPIF operation and investigated an experimental campaign on AA1050-H14 sheets. Honarpisheh et al. (2018) performed the experimental campaign on bimetal sheets during the SPIF process. An error of 9.6% was also recorded between the simulation and experimental values of forming force. Zhai et al. (2020) produced the pyramidal frustums from AA1050-O sheets and studied the impact of punch rotation. A reduction of 40% was noticed in the forming forces by raising the ultrasonic amplitude from 0 to 10 μm. Kumar and Gulati (2018) produced the conical frustums from AA6063 and AA2024 sheets experimentally. A reduction of 14.20% was noticed in axial peak forces when the hemispherical punch of 5.8 mm radius was employed in place of the flat-end punch of the identical radius. Kumar and Gulati (2021) also optimized the SPIF process using the Taguchi method and formed the truncated cone-shaped components.

Figure 1. The principle of SPIF technique

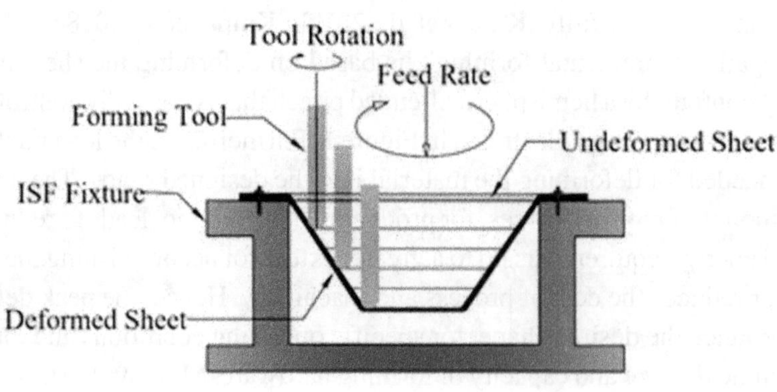

Literature (Kumar et al., 2019b; Oleksik et al., 2010; Fiorentino et al., 2009) also reveals that the axial deforming load (Fz) is higher than the horizontal deforming loads (i.e. Fx and Fy, see Figure 2 for components of loads). Thus, study of peak deforming loads can assure the safe and effective employment of forming machines and auxiliary hardware. Moreover, analysis of the relationship between the input variables and forming forces can open the channel to predict the failures of parts, machinery, and forming processes efficiently. Furthermore, the estimation of forming forces can ensure the prediction of power to be consumed by the forming machine and controlling the process on-line.

Figure 2. The arrangement for recording the deformimg loads

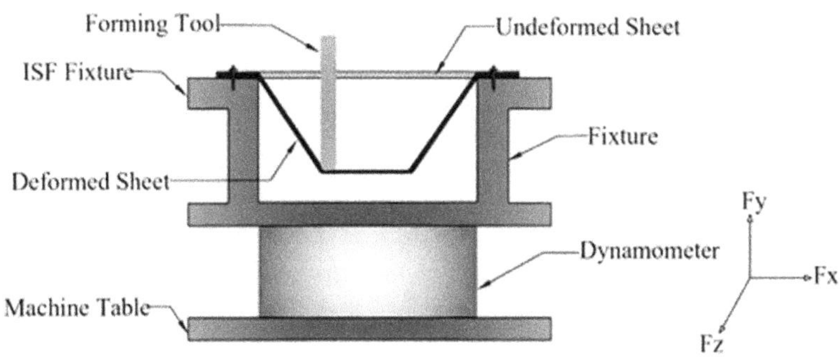

The current study aims at investigating the effects of punch feed rate and forming angle on peak deforming loads. The material, considered for experiments (i.e. AA2024-O), is popularly used for various applications of sheet products due to its suitable characteristics such as lightweight and greater strength. Table 1 delineates the chemical compositions of this alloy. In this work, three levels of forming angle (60°, 64°, 68°) and four levels of punch feed rate (1000, 1500, 2000, 2500 mm/min) have been studied for cheking their impact on deforming loads. The other experimental factors were kept constant as punch radius 5.80 mm, blank thickness 1.0 mm, tool rotation 2000 rpm.

Table 1. Chemical compositions of AA2024-O sheet

Element	Al	Cr	Cu	Fe	Mg	Mn	Si	Ti	Zn
Weight %	91.50	0.10	4.60	0.30	1.70	0.80	0.50	0.10	0.20

2. MATERIALS AND METHODS

An experimental campaign for the SPIF process was explored as represented by Figure 3. Truncated cone-shaped components, of 65 mm upper radius and 75 mm forming depth, were modeled using the Solidworks® software before transferring to Delcam™ software for generating the Computer-Aided Manufacturing (CAM) instructions. The helical tool trajectory was employed to complete the forming of components. The deforming punches were employed, through collet of CNC milling, to accomplish the rotation. The SPIF fixture was mounted on the machine table and the blank was clamped in the SPIF fixture. A relative motion between the blank and deforming punch was produced by moving the machine table. On the other hand, the forming tool was allowed to move along the z-axis (vertical downward direction) to accomplish the step depth according to required process conditions. The axial peak values of forming forces were measured with the help of a dynamometer which was installed between the SPIF fixture and CNC milling table. The exerted loads were measured by utilizing a dynamometer and data acquisition apparatus to save the real-time values.

Figure 3. The sheme for experimental examinations and load measurement

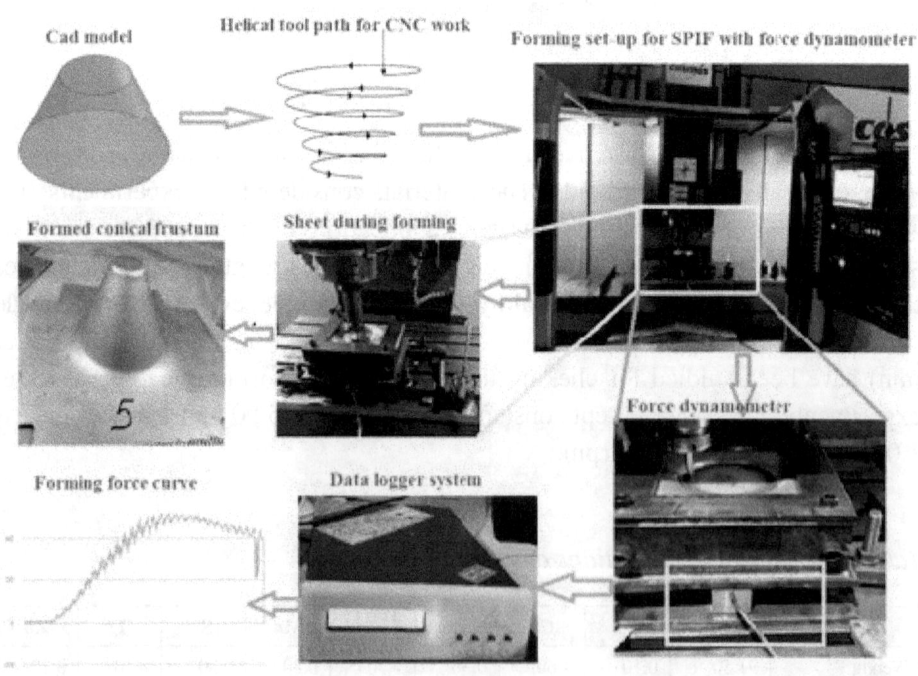

3. EXPERIMENTAL ANALYSIS

The experimentally recorded peak deforming loads (Fz_max.) are represented by **Table 2 and Figure 4.** The rise in forming angle led to sigficant increment in the deforming loads.

Table 2. Experimental values of peak deforming loads

Run	Punch feed rate (mm/min)	Forming angle (Degree)	Experimental Fz_max. (N)
1	1000	60	896
2	1000	64	968
3	1000	68	1059
4	1500	60	908
5	1500	64	1003
6	1500	68	1122
7	2000	60	921
8	2000	64	1034
9	2000	68	1175
10	2500	60	934
11	2500	64	1059
12	2500	68	1202

The greater value of forming angle (68°) resulted in the fracture of specimen at a height of 15.80 mm when a higher feed rate (2500 mm/min) was employed as shown in Figure 5. On the other hand, when the same feed rate was employed with the lower value of wall angle (level 2, 64° in this case), components were formed successfully (forming depth of 70 mm) at a lower forming force (1059 N) as shown in Figure 6. It can also be noticed from the observed values of results that the intensity of increment was also found to raise as the level of wall angle was increased.

Furthermore, the deforming loads was found to increase by 34.16%, 9.41%, and 13.51% when the punch feed rate was raised from level 1 to level 4 for a wall angle of 60°, 64°, and 68° respectively. Besides, deforming loads were raised by 18.20%, 23.57%, 27.58%, and 28.70% as the forming angle was raised from 60° to 68° for level 1, 2, 3 and 4 of the punch feed rate respectively. When the condition of forming changed from the combination of higher forming angle (68°) and greater punch feed rate (2500 mm/min in this case) to a condition of the lower value of wall angle

(60°) and lower feed rate ((1000 mm/min), the forming force was found to decrease by 25.45% which is very significant to save power and energy for producing the components.

Figure 4. Effect of forming angle and punch feed rate on deforming load

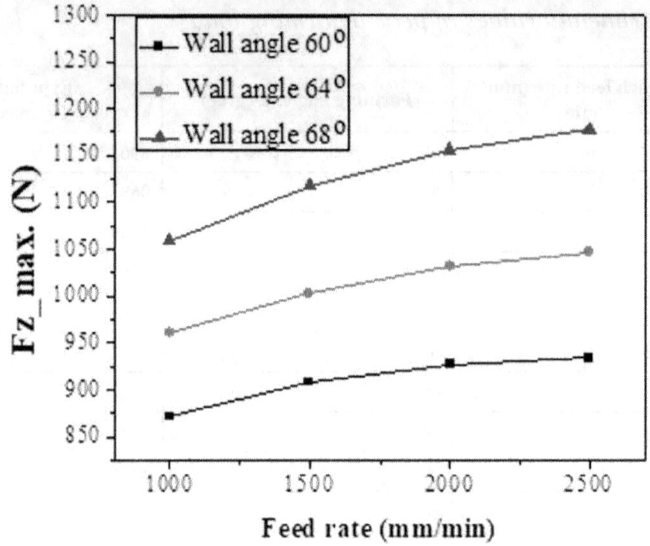

Figure 5. Fracture of specimen during experiment-12 (wall angle = 68°, feed rate = 2500 mm/min, force value = 1202 N, fracture height = 15.80 mm)

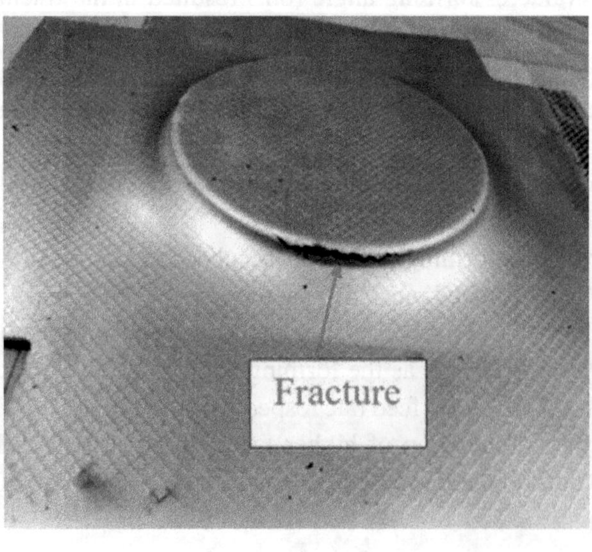

Figure 6. Successfully formed component during experiment-11 (wall angle = 64°, feed rate = 2500 mm/min, force value = 1059 N)

4. CONCLUSION

The current work focused on exploring and predicting the maximal axial forming forces during the SPIF technique to form conical frustums on AA-2024-O sheets. The effects of feed rate and wall angle along with their interactions have also been investigated. The following conclusions have been made:

- The forming angle and puch feed rate proved to be significant factor for influencing the deforming load. The peak deforming loads were raised by 34.16%, 9.41%, and 13.51% by raising the punch feed rate from level 1 to level 4 for the wall angle of 60°, 64°, and 68° respectively.
- Also, deforming loads were raised by 18.20%, 23.57%, 27.58%, and 28.70% as the forming angle was raised from 60° to 68° for the punch feed rate of level 1, 2, 3 and 4 respectively.
- When the condition of forming changed from the combination of maximum forming angle (68°) and maximum punch feed rate (2500 mm/min) to the combination of minimum value of forming angle (60°) and minimum punch feed rate (1000 mm/min), the forming force was found to decrease by 25.45% which is very significant to save power and energy for producing the components.

Furthermore, secure and effective employment of SPIF machine and auxiliary hardware may be assured by measuring the peak deforming loads which offer an alternative economical, easy, and technology-based solution to the cumbersome experimental process of this die-less approach of forming. Future work will aim at the investigation of different materials and input conditions for measuring the forces during the SPIF process.

REFERENCES

Alsamhan, A., Ragab, A. E., Dabwan, A., Nasr, M. M., & Hidri, L. (2019). Prediction of formation force during single-point incremental sheet metal forming using artificial intelligence techniques. *PLoS One*, *14*(8), e0221341. doi:10.1371/journal.pone.0221341 PMID:31437217

Bagudanch, I., Centeno, G., Vallellano, C., & Garcia-Romeu, M. L. (2019). Towards the manufacturing of near net shape medical prostheses in polymeric sheet by incremental sheet forming. *Near Net Shape Manufacturing Processes*, 1-33.

Centeno, G., Bagudanch, I., Martínez-Donaire, A. J., Garcia-Romeu, M. L., & Vallellano, C. (2014). Critical analysis of necking and fracture limit strains and forming forces in single-point incremental forming. *Materials & Design*, *63*, 20–29. doi:10.1016/j.matdes.2014.05.066

Fiorentino, A, Ceretti, E., Attanasio, A., Mazzoni, L. U. C. A., & Giardini, C. (2009). Analysis of forces, accuracy and formability in positive die sheet incremental forming. *International Journal of Material Forming*, *2*(S1), 805–808. doi:10.100712289-009-0467-z

Gulati, V., & Kumar, A. (2018). Investigation of some process parameters on forming force in single point incremental forming. *Int Res J Eng Technol*, *4*(4), 784–791.

Honarpisheh, M., Keimasi, M., & Alinaghian, I. (2018). Numerical and experimental study on incremental forming process of Al/Cu bimetals: Influence of process parameters on the forming force, dimensional accuracy and thickness variations. *Journal of Mechanics of Materials and Structures*, *13*(1), 35–51. doi:10.2140/jomms.2018.13.35

Jeswiet, J., Micari, F., Hirt, G., Bramley, A., Duflou, J., & Allwood, J. (2005). Asymmetric single point incremental forming of sheet metal. *CIRP Annals*, *54*(2), 88–114. doi:10.1016/S0007-8506(07)60021-3

Kumar, A., & Gulati, V. (2018). Experimental investigations and optimization of forming force in incremental sheet forming. *Sadhana*, *43*(10), 1–15. doi:10.100712046-018-0926-7

Kumar, A., & Gulati, V. (2019). Experimental investigation and optimization of surface roughness in negative incremental forming. *Measurement*, *131*, 419–430. doi:10.1016/j.measurement.2018.08.078

Kumar, A., & Gulati, V. (2021). Optimization and investigation of process parameters in single point incremental forming. [IJEMS]. *Indian Journal of Engineering and Materials Sciences*, *27*(2), 246–255.

Kumar, A., Gulati, V., & Kumar, P. (2018a). Investigation of surface roughness in incremental sheet forming. *Procedia Computer Science*, *133*, 1014–1020. doi:10.1016/j.procs.2018.07.074

Kumar, A., Gulati, V., & Kumar, P. (2018b). Effects of process parameters on surface roughness in incremental sheet forming. *Materials Today: Proceedings*, *5*(14), 28026–28032. doi:10.1016/j.matpr.2018.10.043

Kumar, A., Gulati, V., & Kumar, P. (2018c). Investigation of process variables on forming forces in incremental sheet forming. *Int J Eng Technol*, *10*(3), 680–684. doi:10.21817/ijet/2018/v10i3/181003021

Kumar, A., Gulati, V., & Kumar, P. (2019c). Experimental investigation of forming forces in single point incremental forming. In *Advances in Industrial and Production Engineering: Select Proceedings of FLAME 2018* (pp. 423-430). Springer Singapore. 10.1007/978-981-13-6412-9_41

Kumar, A., Gulati, V., Kumar, P., & Singh, H. (2019b). Forming force in incremental sheet forming: A comparative analysis of the state of the art. *Journal of the Brazilian Society of Mechanical Sciences and Engineering*, *41*(6), 1–45. doi:10.100740430-019-1755-2

Kumar, A., Gulati, V., Kumar, P., Singh, H., Singh, V., Kumar, S., & Haleem, A. (2020). Parametric investigation of forming forces in single point incremental forming. *Materials Today: Proceedings*, *24*, 611–617. doi:10.1016/j.matpr.2020.04.315

Kumar, A., Gulati, V., Kumar, P., Singh, V., Kumar, B., & Singh, H. (2019a). Parametric effects on formability of AA2024-O aluminum alloy sheets in single point incremental forming. *Journal of Materials Research and Technology*, *8*(1), 1461–1469. doi:10.1016/j.jmrt.2018.11.001

Manju, V. G., & Kumar, A. (2020). Process parameters and thickness reduction in single point incremental forming. *International Research Journal of Engineering and Technology*, *7*(1), 473–476.

Mittal, R. K. (2020). *Incremental sheet forming technologies: principles, merits, limitations, and applications*. CRC Press.

Oleksik, V., Pascu, A., Gavrus, A., & Oleksik, M. (2010). Experimental studies regarding the single point incremental forming process. *Academic Journal of Manufacturing Engineering*, *8*(2), 51–56.

Suresh, K., Bagade, S. D., & Regalla, S. P. (2015). Deformation behavior of extra deep drawing steel in single-point incremental forming. *Materials and Manufacturing Processes*, *30*(10), 1202–1209. doi:10.1080/10426914.2014.994755

Zhai, W., Li, Y., Cheng, Z., Sun, L., Li, F., & Li, J. (2020). Investigation on the forming force and surface quality during ultrasonic-assisted incremental sheet forming process. *International Journal of Advanced Manufacturing Technology*, *106*(7-8), 2703–2719. doi:10.100700170-019-04870-0

Chapter 7
Microwave-Assisted vs. Conventional Hydrothermal Synthesis, Morphology, Microstructure, and Surface Area Analysis of g-C3N4/MoS2 Nanocomposite

Mukul Saxena
https://orcid.org/0000-0003-0445-5128
Centre for Advanced Studies, Lucknow, India

Anuj Kumar Sharma
Centre for Advanced Studies, Lucknow, India

Ashish Kumar Srivastava
https://orcid.org/0000-0003-2834-1483
G.L. Bajaj Institute of Technology and Management, India

Amit Rai Dixit
Indian Institute of Technology (ISM), Dhanbad, India

Narendra Singh
Indian Institute of Technology, Tirupati, India

Monika Singh
Centre for Advanced Studies, Lucknow, India

DOI: 10.4018/978-1-6684-9224-6.ch007

ABSTRACT

Excellent lubricating characteristics are shown by the solid lubricant MoS2, but due to its spontaneous oxidation and absorption of moisture from the air, it has low wearing resistance and a limited wear life. This study is the first to successfully mix the additive g-C3N4 with MoS2 using sophisticated microwave-assisted synthesis with controllable parameters in addition to traditional hydrothermal synthesis route. The conventional hydrothermal process takes longer than other approaches, such as microwave-assisted synthesis method. There was a lack of comparison between the two synthesis techniques in terms of pore size, morphology, and microstructure. This research evaluated the microstructure, morphology, surface area, and pore diameters of g-C3N4/MoS2 nanocomposite produced employing hydrothermal (g-C3N4/MoS2-HT) and novel microwave methods (g-C3N4/MoS2-MW).

1. INTRODUCTION

Two-dimensional nanomaterial has a large specific area on the surface that provides a significant surface area for absorption onto a substrate surface, decreasing or eliminating friction between the contact surfaces (Vattikuti et al., 2015). Modern composites have been applied in a variety of innovative disciplines (Saxena et al., 2016). Molybdenum disulfide (commonly known to be MoS2) is now being investigated for a possible 2D TMD (transition metal chalcogenide) material due to its physical and chemical endurance in coatings. The lubricant's crystalline lamella structure regulates the rates at which it operates, with the sulphur lamella coupled by a weak VanDerWaal attraction, resulting as for a less friction (Berman et al., 2014). Layers of MoS2 slides with one another efficiently and aligned continuously parallel to the movement during slides, leading as a lubrication action. Lamellar is particularly resistant for asperity penetration because of the intense ionic interaction between sulphur and molybdenum atoms (Dallavalle et al., 2012). Pure MoS2, on the other hand, easily absorbs moisture in any humid atmosphere and can be oxidised in such an environment containing either molecular or atomic oxygen. This causes the coefficient of friction to rapidly increase and the lifetime of surface to decrease. Because of oxidation of MoS2, the frictional coefficient (COF) most often exceeds 0.2 sometimes (Cizaire et al., 2002, Curry et al., 2017, and Donnet et al., 1996). MoS2's practical usage was limited oxidation issue. As a result, the superlubility of MoS2 could be improved at the atomic or molecular scale through proper structural or orientation ordering in a dry setting (humid free) or in a vacuum condition (Berman et al., 2018).

Molybdenum-disulfide is frequently combined with other material commonly grapheme and polyurethane (Bülbül et al., 2007), moreover metal like copper (Yao et al., 2004), chromium (Wang et al., 2018), titanium (Elianov et al., 2018), and aluminium (Lee et al., 2005), in order to overcome these constraints and provide improved properties. Improving the wearing life and COF of MoS2 used as a solid phase lubricant in various domains of applications is becoming a serious concern. Increasing the wear life of molybdenum-disulfide for its use as a solid lubricant in many fields while retaining a low coefficient of friction is currently regarded as being a serious challenge (Li et al., 2017). However, the high expense of the metal additives in MoS2 coatings limits their use. In theory, nonmetal materials such as layered type graphitic-carbon nitride, which is inexpensive, conveniently accessible and ecologically acceptable, can displace metal additives (Zhang et al., 2019 and Flores et al., 2014).

Currently, many disciplines use the CN that has low VanDerWaal forces in between tristriazine unit and its layers (Xu et al., 2018). gC3N4 is commonly added by Kumar et al., 2017 into lubrication media for improving frictional performance. Bonding of gC3N4 with octa-decylamine, for example, responsible in the forming boundary layer on the surface, that increased materials wearing refusal. Duan et al., 2018 chose gC3N4 as an oil ingredient as because it boosts wear resistance substantially. Zhu et al., 2015 created g-C3N4/Polyvinylidene fluoride composites and found that gC3N4 filler increased composite durability. Nanofillers are also widely employed in metal matrix composites (Gupta et al., 2021 and Saxena et al., 2021). Computational approaches also made mechanical and other types of study simpler to analyse (Saxena et al., 2021).

Several experimental techniques to the production of MoS2 nanomaterials have been developed, including sulfurization, laser ablation, the hydrothermal process, chemical vapour deposition, and, in some cases, thermal reduction (Li et al., 2003, Rapoport et al., 2005, Tang et al., 2013, and Gawande et al., 2014). Because of the ease of access to processing equipment, the hydro-thermal technique is extensively used, however it challenges from inadequate heating. In contrast to the typical oven-based hydrothermal technique, the substances could be heated up during the MW-assisted synthesis, resulting in a more constant temperature ramp. Apart from that the vessel made up of Teflon liner which utilised for the reaction is microwave transparent, ensuring that controlled heating was consistent throughout the vessels (Li et al., 2003 and Saxena et al., 2023).

In this study, conventional hydrothermal route and a one-step microwave assisted (MW) process were used to synthesize the g-C3N4 /MoS2 nanocomposite. Morphological, microstructural and surface area analyses were used to compare the quality of synthesized product.

2. MATERIALS AND METHODS

A microwave reactor (Mono-wave450, Anton-Paar) and a Teflon-lined reactor were utilised to synthesise the material, and a drying oven (HTFT1600 Model, Tempsens) employed to complete the drying process. The crystalline phase of synthesized samples in range of 10° to 70° at 4° step was examined using XRD (Ultima-IV with Cu-K, Rigaku). Using a FESEM, the microstructures were examined (Gemini-500, Zeiss with EDS). In order to ascertain the pore volume, specific surface area and pore width of created material in the range of relative pressure of 0 to 1, BrunauerEmmettTeller (BET) analysis was carried out on Belsorp (Model-mini X).

2.1 Reagents and Chemicals

Reagents used for investigation were thioacetamide (CH_3CSNH_2), sodium molybdate dihydrate $Na_2MoO_4.H_2O$, and urea NH_2CONH_2. Since all of the chemicals were of analytical grade quality, additional purification was not necessary. Double DI water was used for all the synthesis.

2.2 Synthesis of Graphitic Carbon Nitride

First, solid reaction was employed to produce graphitic-carbon nitride powder. 20g of urea was placed in a 50mL crucible then dried at roughly 80°C. The crucible was placed in a muffle furnace after being given time to dry before being cover with a lid cap and several layers of foil paper. Following that, it was calcined for four hours at a rate of 3°C per minute at 550°C. g-C_3N_4 powder was made using the grinding method once the temperature was brought down to room temperature. By grinding the material until it reached room temperature, g-C_3N_4 powder has been obtained. $NH_2.CO.NH_2$ was converted into g-C_3N_4 at a rate of about 4.5 percent (Table 1), as calculated from Yield percent = $\left(\frac{M_t - M_c}{M_u}\right)$ X 100, where 'M_t' is catalyst mass, 'M_c' is crucible mass and 'M_u' is initial mass of urea.

2.3 Synthesis of Nanocomposite

As-prepared graphitic carbon nitride, thioacetamide and sodium molybdate dihydrate were mix to a lab beaker that containing 70mL DI, deionized water in a molar ratio as 1:1.91:0.89. Following a three-hour sonication process in which all reagents were fully mix with the DI water, this combination then was subjected to magnetic stirring for an additional hour at RT, room temperature. Schematics of synthesis process is depicted in Figure 1.

Table 1. Yielding calculation

Parameter	Sample weight (g)
M_t	~134
M_c	~133
M_u	20
Yielding %	~ 4.5 percent

Figure 1. Schematics of the synthesis process

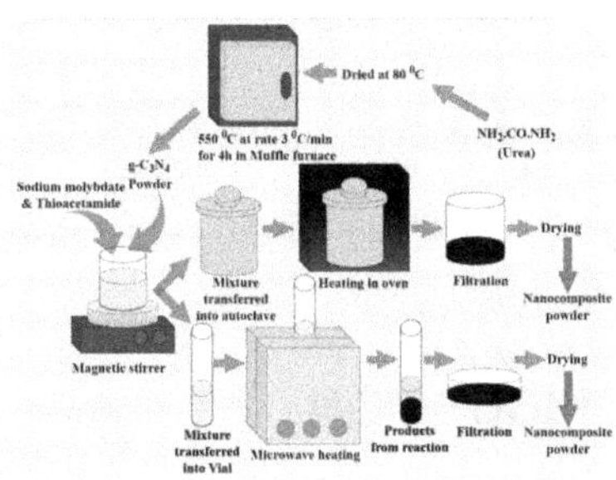

Out of the 70 ml, 60 ml was transferred into an autoclave for conventional hydrothermal synthesis at 180°C for 24 hours, and the remaining 10 ml then transferred in to a 30ml reactor vial, and placed in a microwave (MW) reactor for microwave-assisted hydrothermal process. Figure 2 depicts a microwave-assisted synthesis heat ramp.

After heating reaction was finished, compressed air flow was used to help the mixture reach room temperature. Following that, both reaction products underwent filtration, and gC_3N_4/MoS_2 powder was extracted. Then powder was dried for two hours at 80 °C temperature in an electric oven. In Table 2, nomenclature of nanocomposites are given.

Table 2. Nomenclature of nanocomposite

Nanocomposite	Synthesis method	Heated to	Total duration
$g\text{-}C_3N_4/MoS_2$-HT	Conventional hydrothermal	180 °C	~24 hrs
$g\text{-}C_3N_4/MoS_2$-MW	Microwave-assisted hydrothermal	180 °C	~v45 mins

Figure 2. Heating ramp for MW-assisted synthesis

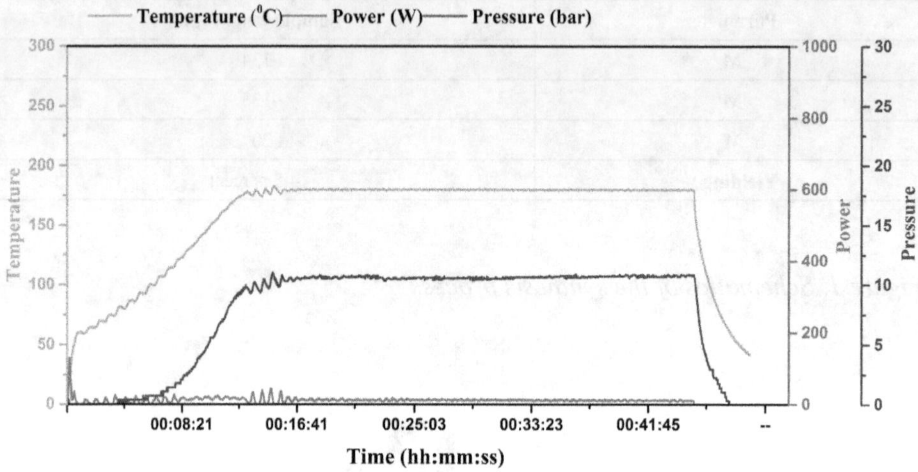

3. RESULTS AND DISCUSSION

3.1 Microstructural and Morphological Characterization

The morphology and micro-structure of as-prepared samples was examined using field emission-SEM. Figure 3 show FE-SEM images of as prepared g-C_3N_4/MoS_2-HT and Figure 4 for g-C_3N_4/MoS_2-MW, respectively.

Figure 3. FESEM image for g-C_3N_4/MoS_2-HT nanocomposite at 200 nm and 100 nm

Figures 3 and 4 show that graphitic carbon nitride has a smooth, layered structure in contrast to flower like structure of MoS_2, which is arranged in form of many nano-sheets. The figures also reveal that MoS_2 and g-C_3N_4 are packed together within the created composite, proving that the components were effectively combined.

Figure 4. FESEM image for g-C_3N_4/MoS_2-MW nanocomposite at 200 nm and 100 nm

To detect dispersed elements in a synthesized nanocomposite, Energy dispersive X-ray analysis was performed. Figure 5 illustrates some of the elements that are involved in forming the composite, including sulfur, carbon, nitrogen, and molybdenum.

Figure 5. EDX results for (a) g-C_3N_4/MoS_2-HT; (b) g-C_3N_4/MoS_2-MW

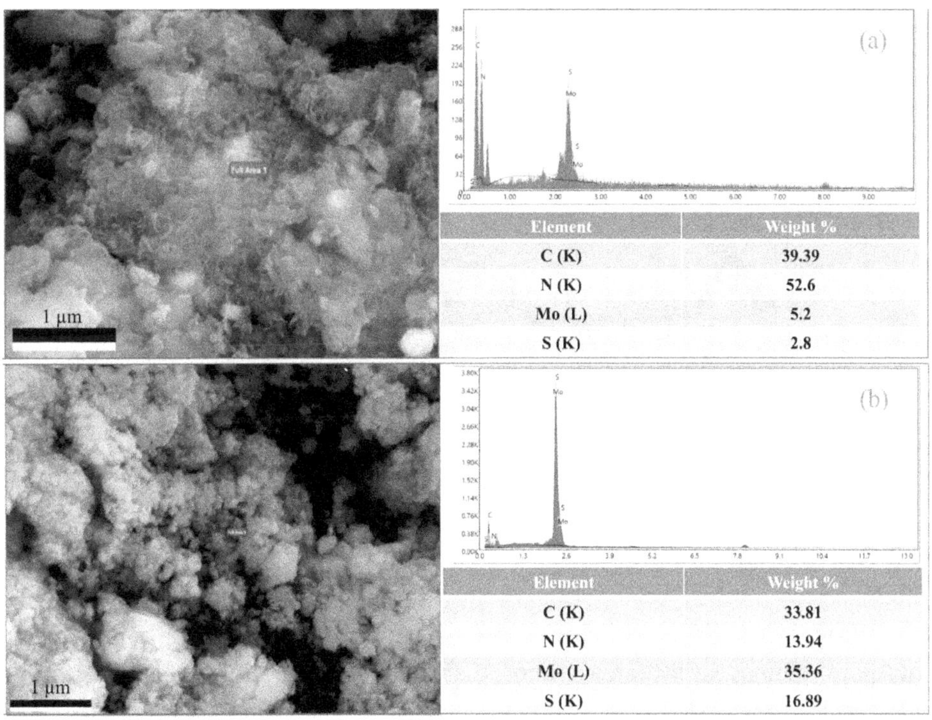

Element	Weight %
C (K)	39.39
N (K)	52.6
Mo (L)	5.2
S (K)	2.8

Element	Weight %
C (K)	33.81
N (K)	13.94
Mo (L)	35.36
S (K)	16.89

3.2 X-Ray Diffraction Analysis

The crystal structures of gC_3N_4/MoS_2-HT and gC_3N_4/MoS_2-MW were investigated using X-ray diffraction. Figure 6 (JCPDS-87-1526) shows peaks for $g\text{-}C_3N_4$ at 13.0° and 27.5°. These values correspond to crystallographic planes (100) and (002), resp. The presence of these two peaks indicate conjugate aromatic system with inter-layer stacking (Saxena et al., 2022). The (002),(100),(102), and (110) planes of the hexagonal MoS_2 are responsible for the extraordinary peaks at 14°, 32.7°, 35.871°, and 58.3°, resp. (JCPDS-75-1539).

Figure 6. XRD plots for g-C_3N_4/MoS_2(CNMS)-HT and g-C_3N_4/MoS_2-MW

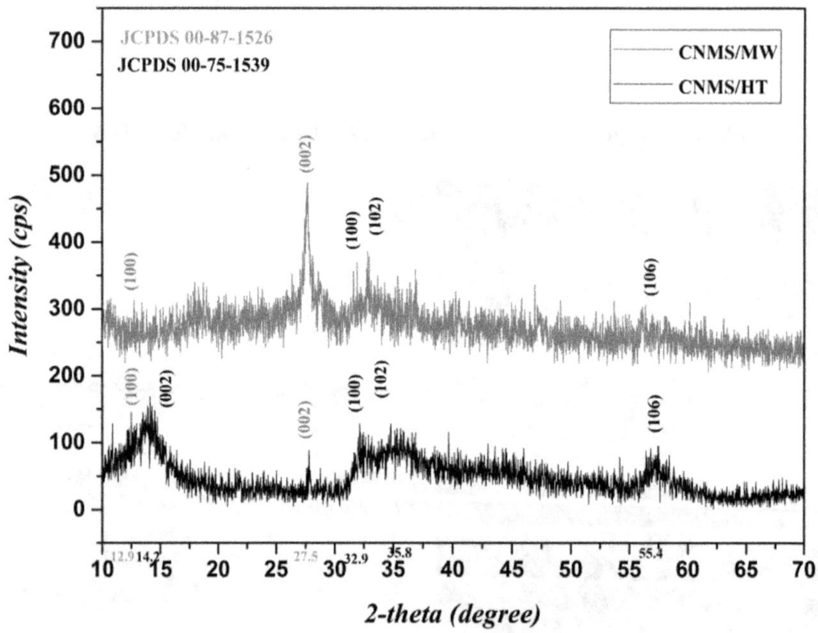

3.3 Surface Area Analysis

The nitrogen (N_2) adsorption and desorption isotherm was utilised for estimating the total pore volumes, surface area and pore diameter of the g-C_3N_4/MoS_2 nanocomposite powders in order to acquire more precise information on their microstructures. At 77 K, the boiling temperature of liquid nitrogen, Brunauer–Emmett–Teller (BET) was carried out utilising nitrogen gas adsorption. Figure 7 shows BET curves between relative pressure (p/p_0) and amount absorbed ($p/V_a(p_0-p)$), where the adsorbtive's pressure is 'p' and the saturation vapour pressure is 'p_0' at the predetermined experimental temperature.

Figure 7. BET surface area plot for (a) g-C_3N_4/MoS_2-HT; (b) g-C_3N_4/MoS_2-MW

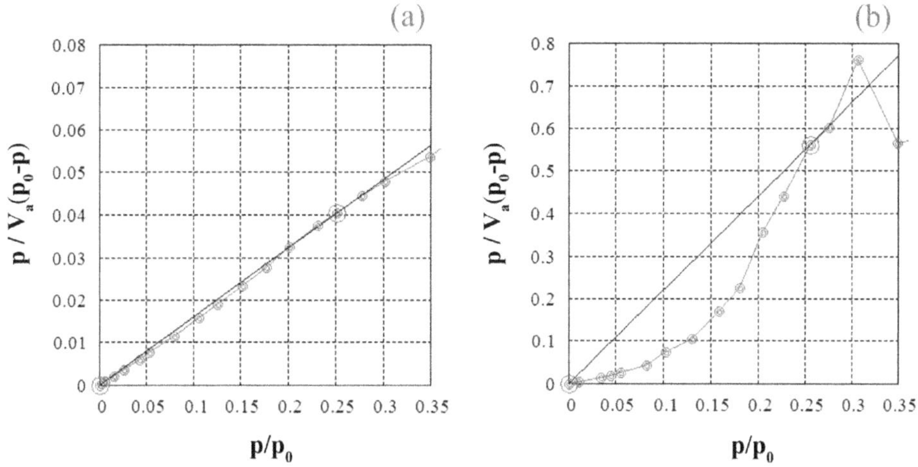

Surface area of a sample can be calculated using the Brunauer–Emmett–Teller equation (1).

$$\frac{p/p_0}{V_a \cdot \left(1 - p/p_0\right)} = \frac{1}{Q_m.K} + \frac{K-1}{Q_m.K} \cdot p/p_0 \quad (1)$$

Where, p/p_0 is relative pressure, 'V_a' is adsorbed gas quantity. 'Q_m' is capacity of monolayer and 'K' is BET constant which can be calculated form equation 2 and 3, respectively.

$$Q_m = \frac{1}{Slope - intercept} \quad (2)$$

$$K = 1 + \frac{Slope}{Intercept} \quad (3)$$

The plot for BET is linearly plotted in the region where the BET model is valid, and the intercept and slope are calculated using a straight line fit. Table 3 displays the outcomes of the BET analysis of the nanocomposites for pore volume, surface area and pore diameter.

Table 3. Brunauer–Emmett–Teller (BET) analyses results

	Slope (Linear)	Intercept (Linear)	Correlation coefficient	a_sBET [m^2/g]	Total pore volume [cm^3/g]	Mean Pore Diameter [nm]
g-C_3N_4/MoS_2 (HT)	0.1611	0	0.9992	27.02	0.072	10.59
g-C_3N_4/MoS_2 (MW)	2.2008	0	0.9392	1.9778	0.021	42.56

In can be seen from Table 3 that the BET surface area for nanocomposite synthesised from hydrothermal route is much larger than microwave assisted synthesized due to larger pore volume that was relatable to mean pore diameter of the pores of nanocomposite powder.

4. CONCLUSION

The g-C_3N_4/MoS_2 nanocomposite was synthesized using traditional hydrothermal and a novel microwave-assisted synthesis approach. XRD, FESEM, and EDS were utilised to characterise the physical and morphological properties. MoS_2 was found to be a flower-like structure, and graphitic carbon nitride had a layered structure. In elemental analysis, EDS (energy dispersive X-ray spectroscopy) was employed, and confirming that the nanocomposite has the necessary elements of MoS_2 and g-C_3N_4. Exploration of the crystal structures in prepared nanocomposite was done using X-ray diffraction. Diffraction peaks of g-C_3N_4 were observed at 13.0° and 27.5°, which correspond to the (100) and (002) crystallographic planes, resp. Peaks for hexagonal MoS_2 were found at 14°, 32.7°, 35.8°, and 58.3°, resp. The pore volume, surface area and pore diameter of the g-C_3N_4/MoS_2-HT nanocomposite were analysed 0.072 cm^3/g, 27.02 m^2/g and 10.59 nm, respectively, while the g-C_3N_4/MoS_2-MW nanocomposite was 0.021 cm^3/g, 1.9778 m^2/g and 42.56 nm. Microwave-assisted synthesis methods took less time than conventional hydrothermal procedures, yet both produced equivalent nanocomposite.

REFERENCES

Berman, D., Erdemir, A., & Sumant, A. V. (2014). Graphene: A new emerging lubricant. *Materials Today*, *17*(1), 31–42. doi:10.1016/j.mattod.2013.12.003

Berman, D., Erdemir, A., & Sumant, A. V. (2018). Approaches for achieving superlubricity in two-dimensional materials. *ACS Nano*, *12*(3), 2122–2137. doi:10.1021/acsnano.7b09046 PMID:29522673

Bülbül, F., Efeoğlu, İ., & Arslan, E. (2007). The effect of bias voltage and working pressure on S/Mo ratio at MoS2–Ti composite films. *Applied Surface Science*, *253*(9), 4415–4419. doi:10.1016/j.apsusc.2006.09.059

Cizaire, L., Vacher, B., Le Mogne, T., Martin, J. M., Rapoport, L., Margolin, A., & Tenne, R. (2002). Mechanisms of ultra-low friction by hollow inorganic fullerene-like MoS2 nanoparticles. *Surface and Coatings Technology*, *160*(2-3), 282–287. doi:10.1016/S0257-8972(02)00420-6

Curry, J. F., Wilson, M. A., Luftman, H. S., Strandwitz, N. C., Argibay, N., Chandross, M., Sidebottom, M. A., & Krick, B. A. (2017). Impact of microstructure on MoS2 oxidation and friction. *ACS Applied Materials & Interfaces*, *9*(33), 28019–28026. doi:10.1021/acsami.7b06917 PMID:28758391

Dallavalle, M., Sändig, N., & Zerbetto, F. (2012). Stability, dynamics, and lubrication of MoS2 platelets and nanotubes. *Langmuir*, *28*(19), 7393–7400. doi:10.1021/la300871q PMID:22530739

Donnet, C., Martin, J. M., Le Mogne, T., & Belin, M. (1996). Super-low friction of MoS2 coatings in various environments. *Tribology International*, *29*(2), 123–128. doi:10.1016/0301-679X(95)00094-K

Duan, C., Yuan, D., Yang, Z., Li, S., Tao, L., Wang, Q., & Wang, T. (2018). High wear-resistant performance of thermosetting polyimide reinforced by graphitic carbon nitride (g-C3N4) under high temperature. *Composites. Part A, Applied Science and Manufacturing*, *113*, 200–208. doi:10.1016/j.compositesa.2018.07.008

Elianov, O., Garusi, S., Rosentsveig, R., Cohen, S. R., Feldman, Y., Pinkas, I., Bendikov, T., Kaplan-Ashiri, I., Moshkovich, A., Perfilyev, V., Rapoport, L., Moshonov, J., Tenne, R., & Shay, B. (2018). Deposition of metal coatings containing fullerene-like MoS2 nanoparticles with reduced friction and wear. *Surface and Coatings Technology*, *353*, 116–125. doi:10.1016/j.surfcoat.2018.08.084

Flores-Ruiz, F. J., Enriquez-Flores, C. I., Chiñas-Castillo, F., & Espinoza-Beltrán, F. J. (2014). Nanotribological performance of fullerene-like carbon nitride films. *Applied Surface Science*, *314*, 193–198. doi:10.1016/j.apsusc.2014.06.168

Gawande, M. B., Shelke, S. N., Zboril, R., & Varma, R. S. (2014). Microwave-assisted chemistry: Synthetic applications for rapid assembly of nanomaterials and organics. *Accounts of Chemical Research*, *47*(4), 1338–1348. doi:10.1021/ar400309b PMID:24666323

Gupta, T. K., Srivastava, A. K., & Srivastava, V. S. (2021). Microstructural and tensile behaviour of hybrid MMC Al7075/SiC/B4C produced by mechanical stir casting. *Materials Today: Proceedings*, *47*, 4107–4113. doi:10.1016/j.matpr.2021.06.430

Kumar, A., Thakre, G. D., Arya, P. K., & Jain, A. K. (2017, December). 2D Structured Nano-Sheets of Octadecylamine Grafted Graphitic-Carbon Nitride (g-C3N4) as Lubricant Additives. In Macromolecular Symposia, 376(1), p. 1700009. doi:10.1002/masy.201700009

Lee, D. B., & Kim, S. K. (2005). High temperature oxidation of a (Cr 2 N– Mo 2 S 3)/Cr double-layered coating deposited on steel by DC magnetron sputtering. *Metals and Materials International*, *11*(3), 227–232. doi:10.1007/BF03027447

Li, H., Wang, J., Gao, S., Chen, Q., Peng, L., Liu, K., & Wei, X. (2017). Superlubricity between MoS2 monolayers. *Advanced Materials*, *29*(27), 1701474. doi:10.1002/adma.201701474 PMID:28497859

Li, W. J., Shi, E. W., Ko, J. M., Chen, Z. Z., Ogino, H., & Fukuda, T. (2003). Hydrothermal synthesis of MoS2 nanowires. *Journal of Crystal Growth*, *250*(3-4), 418–422. doi:10.1016/S0022-0248(02)02412-0

Li, X. L., & Li, Y. D. (2003). Formation of MoS2 inorganic fullerenes (IFs) by the reaction of MoO3 nanobelts and S. *Chemistry (Weinheim an der Bergstrasse, Germany)*, *9*(12), 2726–2731. doi:10.1002/chem.200204635 PMID:12866535

Rapoport, L., Fleischer, N., & Tenne, R. (2005). Applications of WS 2 (MoS 2) inorganic nanotubes and fullerene-like nanoparticles for solid lubrication and for structural nanocomposites. *Journal of Materials Chemistry*, *15*(18), 1782–1788. doi:10.1039/b417488g

Saxena, A., Dwivedi, S. P., Dixit, A., Sharma, S., Srivastava, A. K., & Maurya, N. K. (2021). Computational and experimental investigation on mechanical behavior of zirconia toughened alumina and nickel powder reinforced EN31 based composite material. *Materialwissenschaft und Werkstofftechnik*, *52*(5), 548–560. doi:10.1002/mawe.202000152

Saxena, A., Dwivedi, S. P., Maurya, N. K., & Srivastava, A. K. (2021). Investigation on mechanical properties and deformation behavior of copper-based three-phase metal matrix composite: Experimental and micro-macro-mechanical finite element analysis. *Proceedings of the Institution of Mechanical Engineers, Part L: Journal of Materials: Design and Applications, 235*(8), 1850-1867. 10.1177/14644207211009254

Saxena, M., & Kirtania, S. (2016). Stiffness analysis of symmetric cross-ply laminated composite plates. *ADBU Journal of Engineering Technology, 4*. https://journals.dbuniversity.ac.in/ojs/index.php/AJET/article/view/153/pdf_32

Saxena, M., Sharma, A. K., Srivastava, A. K., Singh, N., & Dixit, A. R. (2023). An Investigation for Minimizing the Wear Loss of Microwave-Assisted Synthesized g-C3N4/MoS2 Nanocomposite Coated Substrate. *Coatings*, *13*(1), 118. doi:10.3390/coatings13010118

Saxena, M., Sharma, A. K., Srivastava, A. K., Singh, R. K., Dixit, A. R., Nag, A., & Hloch, S. (2022). Microwave-Assisted Synthesis, Characterization and Tribological Properties of a g-C3N4/MoS2 Nanocomposite for Low Friction Coatings. *Coatings*, *12*(12), 1840. doi:10.3390/coatings12121840

Tang, G., Wang, Y., Chen, W., Tang, H., & Li, C. (2013). Hydrothermal synthesis and characterization of novel flowerlike MoS2 hollow microspheres. *Materials Letters*, *100*, 15–18. doi:10.1016/j.matlet.2013.02.103

Vattikuti, S. P., & Byon, C. (2015). Synthesis and characterization of molybdenum disulfide nanoflowers and nanosheets: Nanotribology. *Journal of Nanomaterials*, *2015*, 9–9. doi:10.1155/2015/710462

Wang, P., Yue, W., Lu, Z., Zhang, G., & Zhu, L. (2018). Friction and wear properties of MoS2-based coatings sliding against Cu and Al under electric current. *Tribology International*, *127*, 379–388. doi:10.1016/j.triboint.2018.06.028

Xu, Z. D., Xin, J. B., Fan, R. X., Wang, K., & Yang, J. (2018). A simple approach to fabricate CN/MoS2 nanocomposite and its application as a lubricant additive. *Dig. J. Nanomater. Biostruct, 13*, 731-741. https://chalcogen.ro/731_XuZD.pdf

Yao, S. H., Kao, W. H., Su, Y. L., & Liu, T. H. (2004). On the tribology and micro-drilling performance of TiN/AlN nanolayer coatings. *Materials Science and Engineering A*, *386*(1-2), 149–155. doi:10.1016/S0921-5093(04)00919-0

Zhang, Y., Jin, Z., Su, Y., & Wang, G. (2019). Charge separation and electron transfer routes modulated with Co-Mo-P over g-C3N4 photocatalyst. *Molecular Catalysis*, *462*, 46–55. doi:10.1016/j.mcat.2018.10.009

Zhu, L., Wang, Y., Hu, F., & Song, H. (2015). Structural and friction characteristics of g-C3N4/PVDF composites. *Applied Surface Science*, *345*, 349–354. doi:10.1016/j.apsusc.2015.03.197

Chapter 8
Metals, Polymers, Ceramics, Composites Biomaterials Used in Additive Manufacturing for Biomedical Applications

Yusuf Olatunji Waidi
https://orcid.org/0000-0002-5728-1769
Indian Institute of Science, Bangalore, India

Ranjit Barua
https://orcid.org/0000-0003-2236-3876
Indian Institute of Engineering Science and Technology, Shibpur, India

Sudipto Datta
https://orcid.org/0000-0003-2161-6878
Indian Institute of Science, Bangalore, India

ABSTRACT

Additive manufacturing, often called 3D printing, is widely employed in all engineering sectors. Many researchers also refer to 3D printing in the biomedical field as bioprinting. Therefore, the 3D bioprinting methods are used to create patient-specific implants or devices for several tissue engineering needs. In addition, this method is also famous for drug design and targeted drug delivery systems. Nowadays, several researchers have involved smart materials with 3D printing to obtain 4D printing for biomedical applications. In this book chapter, the authors provide a quick overview of the many kinds of 3D printing. They also highlight the popular materials utilized in the technology for biomedical applications. Nanoparticles and smart materials, which are employed in biomedical applications, are also briefly explored. This technology's difficulties and future scope are also discussed. This chapter will give the readers a basic understanding of the materials used for 3D printing biomedical devices.

DOI: 10.4018/978-1-6684-9224-6.ch008

1. INTRODUCTION

Additive manufacturing (AM), often referred to as 3D printing manufacturing technique,has surfaced as a viable advanced fabrication choice for creating complex geometry for application in technology-driven fields, including aerospace (Datta et al., 2020; Barua et al., 2021;Zadpooret al., 2017), automotive (Dämmer et al., 2022), healthcare (Kumar et al., 2022) (Barua et al., 2022), and modern dental science (Sehrawat et al., 2022). AM generates less material waste and scrap than other manufacturing processes like formative and subtractive techniques. In addition, it enables the fabrication of lightweight complex geometry, particularly porous or hollow, needing less energy and material input during production and service. Directed energy deposition, binder jetting, material extrusion, powder bed fusion, material jetting, vat photopolymerization, and sheet lamination are the seven types of AM recognized and specified by the ISO/ASTM 52900 standard [ISO]. However, not every ASTM categorization AM techniques are developed equally and employed for biomedical applications (Bobbert et al., 2017).

In the subsequent sections, we discussed a brief note on the limitations, capabilities, advantages, and disadvantages of traditional techniques and materials generally used for biomaterials fabrication, such as polymers, metals, alloys, and ceramics. Furthermore, several key points are discussed, for example speed of printing process, and the amount of anisotropy, part sizes,the feedstock materials cell encapsulation possibility, achievable resolution, 3D printed support structure, cost, and post-processing necessity. The effectiveness of every 3D printing method is heavily reliant on optimized or appropriate printing and materials parameters (Kumar et al., 2023) inside the limits of the current AM process instruments (Sehrawat et al., 2022).

Besides selecting appropriate AM methods and printing conditions, the biomaterial's structural design is a significant object-making component. For biomedical applications, lattice or porous structures are frequently required. This indicates that the hierarchical pore structures and dimensions of biomaterial should be fully open and linked for oxygen and nutrients to be transported to cells (Wauthle et al., 2015). With the introduction of AM invention, the unprecedented potential for the exact organization of the dimensions and interior architectures of holes at a micro-scale and the production of biomimetic geometries having complex interior structures and pathways have been available (Murr et al., 2010). This dimensional accuracy and flexibilityis one of the most significant advantages of AMcompared to traditional manufacturing processes like moulding and casting (Gokuldoss et al., 2017; Barua et al., 2019), where the manufacturer has little powerover the fine interior features and geometries. Figure 1 shows different types of additive manufacturing process. Figure 2 shows the different steps of AM process, and Table 1 shows their advantages, disadvantages, material resolutions, and biomedical applications. This primary subject goal is to offer anexplicit image of how AM fabrication methods may be used to produce objects with innovative patterns, the obstacles and limits, and the future of thisinvention. We briefly discuss the existing model ideas used in manufacturing AM components for biomedical fields. In addition, we further discuss the use of several AM methods to createceramic, polymeric, and metallic biomaterials. The goal is to encourage the technology's ongoing development and broad deployment to transform design concepts into implants, other healthcare products, and the tissue engineering field.

Table 1. Different types of AM process with advantages, disadvantages, resolutions and biomedical applications

AM Procedure	Process Resolution	Material	Biomedical Application	Advantages	Disadvantages
Material Jetting	20-100 μm	Photopolymer Bioinks	Tissue Engineering, 3D bioprinting.	High resolution and cell viability	Slow
Powder Bed Fusion	100-200 μm	Powder Ceramics, Thermoplastic material	Dental implant, Orthopaedic	Fast processing without any solvent	Costly
Sheet Lamination	-1mm	Thin materials	Anatomical model (Macro)	Support structure not required, Low cost	Slow, Maximum material waste occurs
Binder Jetting	50-400 μm	Metal powder, Sand	Implant	Fast, Colour printing, Support structure not required, Low cost	Less strength
Material Extrusion	100-200 μm	Hydrogel bioink	Soft tissue, scaffold printing	Low cost, Open source design, High cell viability	Slow, Nozzles impart high shear forces on cell
Direct Energy Deposition	250-500 μm	Nylon, Metal	Limited use	Fast procedure	Expensive, Post-processing machining required

Figure 1. Different types of AM process
(Source: Kambale, B., (2021). The Manufacturing Report)

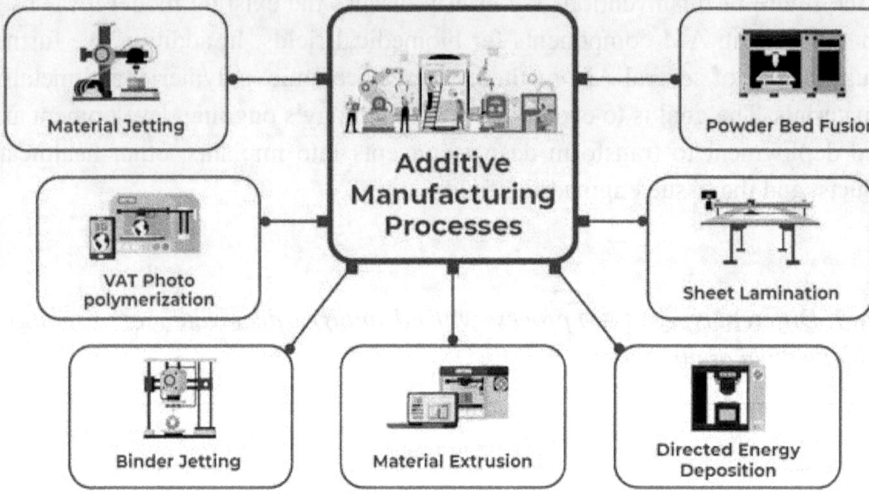

Figure 2. Different steps of AM process
(Source: CTTC)

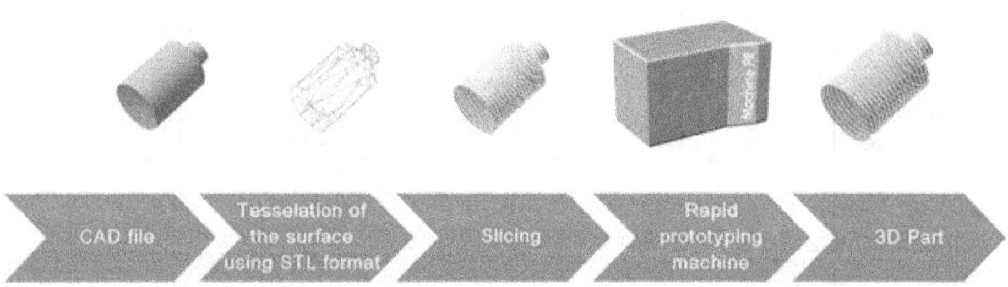

2. METALLIC BIOMATERIALS

2.1. Biodegradable Metals

Biomedical biodegradable implants can be defined as objectswhichgradually degrade in the human body for a certain period. The most popular degradable biomaterials are metal and polymer-based (Datta et al., 2018; Barua et al, 2023). The basic role of metallic biodegradable implants in the body is to be present brieflyand aid in healing before degrading. AM methods can be used to create biodegradable metal components and products. Porous structures made of magnesium alloy (WE43) (Li et al., 2018) and pure iron are two typical examples. Currently,in the market, various medical implants and devices like bone screws (Li et al., 2018) (Windhagen et al., 2013) and cardiovascular stents (Erbel et al., 2007) have taken advantage of Mg alloys as biodegradable materials. Recently, the shape-memory effect was discovered in Fe-Mn-Si alloys containing 30% Mn and 6% Si (Marinopoulou et al., 2022) and seems pretty intriguing for medical and many other industrial fields. The alloy martensitic transition has been known to improve mechanical featureslike hardness, fatigue resistance,and strength (Takhsha Ghahfarokhi et al., 2020). One of the most critical characteristicsof any biodegradable metalis the biodegradation rate or bioabsorbability of biodegradable metal implants in the body. For instance, Zn-based alloys, one of the best-suited biodegradable (Ge et al., 2022, Zhang et al., 2022) metals in vitro degradation rate of about 20-300 m/y (Barua et al., 2022), but Mg-based and Fe-based alloys have greater than 300 m/y and less than 50 m/y, respectively. When pure Mg comes into touch with a physiological chloride-containing condition, it degrades the fastest. The corrosion of Mg resulted in a large amount of hydrogen gas that cannot be handled within the host body. Interestingly, biodegradable Fe-based metals exhibit a substantially low degradation rate. Among all the methods of improving the degradation rate of any metal alloy, an alloying approach seems to be the most efficient.The alloying of Y, Sr, Zn, Zr, and Ca with Mg have shown better degradation resistance thanneat Mg. Aside from having a low degradation rate, they also possess better mechanical features, allowing them to be appropriate for fabricating load-bearing device and implant objects (Hosseini et al., 2021).

Figure 3. Various types of biomanufacturing process

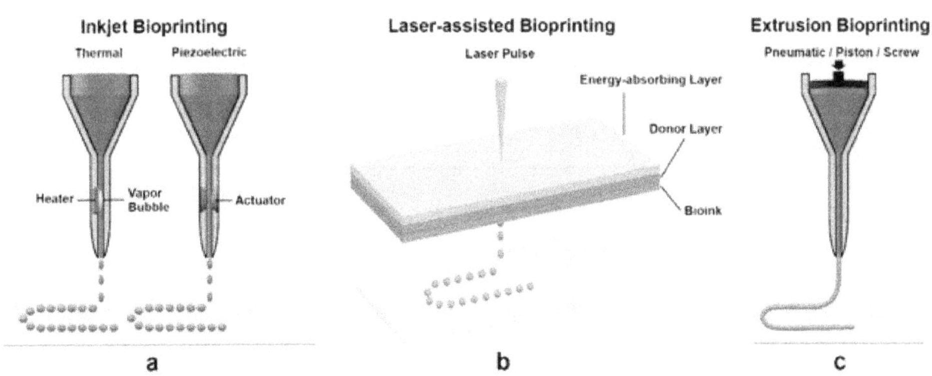

Increasing the surface area of the material can alsobe used to tune the biodegradation rate. As a result, the geometry and amount of porosity are two practical tools that may be employed to modify such material's degradation rate. Additional physical variables, like cyclic mechanical stress, might enhance the rate of biodegradation of porous iron, zinc, or Mg alloy (WE43) constructs coupled with the influence of environmental parameters. However, it is extremely difficult to use AM production method for porous biodegradable metals biomaterials, especially for magnesium and its alloys, which have intense low melting points, great flammability, chemical activity, and less evaporation temperatures. Numerous magnesium alloys may produce crystallisation fractures due to splashing eutectics, huge coefficient stresses of thermal expansion, large deformation, also a wide crystallization temperature range (Lu et al., 2021). As a result, specific safety measures and process adjustments are required for their manufacture. Extrusion-based 3D printingmethods are another method for generating porous biodegradable metals. Such strategies necessitate developing an ink composition compatible with the 3D printing, debinding, and sintering stages processes (Wei et al., 2021). The in vitro degradation data has demonstrated that 3D-printed iron constructs outperformed bulk counterparts (Ji et al., 2021). This may be further managed by developing graded functional porous biodegradable materials. Figure 3 shows the different types of 3D bioprinting process. Notably, using non-biodegradable materials for implants can halt normal bone ingrowth, necessitating another surgery to allow for future bone growth. Hence, biodegradable substances are thus a preferable choice for several device and implant development. Nevertheless, a major problem is thecytotoxicity associated with biodegradablematerials, occurring during the biodegradation process inside that patient's body (Sharma et al., 2021).

Table 2. Different types of bioink use in biomedical applications

Tissue/Cell Types	Bioink	Ref.
Bone/hMSCS; Cartilage/hMSCS	Gelatine	(Byambaa et al., 2017) (Lim et al., 2018)
Chonodrocytes/Cartilage	Alginate	(You et al., 2018)
Skeletal Muscle/hMPCS	Fibrin	(Kim et al., 2018)
Fibroblasts/hMSCS	PEG	(Hockaday et al., 2012)
Liver/Hepatocytes	Collagen	(Mazzocchi et al., 2018)

2.2. Shape-Memory Alloys (SMA)

SMA materials returns to its original form from the deformed form when the external stimuli is removed (Kong et al., 2021). This effect rises because of the transformation in temperature-driven phase in SMAs. These materials are used in various devices in cardiovascular as well as orthopedic implants. The commonly used

SMA is Nitinol which is an alloy of NiTi having equal concentration of Ti and Ni. This effect in SMA materials arises because of the change of austenite to martensite low and high temperatures, correspondingly (Vukojicic et al., 2019). Majority of NiTi having 48 GPa of elastic modulus that is comparaqtively lower than the alloy of Ti has the ability to recover huge strain till 8%. This material is pseudoelastic that means it can regain huge strains during unloading at a specific temperature (Genabai et al., 2015). Because of these advantages, NiTi materials are commonly used in fabricating medical devices like stents, bone healing staples, orthodontic plates and wires, and surgical guides etc. NiTiequiatomic alloy lattice structure is suitable and recognized for biological micro-electro-mechanical systems (bio-MEMS) and bioimplants. Because of the exceptional mixture properties of mechanical and thermal shape memories that is dependent on high corrosion resistance, reversible martensitic phase, biocompatibility, and superelasticity properties (Zhan et al.l, 2021). Further, it is also to be noted that Ni is allergic because of this reason it may distress some of its biomedical applications (Peterson et al., 2022). To overcome this problem, replacement of elements and/or surface modification is required for improving the biocompatibility property (Gorgin Karaji et al., 2017). Alloys like TiNbX, and TiNb shows 4.2% of elastic strains. The SMAs main applications is in implants using 4D printing (Farid et al., 2021) and implants for orthopedic applications (Mercuri et al., 2018). Tsaturyants et al., 2022 has reported that by combining heat treatment and thermal cycling, reduces the martensitic transformation range as well as highly improves the Nitinol alloy's mechanical properties which are processed using L-PBF. In their report they concluded that the heating-cooling combination cycle in temperature range of 350^{0}C and 400^{0}C results in reducing the temperature of martensitic transformation i.e. 10 to 15^{0}C temp can be reduced also it can add one more step for the structure transformation sequence. Also they reported that there is a 10% increment in transformation and dislocation yield stresses as well as 7% increment in dislocation yield stress and maximum stress of the advanced configuration by smearing heating–cooling of 10 cycles.

2.3. In Situ Alloying and Composites

With the use of additive manufacturing technologies, it is potential to dispense ingredients into 3D lattice configurations and place a variety of materials in specific locations throughout the entire structure. Such a capability raises the design difficulty, especially when lattice structures' intricate geometry design is already required. Various feedstock ingredients with various compositions are mixed together and concurrently added to the melt pool as part of the in situ alloying process. The features and functionalities of such a compositional mixture can be customized (Bourell et al., 2017). Amodel of a compositional combination is an in situ Ti-26Nb alloy that has undergone SLM processing for biomedical applications (Fischer et al., 2016). By adding reinforcing particles (mostly ceramics) and in situ alloying, the mechanical properties of metal matrices, such as hardness, stiffness, and strength, can often be greatly increased.The intrinsic qualities, such as toughness and/or electrical and/or thermal conductivity, are also impacted by their combination with metal. During additive manufacturing processes, reinforcing particles can be produced in situ using a metal matrix, alloying components, or ceramic reinforcing particles. Ex situ mixing processes, such ball milling, can also incorporate reinforcing particles. To achieve complete melting of the metal matrix and alloying materials for full contact with the surrounding ex-situ particles as well as a maximum response between the matrix and alloying elements, the laser intensity along with additional process variables of the L-PBF process may be modified (Khorasani et al., 2022). A porous combination of Ti-TiB was produced under specific L-PBF process conditions by an in situ interaction between the Ti matrix and TiB2 reinforcing elements (Attar et al., 2015). Unfortunately, a variety of problems, e.g. fragile interfacial bonding, insufficient responses, interfacial fissures, and uneven diffusion of the extra particles, make it challenging to construct an impeccable combination lattice construction. The types of metallic permeable ingredients are not inadequate to in situ or ex situ combinations. Moreover, using L-PBF procedures, porous metallic glass composites can be made, with the reinforcing elements often being crystalline phases scattered throughout the porous amorphous matrix (Liang et al., 2020). It has been established that L-PBF techniques may produce a variety of metal matrix composites. There are several benefits to producing desired parts (Putra et al., 2020). One of them is the capability to construct cellular constructions that are covered by combined parts at particular points. Due to these benefits, researchers have been able to develop lattice or non-lattice constructions made of FGM (functionally graded materials) for practice in applications of biomedical, which can be taken into consideration for additional research and experimentation (Putra et al., 2020).

2.4. Use of Biomedical Metals and Alloys in Additive Manufacturing

Numerous biomedical applications can make use of metals and their alloys. When most metals interact with biological surroundings, corrosion and ion release occur, which can be harmful to the body. As a result, they must exhibit great in vivo biocompatibility (Wachesk et al., 2021). Titanium (Ti) and the bulk of its alloys, cobalt-based alloys (such as CoCrMo), stainless steel, zirconium (Zr), tantalum (Ta), and niobium (Nb), among others, are some examples of biocompatible metals and alloys. They are outstanding materials for biocompatible applications and have exceptional mechanical and corrosion resistance (Long et al., 1998). Ti and its alloys, such as Ti6Al4V, are arguably the materials that have been investigated the most thoroughly among the numerous biocompatible metals and alloys (Zadpoor et al., 2019). Ti6Al4V is less ductile than pure Ti and therefore significantly more costly. Pure Ti, on the other hand, is regarded as a highly biocompatible metal due to its increased ductility and lower mechanical strength. Stainless steel is relatively biocompatible despite being more affordable than other materials. Stainless steel may be produced readily using laser powder bed fusion (L-PBF) methods, and it has a greater elastic modulus than Ti6Al4V (Zadpoor et al., 2019). The fatigue behaviour of Ti6Al4V is adequate, however its fatigue strength is lower than that of several other metallic materials, such as CoCr (Ahmadi et al., 2018). Porous implants in orthopaedic applications can be made using biomedical metals and alloys. Its elastic moduli are, however, considerably higher than those of the replacement bones. In order to prevent stress shielding at the bone-implant interface, the elastic modulus and yield strength of metallic implants must be carefully regulated. A number of methods, including the creation of graded metallic porosity implants, can be used to improve the mechanical properties of bone and metal interfaces. Another viable option is to add specific elements to the alloys' structure, which lowers the elastic moduli of porous structures. For instance, ß-type Ti alloys with lower elastic moduli than Ti6Al4V can be created by adding ß-phase-stabilizing elements to Ti, such as Ta, Nb, Zr, and Mo. Ti13Nb13Zr (Davidson et al., 1994) and Ti29Nb13Ta4.6Zr (Kuroda et al., 1998) are two examples of such ß-type Ti alloys that enhance the mechanical properties of implants and have elastic moduli of 79 GPa and 55–65 GPa, respectively. Metallic implants' ability to stimulate bone regeneration can be enhanced by surface treatments and coatings (Zhang et al., 2021) (van Hengel et al., 2021) (Fazel et al., 2019) (Barua et al., 2021). It has been suggested that surface modification techniques, such as adding bioactive glass and mesoporous bioactive glass to the surfaces of Ti6Al4V scaffolds, boost the effectiveness of bone tissue regeneration (Zhang et al., 2021). In addition, surface biofunctionalization techniques utilising plasma electrolytic oxidation (PEO) (Fazel et al., 2019) with or without Zn, Ag, or Cu nanoparticles have the probable to reduce implant-associated infections and have immunomodulatory effects (van Hengel et al., 2021) (Razzi et al., 2022) (Singh et al., 2022). Also, by adding nanostructures to the surfaces of metallic implants, the effectiveness of their antibacterial and osteogenic properties can be managed. Making Ti nanostructures via inductively coupled plasma reactive ion etching is one example (Singh et al., 2019) (Ganjian et al., 2019). Another method to give metallic (such as pure titanium (Bosshardt et al., 2017) implants several functionalities simultaneously is layer-by-layer coating biofunctionalization. Both tissue growth factors and antibacterial activity can be improved with this technique.

2.5. Use of Biomedical Polymers in Additive Manufacturing

The earliest materials utilized in AM were polymers. They can be manufactured utilizing AM techniques such extrusion based, PBF based or powder bed fusion process, and vat photo-polymerization because they have lower melting temperatures than ceramics and metals and chemical structures that can be altered (Algamaiah et al., 2021). Polymers for biomedical applications should not only have manufacturing-friendly characteristics but also be well-suited with the host tissue and disintegrate following tissue redevelopment. Figure 4 shows the artificial blood vessel. Therefore, polymers need additional properties, such biocompatibility and biodegradability, to be suitable for implants and other biomedical applications involving natural tissue regeneration, where the polymer is intended to be replaced by the tissue (Datta et al., 2020). Natural and manmade polymers can be broadly categorised as having such properties. Synthetic polymers are more hydrophobic and mechanically more stable than natural polymers due to their slower rates of breakdown. Instead, faster breakdown is preferable for tissue regeneration because the persistence of biomaterials implanted in the host tissue could result in physical harm. This may eventually lead to diminished mechanical strength (Barua et al., 2021). Moreover, 3D-printed polymeric materials' fatigue behaviour is crucial for medical equipment (Datta et al., 2019) (Datta et al., 2023). It is difficult to choose materials and their combinations to attain the desired qualities for specific medical devices. While some specific recent instances are further described, Table 2 addresses regularly utilized bioinks and their applications.

3. CONCLUSION AND FUTURE PROSPECTS

The types, uses, and materials utilized in additive manufacturing are succinctly summarized in this chapter. Metals, polymers, ceramics, and composites are just a few of the materials that may be treated using additive manufacturing, which is flexible in terms of speed, resolution, cost, and energy. To produce the desired product output, there is a new trend of synthesizing new materials using additive manufacturing (AM), such as smart materials, biomaterials, and nanomaterials. Materials are fed into 3D printers in the form of paste, wire, powder, sheets, and inks. The most often used polymers for additive manufacturing (AM) are polymers such as polyamides, PLA, ABS, PC, PS, and photopolymer resins that have been reinforced with fillers, fibers, and nanomaterials to produce composites with improved mechanical properties. Future generations will no longer view additive manufacturing as a separate procedure, but rather as an important part of an intelligent multi-process system. Activities in interdisciplinary research are more likely to expand dramatically. AM is also designed to deal with problems like decreasing waste resources, energy minimization, and workload distribution. Yet it still has issues including expensive costs, a little amount of material input, subpar mechanical performance, and regulatory worries. Also, when using curved parts and geometries, turning a CAD model to a 3D-printed product sometimes results in errors and inconsistencies. Although though additive manufacturing (AM) is a breakthrough method for creating goods for specific markets, it still needs to be improved if it is to compete with conventional methods.

Figure 4. 3D bioprinted artificial blood vessel
(Source: Texas A&M Engineering)

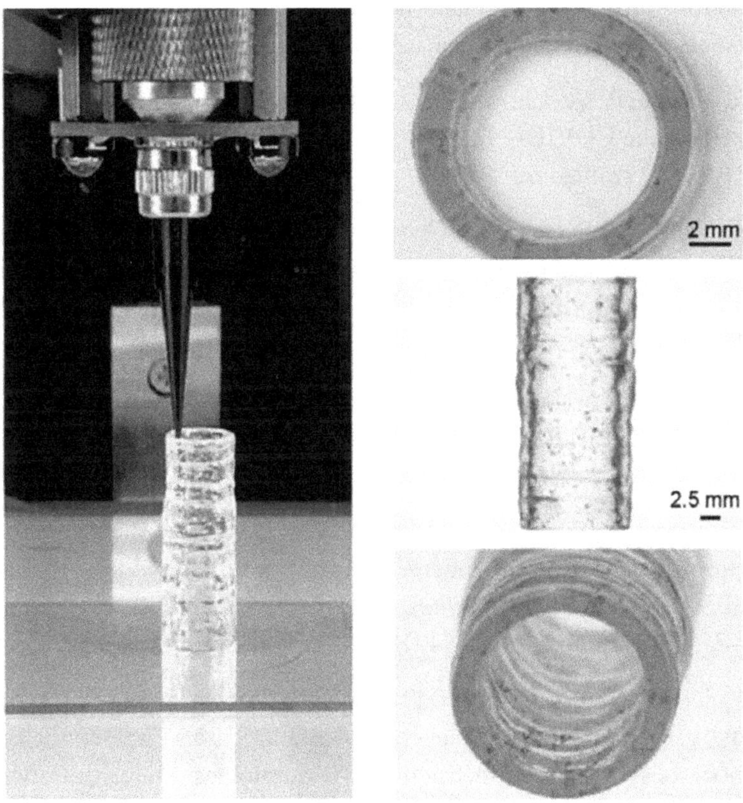

REFERENCES

CTTC. (n.d.). *Additive manufacturing of ceramics*. CTTC. https://www.cttc.fr/en/videos-downloads/additive-manufacturing-of-ceramics/

Ahmadi, S., Hedayati, R., Jain, R. A. K., Li, Y., Leeflang, S., & Zadpoor, A. (2017). Effects of laser processing parameters on the mechanical properties, topology, and microstructure of additively manufactured porous metallic biomaterials: A vector-based approach. *Materials & Design*, *134*, 234–243. doi:10.1016/j.matdes.2017.08.046

Algamaiah, H., Silikas, N., & Watts, D. C. (2021). Polymerization shrinkage and shrinkage stress development in ultra-rapid photo-polymerized bulk fill resin composites. *Dental materials: official publication of the Academy of Dental Materials*, *37*(4), 559–567. doi:10.1016/j.dental.2021.02.012

Attar, H., Löber, L., Funk, A., Calin, M., Zhang, L. C., Prashanth, K. G., Scudino, S., Zhang, Y. S., & Eckert, J. (2015). Mechanical behavior of porous commercially pure Ti and Ti–TiB composite materials manufactured by selective laser melting. *Materials Science and Engineering A*, *625*, 350–356. doi:10.1016/j.msea.2014.12.036

Barua, R., Datta, P., & Chowdhury, A. R. (2021). The Impact of 3D Printing Technology on the COVID-19 Pandemic. *Quality of Healthcare in the Aftermath of the COVID-19 Pandemic*, 135.

Barua, R., Datta, S., Datta, P., & Chowdhury, A. R. (2019). Scaffolds and tissue engineering applications by 3D bio-printing process: a new approach. In Design, Development, and Optimization of Bio-Mechatronic Engineering Products (pp. 78-99). IGI global.

Barua, R., Datta, S., Roychowdhury, A., & Datta, P. (2021). Importance of 3D printing technology in medical fields. In Research Anthology on Emerging Technologies and Ethical Implications in Human Enhancement (pp. 704-717). IGI Global.

Barua, R., Roychowdhury, A., & Datta, P. (2022). Study of Different Additive Manufacturing Processes and Emergent Applications in Modern Healthcare. In *Advanced Manufacturing Techniques for Engineering and Engineered Materials* (pp. 239–259). IGI Global. doi:10.4018/978-1-7998-9574-9.ch014

Barua, R., Sarkar, A., & Datta, S. (2023). Emerging Advancement of 3D Bioprinting Technology in Modern Medical Science and Vascular Tissue Engineering Education. In Handbook of Research on Instructional Technologies in Health Education and Allied Disciplines (pp. 153-175). IGI Global.

Bobbert, F. S. L., & Zadpoor, A. A. (2017). Effects of bone substitute architecture and surface properties on cell response, angiogenesis, and structure of new bone. *Journal of Materials Chemistry. B, Materials for Biology and Medicine*, *5*(31), 6175–6192. doi:10.1039/C7TB00741H PMID:32264433

Bosshardt, D. D., Chappuis, V., & Buser, D. (2017). Osseointegration of titanium, titanium alloy and zirconia dental implants: Current knowledge and open questions. *Periodontology 2000*, *73*(1), 22–40. doi:10.1111/prd.12179 PMID:28000277

Bourell, D., Kruth, J. P., Leu, M., Levy, G., Rosen, D., Beese, A. M., & Clare, A. (2017). Materials for additive manufacturing. *CIRP Annals - Manufacturing Technology*, *66*(2), 659-681. https://doi.org/ doi:10.1016/j.cirp.2017.05.009

Byambaa, B., Annabi, N., Yue, K., Trujillo-de Santiago, G., Alvarez, M. M., Jia, W., Kazemzadeh-Narbat, M., Shin, S. R., Tamayol, A., & Khademhosseini, A. (2017). Bioprinted Osteogenic and Vasculogenic Patterns for Engineering 3D Bone Tissue. *Advanced healthcare materials*, *6*(16). . doi:10.1002/adhm.201700015

Dämmer, G., Bauer, H., Neumann, R., & Major, Z. (2022). Design, additive manufacturing and component testing of pneumatic rotary vane actuators for lightweight robots. *Rapid Prototyping Journal*, *28*(11), 20–32. doi:10.1108/RPJ-03-2021-0052

Datta, S. (2023). Advantage of Alginate Bioinks in Biofabrication for Various Tissue Engineering Applications. *International Journal of Polymer Science*.

Datta, S., Barua, R., & Das, J. (2020). *Importance of alginate bioink for 3D bioprinting in tissue engineering and regenerative medicine*. Alginates—Recent Uses of This Natural Polymer. doi:10.5772/intechopen.90426

Datta, S., Barua, R., Sarkar, R., Barui, A., Chowdhury, A. R., & Datta, P. (2018, September). Design and development of alginate: Poly-l-lysine scaffolds by 3D bio printing and studying their mechanical, structural and cell viability properties. []. IOP Publishing.]. *IOP Conference Series. Materials Science and Engineering*, *402*, 012113. doi:10.1088/1757-899X/402/1/012113

Datta, S., Das, A., Chowdhury, A. R., & Datta, P. (2019). Bioink formulations to ameliorate bioprinting-induced loss of cellular viability. *Biointerphases*, *14*(5), 051006. doi:10.1116/1.5111392 PMID:31610661

Datta, S., Jana, S., Das, A., Chakraborty, A., Chowdhury, A. R., & Datta, P. (2020). Bioprinting of radiopaque constructs for tissue engineering and understanding degradation behavior by use of Micro-CT. *Bioactive Materials*, *5*(3), 569–576. doi:10.1016/j.bioactmat.2020.04.015 PMID:32373763

Davidson, J. A., Mishra, A. K., Kovacs, P., & Poggie, R. A. (1994). New surface-hardened, low-modulus, corrosion-resistant Ti-13Nb-13Zr alloy for total hip arthroplasty. *Bio-Medical Materials and Engineering*, *4*(3), 231–243. doi:10.3233/BME-1994-4310 PMID:7950871

Erbel, R., Di Mario, C., Bartunek, J., Bonnier, J., de Bruyne, B., Eberli, F. R., Erne, P., Haude, M., Heublein, B., Horrigan, M., Ilsley, C., Böse, D., Koolen, J., Lüscher, T. F., Weissman, N., & Waksman, R. (2007). Temporary scaffolding of coronary arteries with bioabsorbable magnesium stents: A prospective, non-randomised multicentre trial. *Lancet*, *369*(9576), 1869–1875. doi:10.1016/S0140-6736(07)60853-8 PMID:17544767

Farid, M. I., Wu, W., Liu, X., & Wang, P. (2021). Additive manufacturing landscape and materials perspective in 4D printing. *International Journal of Advanced Manufacturing Technology*, *115*(9-10), 2973–2988. doi:10.100700170-021-07233-w PMID:34092882

Fazel, M., Salimijazi, H., Shamanian, M., Apachitei, I., & Zadpoor, A. (2019). Influence of hydrothermal treatment on the surface characteristics and electrochemical behavior of Ti-6Al-4V bio-functionalized through plasma electrolytic oxidation. *Surface and Coatings Technology*, *374*, 222–231. doi:10.1016/j.surfcoat.2019.05.088

Fischer, M., Joguet, D., Robin, G., Peltier, L., & Laheurte, P. (2016). In situ elaboration of a binary Ti-26Nb alloy by selective laser melting of elemental titanium and niobium mixed powders. *Materials Science and Engineering C*, *62*, 852–859. doi:10.1016/j.msec.2016.02.033 PMID:26952492

Ganjian, M., Modaresifar, K., Zhang, H., Hagedoorn, P.-L., Fratila-Apachitei, L. E., & Zadpoor, A. A. (2019). Reactive ion etching for fabrication of biofunctional titanium nanostructures. *Scientific Reports*, *9*(1), 18815. doi:10.103841598-019-55093-y PMID:31827149

Ge, Q., Liu, X., Qiao, A., & Mu, Y. (2022). Compressive Properties and Degradable Behavior of Biodegradable Porous Zinc Fabricated with the Protein Foaming Method. *Journal of Functional Biomaterials*, *13*(3), 151. doi:10.3390/jfb13030151 PMID:36135585

Genabai, N. K., Ahmad, S., Zhang, Z., Jiang, X., Gabaldon, C. A., & Gangwani, L. (2015). Genetic inhibition of JNK3 ameliorates spinal muscular atrophy. *Human Molecular Genetics*, *24*(24), 6986–7004. doi:10.1093/hmg/ddv401 PMID:26423457

Gokuldoss, P. K., Kolla, S., & Eckert, J. (2017). Additive Manufacturing Processes: Selective Laser Melting, Electron Beam Melting and Binder Jetting-Selection Guidelines. *Materials (Basel)*, *10*(6), 672. doi:10.3390/ma10060672 PMID:28773031

Gorgin Karaji, Z., Speirs, M., Dadbakhsh, S., Kruth, J. P., Weinans, H., Zadpoor, A. A., & Amin Yavari, S. (2017). Additively Manufactured and Surface Biofunctionalized Porous Nitinol. *ACS Applied Materials & Interfaces*, *9*(2), 1293–1304. doi:10.1021/acsami.6b14026 PMID:28001358

Hockaday, L. A., Kang, K. H., Colangelo, N. W., Cheung, P. Y. C., Duan, B., Malone, E., Wu, J., Girardi, L. N., Bonassar, L. J., Lipson, H., Chu, C. C., & Butcher, J. T. (2012). Rapid 3D printing of anatomically accurate and mechanically heterogeneous aortic valve hydrogel scaffolds. *Biofabrication*, *4*(3), 035005. doi:10.1088/1758-5082/4/3/035005 PMID:22914604

Hosseini, E. S., Dervin, S., Ganguly, P., & Dahiya, R. (2021). Biodegradable Materials for Sustainable Health Monitoring Devices. *ACS Applied Bio Materials*, *4*(1), 163–194. doi:10.1021/acsabm.0c01139 PMID:33842859

ISO. (2015). *IAAF. Additive Manufacturing—General Principles—Terminology*. ISO.

Ji, Y. B., Park, J. Y., Kang, Y., Lee, S., Ju, H. J., Choi, S., Lee, B. Y., & Kim, M. S. (2021). Scaffold printing using biodegradable poly(1,4-butylene carbonate) ink: Printability, *in vivo* physicochemical properties, and biocompatibility. *Materials Today. Bio*, *12*, 100129. doi:10.1016/j.mtbio.2021.100129 PMID:34604731

Kambale, B. (2021). *Which Additive Manufacturing Process Is Right for You? | manufacturing*. REPORT.

Khorasani, M., Ghasemi, A., Leary, M., Sharabian, E., Cordova, L., Gibson, I., Downing, D., Bateman, S., Brandt, M., & Rolfe, B. (2022). The effect of absorption ratio on meltpool features in laser-based powder bed fusion of IN718. *Optics & Laser Technology*, *153*, 108263. doi:10.1016/j.optlastec.2022.108263

Kim, J. H., Seol, Y. J., Ko, I. K., Kang, H. W., Lee, Y. K., Yoo, J. J., Atala, A., & Lee, S. J. (2018). 3D Bioprinted Human Skeletal Muscle Constructs for Muscle Function Restoration. *Scientific Reports*, *8*(1), 12307. doi:10.103841598-018-29968-5 PMID:30120282

Kong, L., Valdivia, D. O., Simon, C. M., Hassinan, C. W., Delestrée, N., Ramos, D. M., Park, J. H., Pilato, C. M., Xu, X., Crowder, M., Grzyb, C. C., King, Z. A., Petrillo, M., Swoboda, K. J., Davis, C., Lutz, C. M., Stephan, A. H., Zhao, X., Weetall, M., & Sumner, C. J. (2021). Impaired prenatal motor axon development necessitates early therapeutic intervention in severe SMA. *Science Translational Medicine*, *13*(578), eabb6871. doi:10.1126citranslmed.abb6871 PMID:33504650

Kumar, A., Kumar, P., Mittal, R. K., & Gambhir, V. (2023). Materials processed by additive manufacturing techniques. In Advances in Additive Manufacturing Artificial Intelligence, Nature-Inspired, and Biomanufacturing (pp. 217-233). Elsevier. doi:10.1016/B978-0-323-91834-3.00014-4

Kumar, A., Mittal, R. K., & Haleem, A. (Eds.). (2022). *Advances in Additive Manufacturing: Artificial Intelligence, Nature-Inspired, and Biomanufacturing*. Elsevier. doi:10.1016/C2020-0-03877-6

Kuroda, D., Niinomi, M., Morinaga, M., Kato, Y., & Yashiro, T. (1998). Design and mechanical properties of new _ type titanium alloys for implant materials. *Materials Science and Engineering A*, *243*(1-2), 244–249. doi:10.1016/S0921-5093(97)00808-3

Li, Y., Jahr, H., Lietaert, K., Pavanram, P., Yilmaz, A., Fockaert, L. I., Leeflang, M. A., Pouran, B., Gonzalez-Garcia, Y., Weinans, H., Mol, J. M. C., Zhou, J., & Zadpoor, A. A. (2018). Additively manufactured biodegradable porous iron. *Acta Biomaterialia*, *77*, 380–393. doi:10.1016/j.actbio.2018.07.011 PMID:29981948

Li, Y., Zhou, J., Pavanram, P., Leeflang, M. A., Fockaert, L. I., Pouran, B., Tümer, N., Schröder, K. U., Mol, J. M. C., Weinans, H., Jahr, H., & Zadpoor, A. A. (2018). Additively manufactured biodegradable porous magnesium. *Acta Biomaterialia*, *67*, 378–392. doi:10.1016/j.actbio.2017.12.008 PMID:29242158

Liang, S. X., Wang, X., Zhang, W., Liu, Y. J., Wang, W., & Zhang, L. C. (2020). Selective laser melting manufactured porous Fe-based metallic glass matrix composite with remarkable catalytic activity and reusability. *Applied Materials Today*, *19*, 100543. doi:10.1016/j.apmt.2019.100543

Lim, K. S., Levato, R., Costa, P. F., Castilho, M. D., Alcala-Orozco, C. R., Van Dorenmalen, K. M., Melchels, F. P. W., Gawlitta, D., Hooper, G. J., Malda, J., & Woodfield, T. B. (2018). Bio-resin for high resolution lithography-based biofabrication of complex cell-laden constructs. *Biofabrication*, *10*(3), 034101. doi:10.1088/1758-5090/aac00c PMID:29693552

Long, M., & Rack, H. (1998). Titanium Alloys in Total Joint Replacement—A Materials Science Perspective. *Biomaterials*, *19*(18), 1621–1639. doi:10.1016/S0142-9612(97)00146-4 PMID:9839998

Lu, Y., Deshmukh, S., Jones, I., & Chiu, Y. L. (2021). Biodegradable magnesium alloys for orthopaedic applications. *Biomaterials Translational*, *2*(3), 214–235. doi:10.12336/biomatertransl.2021.03.005 PMID:35836650

Marinopoulou, E., & Katakalos, K. (2022). Thermomechanical Fatigue Testing on Fe-Mn-Si Shape Memory Alloys in Prestress Conditions. *Materials (Basel)*, *16*(1), 237. doi:10.3390/ma16010237 PMID:36614575

Mazzocchi, A., Devarasetty, M., Huntwork, R., Soker, S., & Skardal, A. (2018). Optimization of collagen type I-hyaluronan hybrid bioink for 3D bioprinted liver microenvironments. *Biofabrication*, *11*(1), 015003. doi:10.1088/1758-5090/aae543 PMID:30270846

Mercuri, E., Finkel, R. S., Muntoni, F., Wirth, B., Montes, J., Main, M., Mazzone, E. S., Vitale, M., Snyder, B., Quijano-Roy, S., Bertini, E., Davis, R. H., Meyer, O. H., Simonds, A. K., Schroth, M. K., Graham, R. J., Kirschner, J., Iannaccone, S. T., Crawford, T. O., & Szlagatys-Sidorkiewicz, A.SMA Care Group. (2018). Diagnosis and management of spinal muscular atrophy: Part 1: Recommendations for diagnosis, rehabilitation, orthopedic and nutritional care. *Neuromuscular Disorders*, *28*(2), 103–115. doi:10.1016/j.nmd.2017.11.005 PMID:29290580

Murr, L. E., Gaytan, S. M., Medina, F., Lopez, H., Martinez, E., Machado, B. I., Hernandez, D. H., Martinez, L., Lopez, M. I., Wicker, R. B., & Bracke, J. (2010). Next-generation biomedical implants using additive manufacturing of complex, cellular and functional mesh arrays. *Philosophical transactions. Series A, Mathematical, physical, and engineering sciences, 368*(1917), 1999–2032. https://doi.org/ doi:10.1098/rsta.2010.0010

Peterson, I. S., Mazzella, A. J., Belter, L. T., Curry, M. A., Cruz, R. E., & Jarecki, J. (2022). The Cure SMA Clinical Trial Experience Survey: A Study of Trial Participant Perspectives on Clinical Trial Management and Patient-Centric Management Practices. *Neurology and Therapy*, *11*(3), 1167–1181. doi:10.100740120-022-00360-w PMID:35635603

Putra, N. E., Leeflang, M. A., Minneboo, M., Taheri, P., Fratila-Apachitei, L. E., Mol, J. M. C., Zhou, J., & Zadpoor, A. A. (2021). Extrusion-based 3D printed biodegradable porous iron. *Acta Biomaterialia*, *121*, 741–756. doi:10.1016/j.actbio.2020.11.022 PMID:33221501

Razzi, F., Lovrak, M., Gruzdyte, D., Den Hartog, Y., Duncker, D. J., van Esch, J. H., van Steijn, V., & van Beusekom, H. M. M. (2022). An Implantable Artificial Atherosclerotic Plaque as a Novel Approach for Drug Transport Studies on Drug-Eluting Stents. *Advanced Healthcare Materials*, *11*(6), e2101570. doi:10.1002/adhm.202101570 PMID:34865315

Sehrawat, S., Kumar, A., Grover, S., Dogra, N., Nindra, J., Rathee, S., Dahiya, M., & Kumar, A. (2022). Study of 3D scanning technologies and scanners in orthodontics. *Materials Today: Proceedings*, *56*, 186–193. doi:10.1016/j.matpr.2022.01.064

Sehrawat, S., Kumar, A., Prabhakar, M., & Nindra, J. (2022). The expanding domains of 3D printing pertaining to the speciality of orthodontics. *Materials Today: Proceedings*, *50*, 1611–1618. doi:10.1016/j.matpr.2021.09.124

Sharma, S., Sudhakara, P., Singh, J., Ilyas, R. A., Asyraf, M. R. M., & Razman, M. R. (2021). Critical Review of Biodegradable and Bioactive Polymer Composites for Bone Tissue Engineering and Drug Delivery Applications. *Polymers*, *13*(16), 2623. doi:10.3390/polym13162623 PMID:34451161

Singh, V., Gupta, M., & Kumar, A. (2019, November). Investigating the morphology and thermophysical properties of Al2O3 and CuO nanofluids. []. IOP Publishing.]. *IOP Conference Series. Materials Science and Engineering*, *577*(1), 012052. doi:10.1088/1757-899X/577/1/012052

Singh, V., Kumar, A., Alam, M., Kumar, A., Kumar, P., & Goyat, V. (2022). A study of morphology, UV measurements and zeta potential of zinc ferrite and Al2O3 nanofluids. *Materials Today: Proceedings*, *59*, 1034–1039. doi:10.1016/j.matpr.2022.02.371

Takhsha Ghahfarokhi, M., Nasi, L., Casoli, F., Fabbrici, S., Trevisi, G., Cabassi, R., & Albertini, F. (2020). Following the Martensitic Configuration Footprints in the Transition Route of Ni-Mn-Ga Magnetic Shape Memory Films: Insight into the Role of Twin Boundaries and Interfaces. *Materials (Basel)*, *13*(9), 2103. doi:10.3390/ma13092103 PMID:32370074

Tsaturyants, M., Sheremetyev, V., Dubinskiy, S., Komarov, V., Polyakova, K., Korotitskiy, A., Prokoshkin, S., Borisov, E., Starikov, K., Kaledina, D., Popovich, A., & Brailovski, V. (2022). Structure and properties of Ti-50.2Ni alloy processed by laser powder bed fusion and subjected to a combination of thermal cycling and heat treatments. *Shape Mem. Superelasticity*, *8*(1), 16–32. doi:10.100740830-022-00363-4

van Hengel, I. A. J., Riool, M., Fratila-Apachitei, L. E., Witte-Bouma, J., Farrell, E., Zadpoor, A. A., Zaat, S. A. J., & Apachitei, I. (2017). Data on the surface morphology of additively manufactured Ti-6Al-4V implants during processing by plasma electrolytic oxidation. *Data in Brief*, *13*, 385–389. doi:10.1016/j.dib.2017.06.015 PMID:28664174

Vukojicic, A., Delestrée, N., Fletcher, E. V., Pagiazitis, J. G., Sankaranarayanan, S., Yednock, T. A., Barres, B. A., & Mentis, G. Z. (2019). The Classical Complement Pathway Mediates Microglia-Dependent Remodeling of Spinal Motor Circuits during Development and in SMA. *Cell Reports*, *29*(10), 3087–3100.e7. doi:10.1016/j.celrep.2019.11.013 PMID:31801075

Wachesk, C. C., Seabra, S. H., Dos Santos, T. A. T., Trava-Airoldi, V. J., Lobo, A. O., & Marciano, F. R. (2021). In vivo biocompatibility of diamond-like carbon films containing TiO_2 nanoparticles for biomedical applications. *Journal of Materials Science. Materials in Medicine*, *32*(9), 117. doi:10.100710856-021-06596-6 PMID:34460018

Wauthle, R., van der Stok, J., Amin Yavari, S., Van Humbeeck, J., Kruth, J. P., Zadpoor, A. A., Weinans, H., Mulier, M., & Schrooten, J. (2015). Additively manufactured porous tantalum implants. *Acta Biomaterialia*, *14*, 217–225. doi:10.1016/j.actbio.2014.12.003 PMID:25500631

Wei, S., Li, Y., Li, K., Kang, A., Zhang, S., Feng, T., Zhang, H., & Zhong, C. (2021). Functional amyloid-chitin hybrid ink coupled with flexible fabrication approaches for diverse macro and micro-structures. *Materials Today. Bio*, *13*, 100179. doi:10.1016/j.mtbio.2021.100179 PMID:34938993

Windhagen, H., Radtke, K., Weizbauer, A., Diekmann, J., Noll, Y., Kreimeyer, U., Schavan, R., Stukenborg-Colsman, C., & Waizy, H. (2013). Biodegradable magnesium-based screw clinically equivalent to titanium screw in hallux valgus surgery: Short term results of the first prospective, randomized, controlled clinical pilot study. *Biomedical Engineering Online*, *12*(1), 62. doi:10.1186/1475-925X-12-62 PMID:23819489

You, F., Chen, X., Cooper, D. M. L., Chang, T., & Eames, B. F. (2018). Homogeneous hydroxyapatite/alginate composite hydrogel promotes calcified cartilage matrix deposition with potential for three-dimensional bioprinting. *Biofabrication*, *11*(1), 015015. doi:10.1088/1758-5090/aaf44a PMID:30524110

Zadpoor, A. A. (2019). Additively manufactured porous metallic biomaterials. *Journal of Materials Chemistry. B, Materials for Biology and Medicine*, *7*(26), 4088–4117. doi:10.1039/C9TB00420C PMID:31701985

Zadpoor, A. A., & Malda, J. (2017). Additive Manufacturing of Biomaterials, Tissues, and Organs. *Annals of Biomedical Engineering*, *45*(1), 1–11. doi:10.100710439-016-1719-y PMID:27632024

Zhan, M., Liu, J., Wang, D., Chen, X., Zhang, L., & Wang, S. (2021). Optimized Neural Network Prediction Model of Shape Memory Alloy and Its Application for Structural Vibration Control. *Materials (Basel)*, *14*(21), 6593. doi:10.3390/ma14216593 PMID:34772118

Zhang, G., Zhao, P., Lin, L., Qin, L., Huan, Z., Leeflang, S., Zadpoor, A. A., Zhou, J., & Wu, L. (2021). Surface-treated 3D printed Ti-6Al-4V scaffolds with enhanced bone regeneration performance: An *in vivo* study. *Annals of Translational Medicine*, *9*(1), 39. doi:10.21037/atm-20-3829 PMID:33553332

Zhang, J., Jiang, Y., Shang, Z., Zhao, B., Jiao, M., Liu, W., Cheng, M., Zhai, B., Guo, Y., Liu, B., Shi, X., & Ma, B. (2021). Biodegradable metals for bone defect repair: A systematic review and meta-analysis based on animal studies. *Bioactive Materials*, *6*(11), 4027–4052. doi:10.1016/j.bioactmat.2021.03.035 PMID:33997491

Chapter 9
Review of Microstructural and Mechanical Properties of Microwave Welding of Lightweight Alloys

Eshan Goswami
G.L. Bajaj Institute of Technology and Management, India

Ashish Kumar Srivastava
https://orcid.org/0000-0003-2834-1483
G.L. Bajaj Institute of Technology and Management, India

Ajay Kumar
School of Engineering and Technology, JECRC University, India

Mohd Salman
G.L. Bajaj Institute of Technology and Management, India

Mohd Zeeshan Choudhary
G.L. Bajaj Institute of Technology and Management, India

Samarthya Shankar Pandey
G.L. Bajaj Institute of Technology and Management, India

ABSTRACT

Manufacturing hard metals and their welding-related operations is still a challenging task for the manufacturer. In this context, many types of welding methods are developed. Among these methods, microwave welding is one of the prominent methods to weld or clad metal plates. The present work is a review chapter, and it highlights the study of microstructural and mechanical properties of the developed components through microwave welding. The present work also highlights the challenges and issues related to microwave welding technology. Finally, the future scope and its commercialization are discussed.

DOI: 10.4018/978-1-6684-9224-6.ch009

1. INTRODUCTION

Due to the rising need for energy worldwide, renewable energy is now thought to be the most popular energy source. The total amount of harvested renewable energy has been greatly boosted by hydroelectric electricity. Stations that produce hydroelectric power frequently use authenticate stainless steel. Additionally, a wide number of industries, including petrochemical, aerospace, defence, shipbuilding, and cryogenic stations, use stainless steel that contains nitrogen (Rajan et al., 2018). The developing method of material processing that is non-intrusive, eco-friendly, and energy-efficient is microwave processing. In this method, materials are processed for a variety of purposes, including boring, moulding, impurity, covering, thawing, and attaching, using electromagnetic energy at specific frequencies within the microwave energy spectrum. There have been numerous studies on microwave casting and drilling of various types of materials (Ajay et al., 2023, Kumar et al., 2023), including metals, glass, Perspex, and, ceramics with reassuring results reported. Similar research has been done and is being done in the field of microwave sintering, including treating different types of electronic and biomaterials materials. Additionally, microwave coating/cladding was reported for employing microwave radiation to process high melting point materials such WC-12Co cermet; the produced clad were stated to have better wear resistance and frictional qualities (Chowdhury et al., 2011).

The ability of a substance to effectively absorb a given amount of electromagnetic (EM) energy determines how well it may be heated. However, it is fairly possible for a variety of materials to heat very fast when employing EM energy than when using standard heating methods. Microwaves with a frequency spectrum of 305 MHz to 305 GHz are discovered to be particularly efficient in the operation of material refining in the EM spectrum. When compared to traditional heating techniques, microwave processing offers the potential to boost material and energy efficiency. Energy movement is not a restricted level of matter in microwave processing; instead, heat can be transferred to all particles, increasing the rate at which heat is transferred with a microwave The inverse heating profile is observed because, during processing, heating is volume-dependent rather than surface-dependent. The efficiency of electromagnetic radiation's conversion to heat is close to 100%. Since energy is delivered using microwaves, which have the quality to break surface layers, a new climate figure exists for warm-up clarification. The microwave thawing method was first used to heat food before expanding its function to include the clarification of a different range of matter, including slip wares, polymaths, and complex; it now has many uses in the processing of fusible matters as well. The sintering, joining, and coating/cladding of fusible matter using oven heat have seen a sharp increase in interest in recent years (Kumar et al., 2022). Growing awareness is one cause of the growing interest. However, because most metals reflect electromagnetic waves under normal circumstances, using microwaves to process metallic materials is a difficult field of research. Now composition has revealed actual little study in the domain of treating fusible material using sintering under certain situations. Additionally, effort has been done on weld particular materials bottom particular circumstances. It has also been documented that bulk metallic objects can be joined in a variety of ways using a home microwave device (Bozkurt et al., 2011).

However, it has been stated in the form of a patent that metallic and non-metallic powders can be coated or clothed on metallic substrates. This presents a chance to make use of the benefits of oven heat for treating fusible metals in order to appropriate the demanding or expanding requirements in several industry functions (Singh et al., 2023). The general effort has successfully completed the connecting of SS-317 to MS in the majority pattern. The basic difference between microwave processing and other methods is that during microwave heating, heat is produced as a result of electromagnetic radiation penetrating the target material due to dielectric losses. As a result, microwave heating produces an inverse temperature profile, meaning the crux is heated to increased heat. Co-related to the objective matters plane. Furthermore, because of their poor absorption coefficient, applying oven heat to process the extent of fusible matter at room heat is difficult. Thus, the microwave hybrid heating (MHH) approach has been employed to perform compound temperature through a susceptible matter such as charcoal (Kumar et al., 2020).

It used Inconel 718 powder as an interface layer and the MHH approach, fusion joining of Inconel 718 and SS-316L has been accomplished in the current study. In general, oven temperature of matter differs from traditional temperature procedures. Materials' dielectric characteristics affect how they interact with microwaves. As opposed to the traditional heating of matter, there is an energy transformation taking place. Through molecular contact, microwave radiation is sent directly to the substance, where it causes volumetric heating. Conventional heating causes poor microstructure because the material surface is heated to an increased heated than the inside. due to the significant heat differential between the boundary and the core inside the core. As a result of warming the undesired locations, a content may drop its expected pattern features when heated another using normal means. Microwaves are used in material processing in an environmentally responsible way. However, there are several challenges when treating metallic materials using microwave energy. At normal temperatures, the interaction of microwaves with fusible matter results in ignition because the bulk metals reflect microwaves because their skin depth is only a few microns. Slip wares and plastics are examples of matter that can be treated with straight exposure to the oven because they readily consume the ambient heat due to increasing dielectric damage (Mishra et al., 2016).

Weld joints, which are defined by the geometry of the metal being considered, are places where two or more metal pieces can be linked together. Advanced as a fusing system for use in increased durability composites that have proven difficult to link with traditional methods, friction stir welding is a hard case welding procedure. Joints come in five different varieties: butt, corner, edge, tee, and lap. Because matter is arranged in the same uniform before joining, which is frequent in motorized joining methods like submerged arc welding butt junctions are the most prevalent and straightforward. Due to their capacity to organically link metal components, TIG and MIG welding methods are ideal for creating butt weld joints (Kandagal et al., 2021).

There are different types of joined plates used to create the join to determine the fuse qualities, including its rust or durability protection (Sireesha et al., 2000). The least defective butt weld has the greatest strength. This is a fillet fuse joint that can be used with a variety of joining processes to join materials of different thicknesses when one of the parts is overlapping the other. Although corner joining joints are the most typical welding joints in production, they are a particular type of junction where the two components come together at a right angle to form the connection (Badiger et al., 2018). When welding sheet metal parts with flanging edges, edge welding joints are a common type of welding groove employed (Zafar & Sharma, 2017), Al6061-6%Gr-SiC hybrid MMC samples were unloaded using an ultrasonic cleaner and an acetone bath, then they were dried to get rid of the impurities. The samples were stored with a welding thickness of roughly 1 mm at the butt joint configuration (Huang, 2010). The inter-facial powder Al6061 was combined in the right amounts along with adhesive resin to create a patch-like slurry. The paste was manually increased across the composite variety interfaces before welding. The Al6061 interface particles were used to bind the pattern with epoxy glue (Adetunji et al., 2015). The specimens were then put in graphite sheets with a thickness of roughly 0.5 mm. The Coke sheet can endure increased heat and keep charcoal powder and specimens separate. The area where the charcoal powder was deposited offers the initial coupling of microwaves (Padhy & Saini, 2008).

2. MATERIALS AND METHODS

Due to the bulk metals' reflection, microwave welding of metals was first difficult. Further study led to the researchers' success in bulk microwave welding of metals and ceramics. Microwave heating with a 2.45 GHz frequency and 700 W of power was used to successfully weld the sintered alumina with 30% zirconia ceramic composites (Aravindan & Krishnamurthy, 1999). Numerous researchers have looked into the welding of Al-based MMCs and discovered that the weld joints feature flaws like process-related flaws, solidification flaws, metallurgical flaws, and the development of dangerous aluminium carbide (Al4C3). Friction stir welding, laser beam welding, electron beam welding, arc welding, etc. are some of the frequently utilized fusion welding techniques for MM Cs. The resultant joint strength is minimal when using alternative welding techniques like adhesive bonding, diffusion bonding, and brazing, which are low temperature processes. Conventional welding techniques not only result in weld joints with flaws, but they also have other drawbacks including pricey setups, lengthy processes, and environmental risks, among others (Kothari & Hung, 2018; Bozkurt et al., 2011). The microstructural analysis revealed a weld connection that was dense and uniform with good tensile strength. Later, using a household microwave oven with a frequency of 2.45 GHz and 900 W, the same researcher successfully welded the incompatible metals (stainless steel to mild steel) (Srinath et al., 2011a). Al6061-6%Gr-SiC hybrid MMC samples were cleaned using an ultrasonic cleaner and an acetone bath, then they were dried to get rid of the impurities. The samples were stored with a welding thickness of around 1 mm at the butt joint configuration. The interfacial powder Al6061 was combined in the right amounts with epoxy resin to create a paste-like slurry. The paste was manually spread across the composite specimen's interfaces before welding. The Al6061 interface particles were used to bond the specimens using epoxy glue. The specimens were then put on graphite sheets with a thickness of around 0.5 mm. The graphite sheets can endure high temperatures and keep charcoal powder and specimens separate. The area where the charcoal powder was deposited offers the first coupling of microwaves (Wong & Gupta, 2015) and as seen in the SEM photos, the homogeneous heating during MHH produced homogenous and thick joint surfaces of hybrid MMCs without any discernible fractures. a greater more Al6061 powder particles were found in the weld zone than SiC and graphite reinforcement. In joints with different reinforcing content, less voids were seen. Al2O3 and graphite particle segregation and cluster formation in inter-dendritic zones were seen. According to SEM pictures (Aravindan & Krishnamurthy, 1999; Kothari & Hung, 2018; Bozkurt et al., 2011), the free carbons were discernible in the weld zone area (Badiger et al., 2015; Singh et al., 2019; Bansal et al., 2016). Energy-disperse spectroscopy (EDS) analysis was used to evaluate the chemical composition of the Al6061-6%Gr-sic composite joints. Along with oxygen, carbon, and sulphur (S) components, the EDS analysis identified aluminium as one of the main constituent elements. The degradation of Sic and the disintegration of the graphite sheet into free silicon and carbon occurred at higher temperatures. Carbon and oxygen components in the EDS study confirm the presence of the reaction products aluminium carbide and aluminium oxide, as revealed by XRD analysis. Numerous researchers utilizing various welding techniques for Al-sic-based MM-Cs obtained similar observations (Ma et al., 2018; Salih et al., 2015; Hassan et al., 2012). Using a Rockwell hardness tester, the hardness was determined throughout the weld zone. For the test, a steel ball indented with a diameter of 1/16 inch and a weight of 100 kgf was employed. The specimens were created in accordance with the ASTM E28 standard. In order to obtain the average value, the hardness was tested at three different points, including the base composite, weld interface, and weld zone. The hardness test results are summarized when the weight % of Si C reinforcement rose, the hardness value increased as well. The basic MMCs were found to have an average hardness of 57 HRB. The weld interface was determined to have a hardness of 62 HRB, which is greater than the weld zone (Srinath et al., 2011a,b; Badiger et al., 2015; Sujan et al., 2012; Badiger et al., 2019; Singla et al., 2009). By segmenting routes into points of welded joints, the impacts of the welded microstructure static performance and the impact of spot-welded joints in cutting-edge steel were thoroughly explored (Afolalu et al., 2019). An impact evaluation and a quasi-static test are used to investigate the impact outcomes of resistance spot-welded joints. Using an instrumented drop-down tower and a universal traction test, quasi-static testing is a sort of test that is carried out (Magudeeswaran et al., 2008). Two distinct tensile samples were arranged to assess the joints' notch tensile strength and notch strength ratio in addition to notch specimens in order to evaluate the transverse tensile properties of the joints.

The most popular metallic materials in the fabrication industry are SS-316 and MS. Commercial quality SS-316 and MS plates with along capacity of 25 mm x 12 mm x 6 mm were used in the experiments. The synthetic makeup or automatic characteristics of the SS-316 and MS employed in the current tests are displayed in Table 1. The joints were constructed as square butts. As an open-faced boundary, a particle made of nickel along a particle size of 40 m was employed. Epoxy resin (Bisphenol-A, Blumer 1450XX) was combined along the particle to create a patch-like slurry (Di Martino et al., 2014). The process of microwave welding is depicted in figure 1.

Table 1 represents the devices and materials which is used in the setup for the development of microwave welding. we have a domestic microwave with the specification of 900W, which works on the frequency of 2.45GHz was used for this experiment. charcoal powder was used as the susceptor material in this experiment. this experimental setup was applicable to both types of materials (similar and dissimilar types of materials). the reason behind using charcoal powder absorbs more heat and doesn't transmit energy throughout the welding process.

Before being cleaned in acetone using an ultrasonic cleaner for 15 minutes and dried with hot air, the 21-4-N nitric steel plates were cut to the desired size and shape. Furthermore, an epoxy resin was mixed with the nickel powder to generate slurry (Bisphenol-A, Blummer 1450XX).

Figure 1. Process of microwave welding

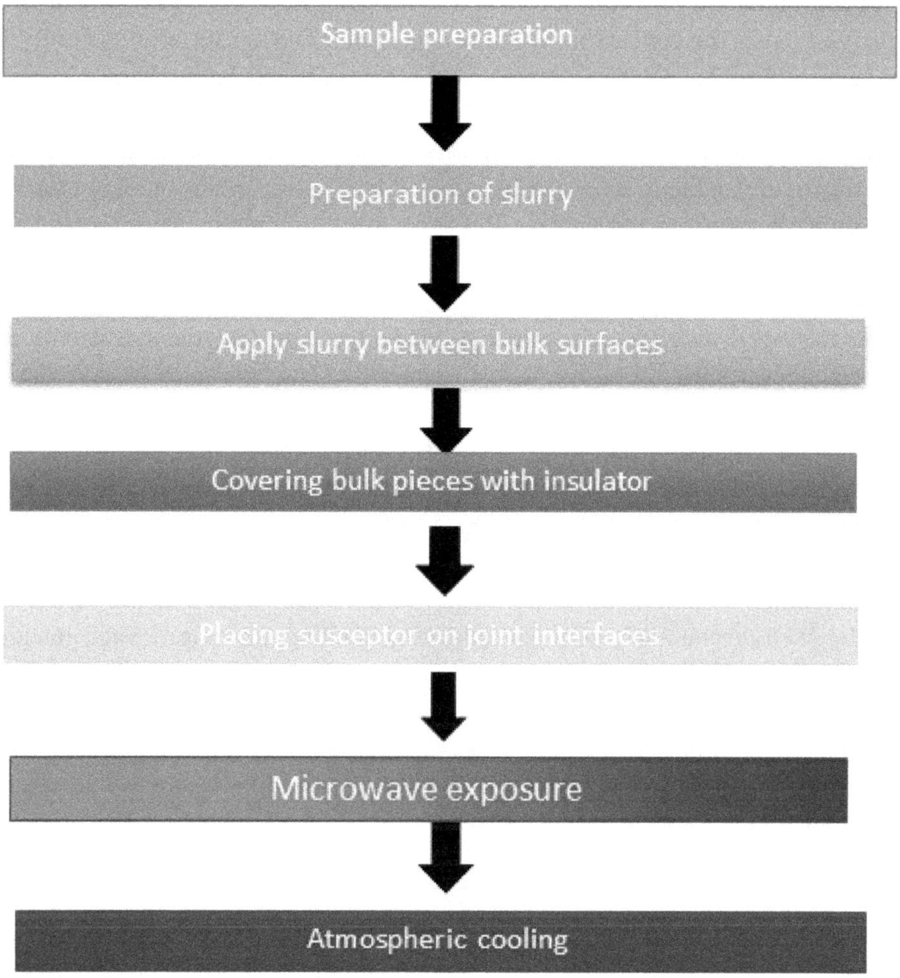

Table 1. Microwave processing parameters

Device	Multi-band
Microwave Range	2.45Ghz
Foundation matter	Commercial applicable SS-316
Accommodate matter	Al-based powder
Sucesptor material	Charcoal
Exposure time	450 s
Exposure powder	900 W

The specimens were maintained with a 0.5 mm space between them in a butt joint configuration. Fire clay served as the masking material, while the produced slurry served as the contact between the specimen's flaying surfaces. After filling the setup with nickel powder, alumina wool was used as insulation. Figure 2 depicts a schematic of the microwave joining procedure. Putting the entire component inside the microwave applicator. The minimal depth of metal penetration made the microwave unreliable at room temperature. It takes place when an alloy under its demanding heat reflects microwaves. Beyond the threshold temperature, metals begin to absorb microwaves, and the crack base importance rises. The base to which an oven can penetrate alloy objects relies on a number of microwave characteristics. These metrics, which are temperature-dependent, are the material's resistivity and magnetic permeability. 24,25In this experiment, tests were run in an oven device under atmospheric circumstances at 950 W impotence and 2.46 GHz range. As a result of the susceptor's aid, the Nickel dust molecule begins to consume the oven during microwave exposure and melt at an increased heat. Consequently, the joint material interface surfaces the creation of a full example of a fused joint. Bulk samples needed 480 s of exposure time at a frequency of 2.45 GHz for complete fusion joining.

The MHH technique's benefits include generating a strong enough junction between lightweight metals that are comparable or dissimilar, which cuts down on melting time and joint surface imperfections. Settings for key parameters and models for heating designs, however, have not been thoroughly investigated for MHH applications. Because of its poor consolidating ability and increase MP at ambient temperature, Nickel-based metallic particle (around 9–10%) exhibits limited ductility. Weldability has been challenged by this issue in a number of production processes. Researchers used microwave hybrid heating (MHH) approaches to accelerate the melting process by increasing the quick Using carbon-based (absorbed) and silicon-based materials, we investigated the effects of heating rate, mechanical parameters (tensile strength and hardness), and microstructure properties (insulator).

Metal powder selection is essential for connecting metals using MHH procedures. Epoxy resin is incorporated into the slurry to get the required metal composition (Srinath et al., 2011b). Fast heating (energy-saving), high control accuracy, selective heating, volumetric heating, environmental friendliness, and quick processing times are some benefits of using the MHH approach (Wong & Gupta, 2015). Microwave radiation's wave energy was equally converted and was able to change the inter-molecular reaction with a narrower power range. The interaction between the absorbed and microwaves is essential to form heat matter that can replace conventional heating techniques due to its effectiveness and rapid heating. Even though the MHH reduces reactivity and offers good mechanical and microstructural qualities.

In conventional heating, first heat the outer surface then gradually heat the inner material so it takes very more time to heat the material in conventional heating there is a heating element present in it. but in microwave welding heating is diffused throughout the material and there is no heating element present in the microwave.

By segmenting routes into points of fused joints, spot welding connections' effects on static performance and the effects of the welded microstructure in cutting-edge steel (Bansal et al., 2016) were thoroughly explored. A shock evaluation or a quasi-static test is used to investigate the shock outcomes to protect spot-fused joints. A universal traction test and an instrumented drop-down tower (Ma et al., 2018) are used in quasi-static testing, which also includes impact testing. It is necessary to examine welded constructions, for instance, the impact joint strength and failure mode under a certain stress rate. This research examines moderately broken weld pieces to determine the precise fracture route under static loading. Utilizing the observed microstructures, a link among motorized parameters or the crack path was constructed. When 2.46 GHz microwaves are applied to the Ti and Ni-Al powder compact, a reaction occurs. and the SHS opinion begins to produce intermetallic compounds. Heat is transported from the freshly created molten Ni-Al to the Ti facial during the exothermic reaction.

An earlier study (Badiger et al., 2015) showed that the delay of the temperature-impacted area is lower than 80 m due to titanium's highly quick reactivity and comparatively low thermal conductivity. Against the unfavourable heat angle (the synthesized substances are much hotter than the environment around them in temperature), the use of microwaves enables the continued delivery of efficiency to

Figure 2. Microwave material processing

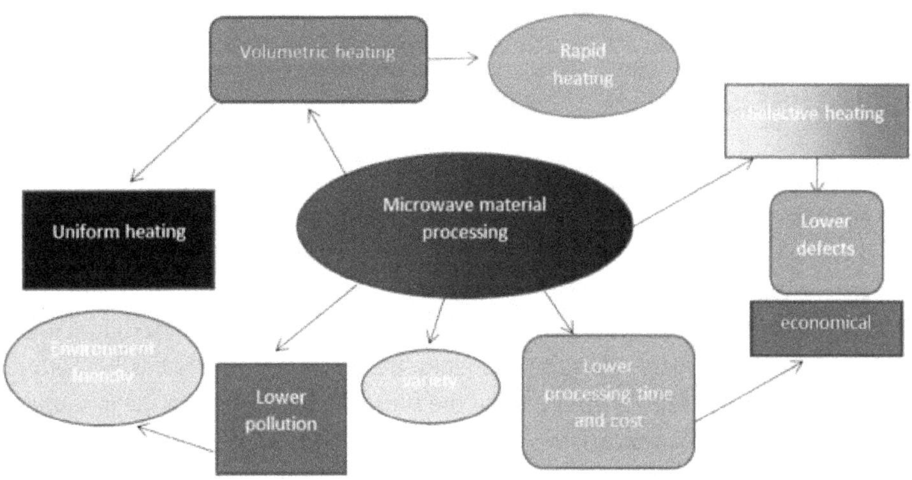

the newly created aspects. This slows the pace of cooling and prolongs the duration that the molten phase is present (Badiger et al., 2019).

Only the microwave's electric field components influence non-magnetic materials. Dipolar losses and power dissipation are the 2 main loss mechanisms for non-magnetic materials like Al, water, Cu, polymers, and porcelain. While dipolar losses dominate in dielectric insulators, conduction losses dominate in metallic and high-conductivity materials. In dielectric insulator materials, whereby dipoles are produced when exposed to an external electric field, the dipole loss is more rapid. These components include food products, water, and ceramic. With dipoles, heat is generated. In the microwave processing of pure metals metallic based materials and semiconductors, such as cu, Al, Si, Fe, Ni, and MMC, this loss is considerable. Most of the thermoplastics have decreased dielectric damage aspect therefore susceptor materials are used to generate a sufficient amount of heat. thermostats start heating after interacting with microwaves but sometimes as polymerization occurs, the heating of thermostats slows down growth of polymeric chains restricts the interaction of thermostats with microwaves.

As part of the mechanical property's characterization of pure magnesium and the manufactured magnesium-based products, microhardness and tensile qualities are measured. The microhardness of polished flat specimens was measured using a Shimadzu HMV automatic digital microhardness tester and a Nickers indented with a load of 25 GF and a dwell time of 15s. The microhardness tests were performed on each sample's three pieces for a minimum of 15 reliable readings. A fully automated servo-hydraulic mechanical testing machine, Model-MTS 810, was used to determine the tensile properties of magnesium-based materials. The test specimens had a gauge length of 25 mm and a diameter of 5 mm, in accordance with ASTM test method E8M-96. The cross-head speed was set at 0.254 mm/min. An Instron 2630-100 was used to take measurements.

3. DISCUSSION

The usual microstructure and morphology at the TA2/5083 contact no significant decontamination or voids were seen. Without an inter-layer, the directly bonded interfacial morphology displayed a "harmonious" condition indicating that the joined connection had an equal format. The detonation velocity of the explosives and the characteristics of the involved metals are related to the creation and design of joined joints for explosive fusion. On the one hand, the explosive link is easily shattered if the detonation velocity is greater than the sensual acceleration of the component alloy. However, the characteristics of the involved metals, particularly their body, reducing, and effective yield durability, have a significant impact on the symmetry and geometry depict a thin micro crack and a locally continuing melting layer close to the bonding zone. The bonding property was thought to degrade as a result of the observed interfacial microstructure. The severe collision under the influence of explosive products led the flyer plate's available kinetic energy to nearly entirely evaporate at the fusion contact (TA2-5083-3.13 MJ m^2). Most of the lost kinetic energy was converted to thermal energy, which quickly raised the temperature and ultimately caused the associated metals to melt and diffuse near the bonding zone, especially for aluminium alloy 5083 with its lower melting point.

Figure 3. How microwave welding differs from other welding

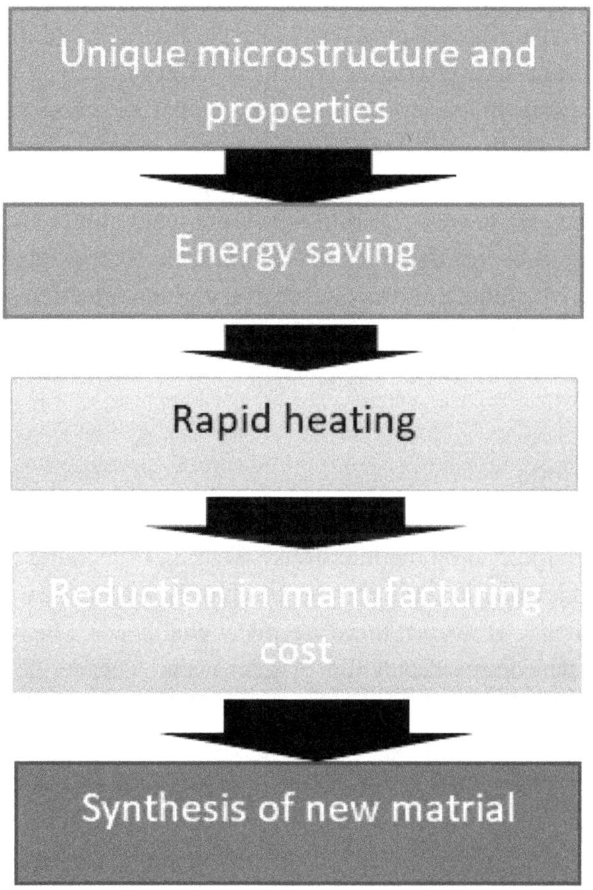

It is highly challenging to join metallic materials in bulk using microwave energy because microwaves at ambient temperature reflect and figure 3 represents the difference between microwave welding and other conventional welding methods. The results of the construction of metallic material joints made of different metals using a microwave hybrid heating process are reported in the current work. It is being done to characterize the joints further. The work lays the groundwork for massive, environmentally friendly, and notably quick bonding of metallic materials. In comparison to as-sprayed Ni coatings, lower porosity and surface roughness, higher microhardness, higher fracture toughness, and stronger abrasive wear resistance were all characteristics of Ni coatings that had undergone microwave post-processing. The microwave post-processed coating layer has lower porosity (8% less) than the as-sprayed coating, higher microhardness (18% higher), higher fracture toughness (35% higher), lower surface roughness (44% less), and lower cumulative weight loss (32% less) than the as-sprayed coating. Due to material withdrawal, the homogenized microstructure results in less material loss and higher wear resistance.

It has been found that the exposure duration is less for similar metals based on SS when using microwave hybrid processing than for dissimilar metals. The situation with Inconel, however, is different. Joint weld-ability and exposure time, respectively it is discovered that microwave joining on composite materials appears to be quite effective. Table 2 shows the summary of the selection of materials with their specifications like material selection, interface material, susceptor material, and exposure time.

4. CONCLUSION

The present work focused on the microwave welding of metallic components. The present work concluded that microwave welding is a prominent method to weld metallic components. However, there are many challenges which are still need to be overcome. The commercialization of microwave welding is still limited due to its compact design and high cycle time which is a major challenge for mass production. However, some futuristic works are going on which can increase the commercialization of microwave welding.

Table 2. Parameters of selected material with their results

Materials	Microwave oven	Interface material	Susceptor	Result	References
Sintered alumina zirconia ceramic 40%	700WAt 2.45GHz	Sodium silicate glass powder	NA	The inter-Diffusion of atomic species promotes the formation of strong joints	Aravindan & Krishnamurthy, 1999
SS316	900W At 2.45GHz	Nickel based powder 40	Charcoal	Evaluation of the tensile properties of the joints showed an ultimate tensile strength of the order 309 MPa with elongation of 11.50%	Srinath et al., 2011c
SS316	900WAt 2.45GHz	Nickel powder	Charcoal	The joint's final tensile strength was determined to be 346.6 MPa with an elongation of 13.58%.	Nandwani et al., 2020
Sisal fiber-reinforced polylactic acid	300W.600W,900W at 2.45 Hz	charcoal	NA	Without addition or acceleration, there was no joining.	Singh et al., 2011
Metal ceramic	900W at 2.45Hz	Nickel powder	charcoal	used-microwave energy to create a metal-ceramic composite that is wrapped in a metallic substrate.	Sharma & Gupta, 2012
Mild Steel (in bulk form)	900W at 2.45Hz	Nickel powder (40 mm)/1450XX	Charcoal	The results of the tensile test revealed that the average joint strength was 250 MPa with an elongation of 6%.	Bansal et al., 2012
Natural Fibres (nettle and grewia optiva)	900W at 2.45Hz	NA	Charcoal	When compared to the strength of joints that were adherently joined, the joint strength attained using microwave joining was greater.	Bajpai et al., 2012
MS-MS	900W at 2.45Hz	Nickel powder (40 mm)/1450XX	Aluminium Plate	The powder and substrate material formed a strong metallurgical connection. 2). Tensile strength of the joint increased by 50% of the strength of the base material.	Bansal et al., 2013

continues on following page

Table 2. Continued

Materials	Microwave oven	Interface material	Susceptor	Result	References
SS	NA	Nickel based EWAC 102ET/NA	Charcoal	Tensile strength of the joint rose as exposure duration and nickel powder fraction increases.	Gupta & Kumar, 2014
Inconel 718 – SS316L	NA	Inconel 718 powder (30 mm)/NA	NA	The joint interface was found to have an average micro-hardness of 230 52. The observed value for the ultimate tensile strength of microwave welds was 517.5 MPa.	Bansal et al., 2016
Aluminium alloy		Aluminium powder/ 1550XX	Silicon Carbide	It was possible to generate a dense, homogenous microstructure with metallurgic connections between the sandwich layer and the interface.	Singh et al., 2015
SS304-SS304	800 W At 2.45 GHz	Nickel based EWAC (50 mm) and 99% pure nickel/1450XX	Charcoal	In terms of weldability, nickel-based EWAC powder performed better than 99% pure nickel powder.	Bagha et al., 2016
Casting Al-7039	1400 W At 2.44 GHz	NA/Silicon Carbide Plates	Ceramic Crucible	The simulation results demonstrated that at higher power levels and at particular areas within the mould assembly, the heating of the bulk metal was quicker and more uniform.	Mishra & Sharma, 2016
Inconel-625	900 W At 2.45 GHz	Inconel-625/1450XX	Graphite sheet	Joints' ultimate tensile and flexural strengths were calculated to be 375 and 377 MPa, respectively.	Nandwani et al., 2020
SS304-SS304	800 W At 2.45 GHz	Nickel powder (50 mm, 40 mm, 30 mm, 20 mm)/ 1400XX	Graphite sheet	Greater tensile strength of the joint and tougher welding beads was produced using nickel powder with smaller grain sizes, while ductility was decreased.	Bagha et al., 2017

continues on following page

Table 2. Continued

Materials	Microwave oven	Interface material	Susceptor	Result	References
Al 6061-T6 – AZ31B	700 W At 2.45 GHz	Braze/NA	Charcoal, Graphite	The intermetallic's hardness toward the parent material, aluminium, was high (4.31 GPa). 2.) The intermetallic compounds formed, including aluminium, magnesium, and silver, were to blame for the weak strength.	Tamang & Arvindan, 2017
Casting of Copper	1400 W At 2.44 GHz	NA/Silicon Carbide Plates	Ceramic Crucible	The cast was determined to have a uniform and thick structure. The presence of copper in various orientations, with (1 1 1) being the dominant orientation, was shown by the X-ray diffraction pattern.	Mishra & Sharma, 2018
SS316-SS316	800 W At 2.45 GHz	Nickel based nano powder (40 mm)/ 1400XX	Graphite	The joint's tensile strength was 24% greater than that of the micro powder.	Soni et al., 2018
Cast Iron- cast Iron	900 W At 2.45 GHz	Nickel-based powder/ NA	Charcoal	1. A 400-second joint generated without errors. 2.) At 440 seconds, the joint began to twist. 3.) The deformation lasted 480 seconds.	Singh et al., 2019

REFERENCES

Adetunji, O. R., Adegbola, A. O., & Afolalu, S. A. (2015). Comparative study of case-hardening and water-quenching of mild steel rod on its mechanical properties. *International Journal of Advanced Research*, *3*(6), 1–9.

Afolalu, S. A., Soetan, S. B., Ongbali, S. O., Abioye, A. A., & Oni, A. S. (2019, November). Impact of activated–flux tungsten inert gas (a-tig) welding on weld joint of a metal–Review. [). IOP Publishing.]. *IOP Conference Series. Materials Science and Engineering*, *640*(1), 012064. doi:10.1088/1757-899X/640/1/012064

Ajay, P. Ahmad, S., Sharma, J., & Gambhir, V. (Eds.). (2023). Handbook of Sustainable Materials: Modelling, Characterization, and Optimization (1st ed.). CRC Press. doi:10.1201/9781003297772

Aravindan, S., & Krishnamurthy, R. (1999). Joining of ceramic composites by microwave heating. *Materials Letters*, *38*(4), 245–249. doi:10.1016/S0167-577X(98)00166-9

Badiger, R. I., Narendranath, S., & Srinath, M. S. (2015). Joining of Inconel-625 alloy through microwave hybrid heating and its characterization. *Journal of Manufacturing Processes*, *18*, 117–123. doi:10.1016/j.jmapro.2015.02.002

Badiger, R. I., Narendranath, S., & Srinath, M. S. (2018). Microstructure and mechanical properties of Inconel-625 welded joint developed through microwave hybrid heating. *Proceedings of the Institution of Mechanical Engineers. Part B, Journal of Engineering Manufacture*, *232*(14), 2462–2477. doi:10.1177/0954405417697350

Badiger, R. I., Narendranath, S., Srinath, M. S., & Hebbale, A. M. (2019). Effect of power input on metallurgical and mechanical characteristics of Inconel-625 welded joints processed through microwave hybrid heating. *Transactions of the Indian Institute of Metals*, *72*(3), 811–824. doi:10.100712666-018-1537-z

Bagha, L., Sehgal, S., Thakur, A., & Kumar, H. (2017). Effects of powder size of interface material on selective hybrid carbon microwave joining of SS304–SS304. *Journal of Manufacturing Processes*, *25*, 290–295. doi:10.1016/j.jmapro.2016.12.013

Bagha, L., Shegal, S., & Thakur, A. (2016). Comparative analysis of microwave based joining/welding of SS304-SS304 using different interfacing materials. In *MATEC Web of Conferences* (*Vol. 57*, p. 03001). EDP Sciences.

Bajpai, P. K., Singh, I., & Madaan, J. (2012). Joining of natural fiber reinforced composites using microwave energy: Experimental and finite element study. *Materials & Design*, *35*, 596–602. doi:10.1016/j.matdes.2011.10.007

Bansal, A., Sharma, A. K., & Das, S. (2013). Metallurgical and mechanical characterization of mild steel-mild steel joint formed by microwave hybrid heating process. *Sadhana, 38*(4), 679–686. doi:10.100712046-013-0142-4

Bansal, A., Sharma, A. K., Das, S., & Kumar, P. (2016). On microstructure and strength properties of microwave welded Inconel 718/stainless steel (SS-316L). *Proceedings of the Institution of Mechanical Engineers, Part L: Journal of Materials: Design and Applications, 230*(5), 939-948.

Bansal, A., Sharma, A. K., Kumar, P., & Das, S. (2012). Joining of mild steel plates using microwave energy. *Advanced Materials Research, 585*, 465–469. doi:10.4028/www.scientific.net/AMR.585.465

Bozkurt, Y., Uzun, H., & Salman, S. (2011). Microstructure and mechanical properties of friction stir welded particulate reinforced AA2124/SiC/25p–T4 composite. *Journal of Composite Materials, 45*(21), 2237–2245. doi:10.1177/0021998311401067

Chowdhury, S. M., Chen, D. L., Bhole, S. D., Powidajko, E., Weckman, D. C., & Zhou, Y. (2011). Microstructure and mechanical properties of fiber-laser-welded and diode-laser-welded AZ31 magnesium alloy. *Metallurgical and Materials Transactions. A, Physical Metallurgy and Materials Science, 42*(7), 1974–1989. doi:10.100711661-010-0574-y

Di Martino, S. F., Faulkner, R. G., Hogg, S. C., Vujic, S., & Tassa, O. (2014). Characterisation of microstructure and creep properties of alloy 617 for high-temperature applications. *Materials Science and Engineering A, 619*, 77–86. doi:10.1016/j.msea.2014.09.046

Gupta, P., & Kumar, S. (2014). Investigation of stainless steel joint fabricated through microwave energy. *Materials and Manufacturing Processes, 29*(8), 910–915. doi:10.1080/10426914.2014.892975

Hassan, A. M., Almomani, M., Qasim, T., & Ghaithan, A. (2012). Effect of processing parameters on friction stir welded aluminum matrix composites wear behavior. *Materials and Manufacturing Processes, 27*(12), 1419–1423. doi:10.1080/10426914.2012.700156

Huang, H. Y. (2010). Effects of activating flux on the welded joint characteristics in gas metal arc welding. *Materials & Design (1980-2015), 31*(5), 2488-2495.

Kandagal, Z. B., Akkimardi, V. G., & Sonnad, A. N. (2021). Investigation on microstructural and mechanical properties of microwave welded Al6061-SiC-graphite hybrid metal matrix composites. *Metallography, Microstructure, and Analysis, 10*(3), 334–346. doi:10.100713632-021-00745-7

Kothari, M., & Hung, W. N. (2018). Suppressing aluminum carbide in welding aluminum silicon carbide composite. *International journal of engineering materials and manufacture, 3*(1), 41-54.

Kumar, A., Kumar, P., Mittal, R. K., & Gambhir, V. (2023). Materials processed by additive manufacturing techniques. In Advances in Additive Manufacturing Artificial Intelligence, Nature-Inspired, and Biomanufacturing (pp. 217-233). Elsevier. doi:10.1016/B978-0-323-91834-3.00014-4

Kumar, A., Mittal, R. K., & Haleem, A. (Eds.). (2022). *Advances in Additive Manufacturing: Artificial Intelligence, Nature-Inspired, and Biomanufacturing*. Elsevier. doi:10.1016/C2020-0-03877-6

Kumar, N., Arora, N., & Goel, S. K. (2020). Effect of base metal solution annealing on mechanical and metallurgical properties of GMA welded nitronic steel. *Materials Science and Engineering A, 771*, 138542. doi:10.1016/j.msea.2019.138542

Ma, B., Wang, J., Lee, T. H., Dorris, S. E., Wen, J., & Balachandran, U. (2018). Microstructural characterization of Al 4 C 3 in aluminum–graphite composite prepared by electron-beam melting. *Journal of Materials Science, 53*(14), 10173–10180. doi:10.100710853-018-2336-y

Magudeeswaran, G., Balasubramanian, V., Reddy, G. M., & Balasubramanian, T. S. (2008). Effect of welding processes and consumables on tensile and impact properties of high strength quenched and tempered steel joints. *Journal of Iron and Steel Research International, 15*(6), 87–94. doi:10.1016/S1006-706X(08)60273-3

Mishra, R. R., & Sharma, A. K. (2016). On mechanism of in-situ microwave casting of aluminium alloy 7039 and cast microstructure. *Materials & Design, 112*, 97–106. doi:10.1016/j.matdes.2016.09.041

Mishra, R. R., & Sharma, A. K. (2018, April). Experimental investigation on in-situ microwave casting of copper. [). IOP Publishing.]. *IOP Conference Series. Materials Science and Engineering, 346*(1), 012052. doi:10.1088/1757-899X/346/1/012052

Nandwani, S., Vardhan, S., & Bagha, A. K. (2020). A literature review on the exposure time of microwave based welding of different materials. *Materials Today: Proceedings, 27*, 2526–2528. doi:10.1016/j.matpr.2019.10.056

Padhy, M. K., & Saini, R. P. (2008). A review on silt erosion in hydro turbines. *Renewable & Sustainable Energy Reviews, 12*(7), 1974–1987. doi:10.1016/j.rser.2007.01.025

Rajan, P. K., Ganesan, V., Dennison, M. S., & Kalishwaran, P. (2018). Fabrication and characterisation of hybrid aluminium metal matrix composite with welding slag and flux by using stir casting technique. *International Research Journal of Engineering and Technology*, *25*, 24–27.

Salih, O. S., Ou, H., Sun, W., & McCartney, D. G. (2015). A review of friction stir welding of aluminium matrix composites. *Materials & Design*, *86*, 61–71. doi:10.1016/j.matdes.2015.07.071

Sharma, A. K., & Gupta, D. (2012). On microstructure and flexural strength of metal–ceramic composite cladding developed through microwave heating. *Applied Surface Science*, *258*(15), 5583–5592. doi:10.1016/j.apsusc.2012.02.019

Singh, I., Bajpai, P. K., Malik, D., Sharma, A. K., & Kumar, P. (2011). Feasibility Study on Microwave Joining of 'green composites'. *Akademeia, 1*(1), ea0101.

Singh, S., Belokar, R. M., & Suri, N. M. (2019). XRD characterization of aluminium joint processed by microwave hybrid heating. *International Journal of Scientific Research and Reviews*, *7*(2).

Singh, S., Singh, P., Gupta, D., Jain, V., Kumar, R., & Kaushal, S. (2019). Development and characterization of microwave processed cast iron joint. *Engineering Science and Technology, an International Journal, 22*(2), 569-577.

Singh, S., Suri, N. M., & Belokar, R. M. (2015). Characterization of joint developed by fusion of aluminum metal powder through microwave hybrid heating. *Materials Today: Proceedings*, *2*(4-5), 1340–1346. doi:10.1016/j.matpr.2015.07.052

Singh, V., Gupta, M., Kumar, A., Luthra, S., & Kumar, A. (2023). Experimental investigations of thermo physical properties and convective heat transfer of Al2O3 and CuO nanofluids in a copper tube: Proposing New Correlations. *Biointerface Research in Applied Chemistry*, *13*(3), 229.

Singla, M., Dwivedi, D. D., Singh, L., & Chawla, V. (2009). Development of aluminium based silicon carbide particulate metal matrix composite. *Journal of Minerals & Materials Characterization & Engineering*, *8*(06), 455–467. doi:10.4236/jmmce.2009.86040

Sireesha, M., Albert, S. K., Shankar, V., & Sundaresan, S. (2000). A comparative evaluation of welding consumables for dissimilar welds between 316LN austenitic stainless steel and Alloy 800. *Journal of Nuclear Materials*, *279*(1), 65–76. doi:10.1016/S0022-3115(99)00275-5

Soni, P., Sehgal, S., Kumar, H., & Singh, A. P. (2018). Joining of SS316-SS316 through microwave hybrid heating by using Nickel nano-powder. *International Journal of Applied Engineering Research: IJAER*, *8*(8), 6446–6449.

Srinath, M. S., Sharma, A. K., & Kumar, P. (2011a). A novel route for joining of austenitic stainless steel (SS-316) using microwave energy. *Proceedings of the Institution of Mechanical Engineers. Part B, Journal of Engineering Manufacture*, *225*(7), 1083–1091. doi:10.1177/2041297510393451

Srinath, M. S., Sharma, A. K., & Kumar, P. (2011b). Investigation on microstructural and mechanical properties of microwave processed dissimilar joints. *Journal of Manufacturing Processes*, *13*(2), 141–146. doi:10.1016/j.jmapro.2011.03.001

Srinath, M. S., Sharma, A. K., & Kumar, P. (2011c). A new approach to joining of bulk copper using microwave energy. *Materials & Design*, *32*(5), 2685–2694. doi:10.1016/j.matdes.2011.01.023

Sujan, D., Yeo, C. W., Rahman, M. E., Reddy, M. M., Maleque, M. A., & Mohammad, Y. A. (2012). Aluminum-silicon carbide composites for enhanced physio-mechanical properties. *Advanced Materials Research*, *576*, 370–373. doi:10.4028/www.scientific.net/AMR.576.370

Tamang, S., & Aravindan, S. (2017). An investigation on joining of Al6061-T6 to AZ31B by microwave hybrid heating using active braze alloy as an interlayer. *Journal of Manufacturing Processes*, *28*, 94–100. doi:10.1016/j.jmapro.2017.05.027

Wong, W. L. E., & Gupta, M. (2015). Using microwave energy to synthesize light weight/energy saving magnesium based materials: A review. *Technologies*, *3*(1), 1–18. doi:10.3390/technologies3010001

Zafar, S., & Sharma, A. K. (2017). Microstructure and mechanical properties of microwave post-processed Ni coating. *Journal of Materials Engineering and Performance*, *26*(3), 1382–1390. doi:10.100711665-017-2540-y

Chapter 10
Wilson's Matrix Method for Calculating Molecular Vibration Frequencies and Normal Modes for Smart Materials

Ruby Jindal
Department of Physics, School of Basic and Applied Sciences, K.R. Mangalam University, Gurugram, India

Archana Tripathi
Physics Department, Gargi College, University of Delhi, New Delhi, India

Neenu Saini
https://orcid.org/0000-0001-6913-8911
Department of Physics, Pt. Jawaharlal Nehru Government College, Faridabad, India

ABSTRACT

This book chapter provides an in-depth exploration of Wilson's matrix method, a computational technique used to calculate molecular vibration frequencies and normal modes in smart materials. The chapter begins with an introduction to molecular vibrations and normal modes, emphasizing their importance in understanding material behavior at the molecular level. It then delves into the theoretical foundation and practical steps involved in Wilson's matrix method, highlighting its use of force constants and eigenvalue calculations to determine vibrational frequencies and

DOI: 10.4018/978-1-6684-9224-6.ch010

associated modes. The chapter further examines the application of Wilson's matrix method to smart materials, emphasizing their unique vibrational characteristics and the role of molecular vibrations in shaping their functionality. The advancements in the method, including the inclusion of anharmonicity and improved force constant determination, are discussed, along with the integration of enhanced computational techniques and increased computational power.

1. INTRODUCTION

Wilson's matrix method is a powerful computational technique used to calculate molecular vibration frequencies and normal modes in smart materials. It provides valuable insights into the dynamic behavior of molecules and enables the prediction and design of materials with desired vibrational properties. The method is based on the concept of force constants, which describe the strength of the bonds between atoms in a molecule. By constructing a symmetric Wilson matrix using the force constants and solving the corresponding eigenvalue problem, the vibrational frequencies and associated normal modes of the molecules can be determined (Wilson, 1939).

The Wilson matrix method is particularly relevant for smart materials, which exhibit unique vibrational characteristics that underpin their functionality. These materials often undergo complex structural changes or exhibit dynamic responses to external stimuli such as temperature, stress, or electric fields. By accurately calculating the molecular vibration frequencies and normal modes, the method helps in understanding the intricate interplay between molecular motions and macroscopic material behavior.

Moreover, the method has seen advancements over the years, including the incorporation of anharmonicity and improved force constant determination. Anharmonicity accounts for the deviation of molecular vibrations from simple harmonic oscillations, which is crucial for accurately predicting the behavior of smart materials under varying conditions. Furthermore, the development of enhanced computational techniques and increased computational power has expanded the applicability of the method to larger and more complex systems.

The insights obtained from Wilson's matrix method (Shimanouchi et al., 1961) can be leveraged in the design and optimization of smart materials. By manipulating the force constants or modifying the molecular structure, researchers can tailor the vibrational properties of materials to achieve specific functionalities such as shape memory behavior, sensing capabilities, or adaptive responses. This fine-tuning of vibrational characteristics enables the development of smart materials with improved performance and functionality in a wide range of applications.

2. FUNDAMENTALS OF MOLECULAR VIBRATIONS

2.1 Molecular Vibrations and Normal Modes

Molecular vibrations refer to the periodic motions of atoms within a molecule. These vibrations occur due to the interplay between interatomic forces, such as chemical bonds, and the distribution of mass within the molecule. Understanding molecular vibrations is essential for comprehending the properties and behavior of molecules and materials (Shimanouchi et al., 1961).

Normal modes are the fundamental vibrations of a molecule, each associated with a specific frequency and pattern of atomic motion. They represent independent modes of motion that are orthogonal to each other. In other words, each normal mode is a unique combination of atomic displacements that the molecule can undergo while conserving energy (Wilson et al., 1955, 1980).

Molecular vibrations and normal modes are typically analyzed using the concepts of harmonic oscillation, potential energy surfaces, and quantum mechanics. The harmonic approximation assumes that the interatomic forces can be approximated by simple harmonic potentials. While this approximation may not capture the full complexity of molecular vibrations, it provides a useful starting point for understanding the overall behavior of the system (Hargittai & Hargittai, 1995).

In the context of molecular vibrations, the potential energy surface represents the energy of the molecule as a function of atomic positions. By calculating the curvature of the potential energy surface around the equilibrium positions, the force constants of the molecule can be determined. Force constants describe the strength of the interatomic forces and play a crucial role in characterizing the vibrational behavior of the molecule.

To analyze molecular vibrations and normal modes (Rempe & Jonsson, 1998), the Wilson matrix method is often employed. This method involves constructing a symmetric Wilson matrix using the force constants, which are then diagonalized to obtain the vibrational frequencies and corresponding normal modes (Wilson et al., 1955). The diagonalization process essentially separates the coupled vibrational motions into independent normal modes.

The molecular vibrations and normal modes provide valuable information about the dynamic behavior of a molecule. Vibrational frequencies determine the energy required to induce a particular vibrational motion, while the normal modes (Radhikaa, 2020) describe the specific patterns of atomic displacements associated with each vibrational frequency. These vibrational characteristics influence various properties of molecules, such as their spectroscopic features (e.g., infrared and Raman spectra) and their reactivity in chemical reactions.

So, molecular vibrations and normal modes play a fundamental role in understanding the behavior of molecules and materials. They are determined by the interplay of interatomic forces and the distribution of mass within the molecule (Wilson et al., 1980). The Wilson matrix method is a valuable tool for calculating the vibrational frequencies and normal modes, providing insights into the dynamic behavior of molecules and facilitating the design and optimization of materials with desired vibrational properties.

2.2 Anharmonicity and Nonlinear Vibrations

While simple harmonic motion often provides a good approximation for molecular vibrations, anharmonic effects become significant in many cases. Anharmonicity (Pour et al., 2017) arises from deviations of the potential energy surface from a purely quadratic form and leads to deviations from the simple harmonic oscillator behavior. In anharmonic systems, higher-order terms in the potential energy expansion become important, allowing for coupling between different vibrational modes.

Nonlinear vibrations occur when the anharmonic terms dominate the vibrational motion. In nonlinear vibrations, the amplitude of the motion can affect the vibrational frequency and mode shape, leading to complex and intricate behavior. Nonlinear effects are particularly relevant in smart materials where large deformations and strong intermolecular interactions can give rise to nonlinear vibrational behavior.

Accounting for anharmonicity and nonlinear vibrations in the analysis of molecular vibrations is crucial for accurately predicting the vibrational spectrum and understanding the dynamic behavior of smart materials. Wilson's matrix method provides a powerful approach to handling these complexities and calculating the molecular vibration frequencies and corresponding normal modes, enabling a deeper understanding of the vibrational properties of smart materials (Cornaton et al., 2016).

3. WILSON'S MATRIX METHOD

3.1 Overview of Wilson's Matrix Method

Wilson's matrix method, also known as the Wilson GF (geometric factor) matrix method, is a widely used approach for calculating molecular vibration frequencies and normal modes. It was developed by Edwin Bright Wilson in the early 20th century and has since become a cornerstone of vibrational analysis in chemistry and materials science (Wilson et al., 1980).

The method is based on the concept of the Hessian matrix, which describes the second derivatives of the molecular potential energy for atomic displacements. The

Hessian matrix provides crucial information about the curvature of the potential energy surface and is essential for characterizing molecular vibrations (Wilson, 1939).

3.2 Formulation of Wilson's Matrix Method

Wilson's matrix method provides a systematic approach to calculating the vibrational frequencies and normal modes of a molecule based on its potential energy surface.

The basic idea behind Wilson's matrix method is to express the potential energy function of a molecule as a Taylor series expansion around its equilibrium geometry. The potential energy surface is expanded up to the second-order terms, which takes into account the anharmonic behavior of the molecule.

The formulation of Wilson's matrix method involves several steps:

Coordinate system: A set of internal coordinates is chosen to describe the molecule's geometry. These coordinates should be capable of representing the molecule's internal motions, such as bond stretching, angle bending, and torsional rotations. Common choices include bond lengths, bond angles, and dihedral angles (Shimanouchi et al., 1961; Wilson, 1939).

Equilibrium geometry: The equilibrium values of the chosen internal coordinates are determined. These values correspond to the geometry of the molecule at the potential energy minimum.

Expansion of potential energy: The potential energy function of the molecule is expanded as a Taylor series around the equilibrium geometry. The expansion includes terms up to the second order, taking into account quadratic and cross-product terms. This expansion provides a local approximation of the potential energy surface around the equilibrium geometry.

Force constants: The force constants are calculated by differentiating the potential energy expansion with respect to the internal coordinates. The force constants describe the strength of the interactions and provide information about the molecular bond strengths (Saini et al., 2022, 2023a).

Wilson's matrix: The force constants are used to construct a symmetric matrix known as the Wilson matrix (G matrix). The elements of the matrix represent the second derivatives of the potential energy with respect to pairs of internal coordinates. The diagonal elements contain the force constants for individual internal coordinates, while the off-diagonal elements represent the coupling between different internal coordinates.

Diagonalization: The Wilson matrix is diagonalized to obtain eigenvalues and eigenvectors. The eigenvalues correspond to the vibrational frequencies of the molecule, while the eigenvectors represent the normal modes of vibration. Each normal mode describes a specific pattern of atomic displacements that occur during a vibration (Wilson, 1939).

Calculation of frequencies: The vibrational frequencies are obtained by converting the eigenvalues, which are typically in units of force constants, into frequency units using appropriate conversion factors.

By following these steps, Wilson's matrix method provides a systematic approach to analyzing molecular vibrations and obtaining vibrational frequencies and normal modes. It has been widely used in theoretical and computational studies of molecular spectroscopy and vibrational dynamics.

3.3 Solving the Wilson Matrix

Once the Wilson matrix is constructed, it is diagonalized to obtain the eigenvalues and eigenvectors. The eigenvalues represent the vibrational frequencies of the normal modes, while the eigenvectors describe the corresponding displacements of the atoms. The eigenvalues provide insights into the energy distribution among the different modes, and the eigenvectors help visualize the vibrational patterns.

Solving the Wilson matrix involves numerical methods, such as matrix diagonalization algorithms, to obtain accurate vibrational frequencies and normal modes. Modern computational techniques have significantly enhanced the efficiency and accuracy of the calculations, allowing for the analysis of complex molecular systems (Shimanouchi et al., 1961; Wilson et al., 1980).

Wilson's matrix method is particularly well-suited for analyzing medium-sized molecules and solid-state systems. It provides a systematic and efficient approach to calculating the vibrational frequencies and normal modes, making it an invaluable tool for understanding the dynamic behavior of smart materials.

By applying Wilson's matrix method to smart materials, researchers can gain insights into the vibrational characteristics that influence their mechanical, thermal, and optical properties. This knowledge facilitates the design and optimization of smart materials for various applications, such as sensing, actuation, energy harvesting, and drug delivery systems (Shimanouchi et al., 1961; Wilson, 1939; Wilson et al., 1955, 1980).

4. APPLICATION TO SMART MATERIALS

4.1 Smart Materials and Their Molecular Vibrations

Smart materials are a class of advanced materials that exhibit dynamic and responsive behavior in response to external stimuli, such as temperature, light, pressure, or electric fields. The unique properties of smart materials arise from their molecular structure and the interactions between atoms and molecules within them. The molecular vibrations within smart materials play a crucial role in determining their dynamic behavior and functional properties.

Molecular vibrations refer to the oscillatory motion of atoms within a molecule. These vibrations occur due to the interplay of the bonding and non-bonding interactions between atoms. Each molecular vibration can be characterized by a specific vibrational frequency, which corresponds to the energy associated with the motion. The collective set of molecular vibrations in a molecule is referred to as the vibrational spectrum.

In smart materials (Bahl et al., 2020), molecular vibrations influence the material's mechanical, thermal, and optical properties, as well as its response to external stimuli. By understanding and manipulating these vibrations, researchers can tailor the material's response to specific stimuli and design materials with enhanced functionality. The study of molecular vibrations in smart materials involves analyzing the vibrational spectrum and investigating the corresponding normal modes.

Normal modes are the fundamental vibrational motions of a molecule. They represent the coordinated motion of all atoms in the molecule and can be visualized as distinct patterns of atomic displacements. Each normal mode is associated with a unique vibrational frequency and contributes to the overall vibrational spectrum of the molecule (Wilson et al., 1980). The number of normal modes is determined by the number of atoms in the molecule, minus the degrees of freedom associated with overall translation and rotation.

The molecular vibrations and normal modes in smart materials are crucial for several reasons:

Functional Properties: The vibrational properties of smart materials directly impact their functional properties. For example, in shape-memory materials, the molecular vibrations contribute to the material's ability to undergo large deformations and recover its original shape upon exposure to certain triggers, such as temperature. Understanding the vibrational modes associated with shape memory behavior helps in designing materials with tailored thermomechanical properties (Kumar et al., 2023).

Sensing and Actuation: Smart materials often exhibit changes in their molecular vibrations upon exposure to external stimuli. These changes can be exploited for sensing applications. By monitoring the vibrational spectrum or specific vibrational modes, smart materials can be used as sensors to detect changes in temperature, pressure, or other environmental parameters. Additionally, the controlled manipulation of molecular vibrations can lead to actuation, where materials undergo mechanical deformations in response to external stimuli (Qader et al., 2019; Sehrawat, Kumar, Grover, et al., 2022; Sehrawat, Kumar, Prabhakar, et al., 2022).

Optoelectronic Properties: The vibrational properties of smart materials can affect their optical and electronic behavior. Vibrational modes can couple with electronic transitions, leading to phenomena such as Raman scattering or infrared absorption. By studying the vibrational modes, researchers can gain insights into the material's optical properties, which is crucial for applications in optoelectronics, sensing, and energy harvesting (Joshi et al., 2021; Maheswari, 2022).

Material Design: The knowledge of molecular vibrations and normal modes enables researchers to design smart materials with desired properties. By tailoring the vibrational spectrum and controlling specific vibrational modes, materials can be engineered to exhibit specific responses to external stimuli. This opens up opportunities for developing new materials with improved functionality for various applications, including biomedical devices, energy storage systems, and adaptive structures (Gambhir, 2023).

So, the molecular vibrations and normal modes within smart materials provide a deeper understanding of their dynamic behavior and underpin their unique properties. By studying and manipulating these vibrations, researchers can design and optimize smart materials with enhanced functionality and performance in a wide range of applications

4.2 Case Studies

The application of Wilson's matrix method to smart materials has led to significant advancements in various fields. Here are a few examples of its application:

4.2.1 Shape Memory Polymers

Shape memory polymers (SMPs) exhibit the ability to undergo large deformations and recover their original shape upon exposure to certain triggers, such as temperature. Wilson's matrix method can be employed to analyze the vibrational modes of SMPs and understand the mechanisms behind their shape memory behavior. This knowledge aids in designing SMPs with tailored thermomechanical properties for applications like biomedical devices, aerospace engineering, and textiles (Bahl et al., 2020; Dayyoub et al., 2022).

4.2.2 Ferroelectric Materials

Ferroelectric materials possess a spontaneous electric polarization that can be reversed under an external electric field. The molecular vibrations within these materials contribute to their ferroelectric behavior. Wilson's matrix method enables the calculation of vibrational frequencies and normal modes, shedding light on the vibrational modes associated with the ferroelectric transition and assisting in the design of new ferroelectric materials with enhanced properties for applications like data storage and sensors (Bussmann-Holder, 2020).

4.2.3 Photomechanical Materials

Photomechanical materials undergo shape changes or mechanical responses upon exposure to light. Wilson's matrix method has been utilized to study the vibrational behavior of these materials and understand the underlying mechanisms. By analyzing the vibrational frequencies and normal modes, researchers can gain insights into the structural changes induced by light and the resulting mechanical response. This knowledge helps in the development of photomechanical materials for applications such as light-driven actuators, artificial muscles, and responsive surfaces (Kuzyk & Dawson, 2020; Singh et al., 2022).

4.2.4 Piezoelectric Materials

Piezoelectric materials exhibit the ability to generate an electric charge in response to mechanical stress or deformations, and vice versa. Wilson's matrix method has been applied to analyze the vibrational behavior of piezoelectric materials, such as piezoceramics and piezo composites. By considering the anharmonic and nonlinear effects, this method allows for the prediction of piezoelectric response under various excitation conditions and the optimization of material properties for sensor and actuator applications (Meng et al., 2022).

4.2.5 Electrostrictive Polymers

Electrostrictive polymers are a type of smart material that undergo significant mechanical deformation under an applied electric field. Wilson's matrix method has been used to investigate the vibrational properties and nonlinear behavior of electrostrictive polymers (Su & Tajitsu, 2016). By analyzing the Wilson matrix, researchers have gained insights into the coupling between different vibrational modes and the effect of anharmonicity on the material's response. This knowledge has contributed to the development of improved electrostrictive polymer materials for applications in artificial muscles, energy harvesting, and adaptive structures.

4.2.6 Organic Molecular Crystals

Wilson's matrix method has found applications in studying the vibrational properties of organic molecular crystals, which are used in various organic electronic devices such as organic light-emitting diodes (OLEDs) and organic photovoltaics (OPVs). By employing Wilson's matrix method, researchers can investigate the anharmonic vibrations and phonon dispersion in these materials, providing insights into their thermal transport, stability, and optical properties (Silinsh, 1980). This knowledge aids in the design of organic electronic devices with enhanced performance and efficiency.

These case studies demonstrate the versatility of Wilson's matrix method in studying the molecular vibrations of smart materials. By unraveling the vibrational behavior, researchers can gain insights into the mechanisms underlying their unique properties and engineer materials with improved functionality and performance.

5. ADVANCEMENTS AND CHALLENGES

5.1 Advancements in Wilson's Matrix Method

Wilson's matrix method has been a fundamental tool for calculating molecular vibration frequencies and normal modes. Over the years, advancements in computational techniques and experimental methods have contributed to the improvement and expansion of Wilson's matrix method. Here are some key advancements in detail:

5.1.1 Improved Force Constants

Force constants play a crucial role in Wilson's matrix method as they determine the stiffness of molecular bonds and govern the vibrational behavior of molecules. Initially, force constants were estimated using empirical relationships or simple models. However, advancements in quantum mechanical calculations have allowed for more accurate determination of force constants. Density functional theory (DFT) and ab initio methods (Chrétien & Salahub, 2001) have been employed to calculate force constants with higher precision, considering the electronic structure and molecular interactions more accurately.

5.1.2 Inclusion of Anharmonicity

Anharmonic effects, which deviate from the simple harmonic oscillator behavior, become significant in many molecular systems. Early applications of Wilson's matrix method focused primarily on harmonic vibrations. However, researchers have developed strategies to incorporate anharmonicity into the analysis. Higher-order terms in the potential energy expansion, beyond the quadratic terms, are considered to capture anharmonic effects. This allows for more accurate prediction of vibrational frequencies and mode shapes, especially in systems with large-amplitude motions or strong intermolecular interactions.

5.1.3 Hybrid Approaches

To address the limitations of purely harmonic approximation, researchers have developed hybrid approaches that combine Wilson's matrix method with other computational tools. For example, molecular dynamics simulations can be used to generate anharmonic force fields, which are then incorporated into the Wilson matrix analysis. This integration of different computational techniques allows for a more comprehensive treatment of anharmonicity and provides a better description of the vibrational properties of complex systems.

5.1.4 Computational Power

Advancements in computational power have significantly enhanced the efficiency and applicability of Wilson's matrix method. The ability to perform large-scale calculations has expanded the scope of the method to larger and more complex molecular systems. High-performance computing and parallelization techniques have accelerated the diagonalization of large matrices (Sunderland et al., 2009), allowing for faster and more accurate determination of vibrational frequencies and normal modes.

5.1.5 Experimental Techniques

Experimental methods for collecting vibrational data have also advanced, complementing the computational aspects of Wilson's matrix method. Techniques such as infrared spectroscopy, Raman spectroscopy, and neutron scattering have become more refined, providing accurate measurements of vibrational frequencies and intensities. These experimental data serve as valuable inputs for validating and refining the computational predictions obtained from Wilson's matrix method (Wang et al., 2018).

5.1.6 Integration With Other Analysis Tools

Wilson's matrix method has been integrated with other analysis tools to enhance its capabilities. For example, it can be combined with molecular visualization software to visualize and interpret the vibrational modes in a molecule or solid-state system. This integration allows researchers to gain intuitive insights into vibrational behavior and understand the relationship between structure and function in smart materials.

5.2 Challenges and Limitations

While Wilson's matrix method is a valuable tool for calculating molecular vibration frequencies and normal modes, it does have some challenges and limitations that should be considered. Here are some of the key challenges and limitations:

5.2.1 Harmonic Approximation

Wilson's matrix method assumes a harmonic potential energy surface, where the potential energy is approximated as a quadratic function around the equilibrium geometry. This approximation neglects higher-order anharmonic effects, which can become significant in certain molecular systems, particularly those involving large-amplitude motions or strong intermolecular interactions. Neglecting anharmonicity may lead to inaccuracies in predicting vibrational frequencies and mode shapes, especially for highly excited or low-frequency vibrations.

5.2.2 Computational Complexity

The computational complexity of Wilson's matrix method increases with the size and complexity of the molecular system. As the number of atoms and degrees of freedom increases, the size of the Wilson matrix grows rapidly, making the diagonalization process computationally demanding. Diagonalizing large matrices can be time-consuming and requires significant computational resources. Consequently, the method may face limitations when applied to very large or highly flexible systems (Sunderland et al., 2009).

5.2.3 Force Constant Determination

Accurate determination of force constants is essential for the success of Wilson's matrix method. Experimentally measuring force constants can be challenging, especially for novel or highly flexible molecules. Obtaining accurate force constants often requires sophisticated experimental techniques or reliable quantum mechanical calculations. The quality of force constants directly affects the accuracy of the calculated vibrational frequencies and normal modes (Wilson et al., 1955).

5.2.4 Anharmonic Effects and Coupling

While efforts have been made to incorporate anharmonic effects into Wilson's matrix method, the treatment of anharmonicity can still be challenging. The inclusion of higher-order terms in the potential energy expansion adds complexity to the calculations. Additionally, the coupling between different vibrational modes, which becomes more prominent in anharmonic systems, is not explicitly accounted for in the method. Neglecting or approximating these couplings may affect the accuracy of the calculated vibrational frequencies and mode shapes (Pour et al., 2017).

5.2.5 Experimental Validation

Validating the results obtained from Wilson's matrix method with experimental data can be crucial for confirming the accuracy of the calculations. However, experimental measurements of vibrational frequencies and mode shapes can be challenging and subject to uncertainties. Differences between experimental and calculated results may arise due to limitations in force constants, experimental conditions, or approximations made in the calculations (Wilson et al., 1980).

6. FUTURE PERSPECTIVES

6.1 Enhanced Computational Techniques

Enhanced computational techniques play a crucial role in improving the accuracy and efficiency of Wilson's matrix method for calculating molecular vibration frequencies and normal modes. Here are some key advancements in computational techniques that have contributed to the enhancement of the method:

6.1.1 Increased Computing Power

Advancements in computing power have significantly accelerated the calculations involved in Wilson's matrix method. High-performance computing (HPC) facilities and parallelization techniques allow for the efficient diagonalization of large matrices, which is a computationally demanding step in the method. Faster computing speeds enable researchers to handle more complex systems and perform larger-scale calculations, expanding the applicability of the method.

6.1.2 Improved Algorithms

Developments in algorithms and numerical methods have contributed to the efficiency and accuracy of Wilson's matrix method. Efficient matrix diagonalization algorithms, such as the Lanczos algorithm or the Davidson algorithm, have been employed to solve the eigenvalue problem and obtain the vibrational frequencies and normal modes. These algorithms take advantage of the sparsity and symmetry of the matrices involved, leading to faster and more accurate results.

6.1.3 Quantum Mechanical Calculations

Advancements in quantum mechanical calculations, particularly density functional theory (DFT), have improved the accuracy of force constant determination, which is a critical input in Wilson's matrix method. DFT provides a more realistic description of electronic structure and molecular interactions, allowing for a more accurate prediction of force constants. Higher-level quantum mechanical methods, such as wavefunction-based approaches, can also be employed for more accurate calculations of force constants, albeit at a higher computational cost (Wang et al., 2018).

6.1.4 Hybrid Approaches

Hybrid approaches that combine Wilson's matrix method (Saini et al., 2023c, 2023b) with other computational techniques have been developed to overcome the limitations of purely harmonic approximation. Molecular dynamics (MD) simulations, for example, can generate anharmonic force fields that capture the effects of anharmonicity on molecular vibrations. These anharmonic force fields can then be incorporated into the Wilson matrix analysis, allowing for a more comprehensive treatment of vibrational properties (Phillies, 2013).

6.1.5 Multiscale Modeling

Multiscale modeling approaches integrate different computational techniques and simulation methods (Odegard, 2018) to study molecular vibrations in smart materials. By combining quantum mechanical calculations, molecular mechanics, and continuum mechanics, researchers can capture the effects of various lengths and time scales on vibrational behavior. This allows for a more realistic and accurate description of molecular vibrations in complex systems and interfaces.

6.1.6 Data-Driven Approaches

Advancements in data-driven approaches, such as machine learning and data mining, have the potential to enhance Wilson's matrix method. These approaches can be used to extract patterns and correlations from large datasets of experimental and computational vibrational data. Machine learning algorithms can aid in the prediction of force constants, vibrational frequencies, and mode shapes, reducing the computational cost and improving the efficiency of the method.

6.2 Multiscale Modeling and Design

Multiscale modeling and design approaches have emerged as powerful tools for understanding and optimizing the vibrational behavior of smart materials. These approaches integrate different computational techniques and simulation methods to bridge the gap between molecular-level insights and macroscopic material behavior. By considering multiple lengths and time scales, multiscale modeling (Odegard, 2018) enables the prediction and design of smart materials with desired vibrational characteristics, leading to enhanced functionality and performance in various applications.

At the molecular level, quantum mechanical calculations, such as density functional theory (DFT), provide detailed information about the force constants and vibrational frequencies of individual molecules (Baroni et al., 2001; Ditler & Luber, 2022). These calculations capture the electronic structure and molecular interactions, allowing for accurate determination of the potential energy surface and vibrational properties. The results from quantum mechanical calculations serve as inputs for Wilson's matrix method, which analyzes the collective vibrational behavior of the molecules and calculates the normal modes and frequencies.

However, the behavior of smart materials is not solely governed by molecular vibrations. The macroscopic properties and responses of these materials depend on their microstructure, mesoscale organization, and interactions with the environment. To capture these effects, multiscale modeling approaches combine molecular-level information with higher-level simulation methods. Molecular mechanics and coarse-grained models describe the mechanical behavior of materials at the mesoscale, accounting for the collective motions of molecules and the interactions between them. Continuum mechanics models capture the macroscopic behavior of materials, considering factors such as strain, stress, and boundary conditions.

By integrating these different scales, multiscale modeling enables the prediction of material properties, such as mechanical response, thermal conductivity, and optical behavior, based on the underlying molecular vibrations. This integration allows researchers to optimize the design of smart materials by tailoring the vibrational characteristics to achieve specific functionalities. For example, by adjusting the force constants or modifying the microstructure, materials with desired thermomechanical properties, shape memory behavior, or specific vibrational modes can be designed.

Multiscale modeling also facilitates the understanding of how external stimuli, such as temperature changes, electric fields, or mechanical stress, affect the vibrational behavior of smart materials. By simulating the response of materials at different scales, researchers can gain insights into the mechanisms behind the material's dynamic behavior and optimize its response to specific stimuli. This understanding is crucial for developing smart materials with enhanced sensing, actuation, or adaptive properties.

Overall, multiscale modeling and design approaches provide a comprehensive framework for studying molecular vibrations in smart materials. By considering multiple lengths and time scales, these approaches enable the prediction, optimization, and tailoring of vibrational properties to achieve desired functionality and performance. The integration of molecular-level insights with macroscopic material behavior bridges the gap between fundamental understanding and practical applications, accelerating the development of advanced smart materials in diverse technological domains.

7. CONCLUSION

In conclusion, this chapter has focused on Wilson's matrix method for calculating molecular vibration frequencies and normal modes, particularly in the context of smart materials. The chapter began by introducing the concepts of molecular vibrations and normal modes, highlighting their significance in understanding the properties and behavior of materials at the molecular level.

The discussion then delved into the details of Wilson's matrix method, explaining its theoretical foundation and the steps involved in the calculation process. The method's reliance on force constants and the diagonalization of the Wilson matrix were emphasized as key components for obtaining accurate vibrational frequencies and normal modes.

Furthermore, the chapter explored the application of Wilson's matrix method to smart materials. The unique vibrational characteristics of smart materials, such as shape memory behavior, sensing capabilities, and adaptive responses, make the accurate prediction and understanding of their molecular vibrations crucial. Wilson's matrix method provides a valuable tool for investigating these vibrational properties, aiding in the design and optimization of smart materials with tailored functionalities.

The advancements in Wilson's matrix method were also discussed. Improved force constant determination, the inclusion of anharmonicity, hybrid approaches combining different computational techniques, increased computational power, and integration with experimental techniques were identified as notable advancements that have enhanced the accuracy and applicability of the method. These advancements have expanded the scope of the method to tackle more complex systems and have contributed to a deeper understanding of molecular vibrations in smart materials.

However, it is important to acknowledge the challenges and limitations of Wilson's matrix method. The harmonic approximation, computational complexity, accurate determination of force constants, treatment of anharmonic effects and couplings, experimental validation, and limited treatment of solvation effects were highlighted as areas that require further attention and improvement.

Overall, Wilson's matrix method, combined with enhanced computational techniques and advancements in experimental methods, provides valuable insights into the molecular vibrations and normal modes of smart materials. It serves as a powerful tool for the design and optimization of smart materials with tailored vibrational properties and functionalities. Future research efforts should continue to address the method's limitations, further refine its accuracy, and explore its applications in a wide range of smart materials systems.

REFERENCES

Bahl, S., Nagar, H., Singh, I., & Sehgal, S. (2020). Smart materials types, properties and applications: A review. *Materials Today: Proceedings*, *28*, 1302–1306. doi:10.1016/j.matpr.2020.04.505

Baroni, S., de Gironcoli, S., Dal Corso, A., & Giannozzi, P. (2001). Phonons and related crystal properties from density-functional perturbation theory. *Reviews of Modern Physics*, *73*(2), 515–562. doi:10.1103/RevModPhys.73.515

Bussmann-Holder, A. (2020). Ferroelectrics: Past, present and future. *Ferroelectrics*, *569*(1), 1–8. doi:10.1080/00150193.2020.1791652

Chrétien, S., & Salahub, D. R. (2001). Density Functional Theory, Methods, Techniques, and Applications. In C. Guet, P. Hobza, F. Speigelman, & F. David (Eds.), *Atomic clusters and nanoparticles. Agregats atomiques et nanoparticules: Les Houches Session LXXIII 2–28 July 2000* (pp. 105–160). Springer. doi:10.1007/3-540-45621-X_4

Cornaton, Y., Ringholm, M., Louant, O., & Ruud, K. (2016). Analytic calculations of anharmonic infrared and Raman vibrational spectra. *Physical Chemistry Chemical Physics*, *18*(5), 4201–4215. doi:10.1039/C5CP06657C PMID:26784673

Dayyoub, T., Maksimkin, A. V., Filippova, O. V., Tcherdyntsev, V. V., & Telyshev, D. V. (2022). Shape Memory Polymers as Smart Materials: A Review. *Polymers*, *14*(17), 3511. doi:10.3390/polym14173511 PMID:36080587

Ditler, E., & Luber, S. (2022). Vibrational spectroscopy by means of first-principles molecular dynamics simulations. *Wiley Interdisciplinary Reviews. Computational Molecular Science*, *12*(5), e1605. doi:10.1002/wcms.1605

Gambhir, A. Parveen, Sharif Ahmad, Jyotsna Sharma, Victor (Ed.). (2023). Handbook of Sustainable Materials: Modelling, Characterization, and Optimization. CRC Press. doi:10.1201/9781003297772

Joshi, G. P., Tyagi, A., Meena, S. L., Chourasia, S., Saxena, N. S., & Kumar, A. (2021). Optical Band Gap Determination of Ni-Al Doped Polyaniline at Room Temperature and Different Annealing Temperatures. *Journal of Physics: Conference Series*, *1950*(1), 012091. doi:10.1088/1742-6596/1950/1/012091

Kumar, A., Kumar, P., Mittal, R. K., & Gambhir, V. (2023). Materials processed by additive manufacturing techniques. In A. Kumar, R. K. Mittal, & A. Haleem (Eds.), *Advances in Additive Manufacturing* (pp. 217–233). Elsevier. doi:10.1007/978-3-031-34563-0

Kuzyk, M., & Dawson, N. (2020). Photomechanical materials and applications: A tutorial. *Advances in Optics and Photonics*, *12*(4), 847. doi:10.1364/AOP.387366

Maheswari, A. U. (2022). Smart Materials- Types & Applications. *International Journal for Research in Applied Science and Engineering Technology*, *10*(1), 1752–1755. doi:10.22214/ijraset.2022.40147

Meng, Y., Chen, G., & Huang, M. (2022). Piezoelectric Materials: Properties, Advancements, and Design Strategies for High-Temperature Applications. *Nanomaterials (Basel, Switzerland)*, *12*(7), 1171. doi:10.3390/nano12071171 PMID:35407289

Springer. (1995). Molecular Vibrations. In I. Hargittai & M. Hargittai (Eds.), Symmetry through the Eyes of a Chemist (pp. 207–225). Springer US. doi:10.1007/978-0-585-31234-7_5

Odegard, G. M. (2018). 6.2 Computational Multiscale Modeling – Nanoscale to Macroscale. In P. W. R. Beaumont & C. H. Zweben (Eds.), *Comprehensive Composite Materials II* (pp. 28–51). Elsevier. doi:10.1016/B978-0-12-803581-8.10021-9

Phillies, G. D. J. (2013). *The VW transformation—A Simple Alternative to the Wilson GF Method* (arXiv:1306.5499). arXiv. https://doi.org//arXiv.1306.5499 doi:10.48550

Pour, A. G., Lincoln, C. N., Perlík, V., Šanda, F., & Hauer, J. (2017). Anharmonic vibrational effects in linear and two-dimensional electronic spectra. *Physical Chemistry Chemical Physics*, *19*(36), 24752–24760. doi:10.1039/C7CP05189A PMID:28868559

Qader, I., Kök, M., Dagdelen, F., & Aydogdu, Y. (2019). A Review of Smart Materials Researches and Applications. *El-Cezeri Fen ve Mühendislik Dergisi*, *6*, 755–788. doi:10.31202/ecjse.562177

Radhikaa, C. (2020). A Study on various modes of vibration in spectroscopy. *Journal of Matematik,* (2), 1103-1105. https://doi.org/ doi:10.26637/MJM0S20/0285

Rempe, S. B., & Jonsson, H. (1998). A Computational Exercise Illustrating Molecular Vibrations and Normal Modes. *Chemical Educator (Boise, Idaho)*, *3*(4), 1. doi:10.100700897980231a

Saini, N., Jindal, R., & Tripathi, A. (2022). Effect of M site cation ordering on the lattice dynamical properties of the tetragonal Sr2MO4 (M=Cr, Fe, Co, Mo, and Sn) crystals. *Philosophical Magazine*, *102*(11), 1055–1068. doi:10.1080/14786435.2022.2035006

Saini, N., Jindal, R., & Tripathi, A. (2023a). Influence of B-site cation in lattice dynamics of bilayered Ruddlesden–Popper Sr3B2O7 (B = Zr, Mo, Sn, Hf) compounds. *Journal of Raman Spectroscopy : JRS*, jrs.6561. doi:10.1002/jrs.6561

Saini, N., Jindal, R., & Tripathi, A. (2023b). Normal coordinate analysis to study the optical phonons in B-site cation disordered Ruddlesden-Popper Sr3B2O7 (B = V, Cr, Fe, Co) crystals. *Solid State Sciences*, *107209*, 107209. doi:10.1016/j.solidstatesciences.2023.107209

Saini, N., Jindal, R., & Tripathi, A. (2023c). Study of lattice dynamics of bilayered tetragonal Ruddlesden-Popper compounds (Ca, Sr)3Mn2O7. *Materials Today. Communications*, *34*, 105190. doi:10.1016/j.mtcomm.2022.105190

Sehrawat, S., Kumar, A., Grover, S., Dogra, N., Nindra, J., Rathee, S., Dahiya, M., & Kumar, A. (2022). Study of 3D scanning technologies and scanners in orthodontics. *Materials Today: Proceedings*, *56*, 186–193. doi:10.1016/j.matpr.2022.01.064

Sehrawat, S., Kumar, A., Prabhakar, M., & Nindra, J. (2022). The expanding domains of 3D printing pertaining to the speciality of orthodontics. *Materials Today: Proceedings*, *50*, 1611–1618. doi:10.1016/j.matpr.2021.09.124

Shimanouchi, T., Tsuboi, M., & miyazawa, T. (1961). Optically Active Lattice Vibrations as Treated by the GF-Matrix Method. *The Journal of Chemical Physics*, *35*(5), 1597–1612. doi:10.1063/1.1732116

Silinsh, E. A. (1980). Introduction: Characteristic Features of Organic Molecular Crystals. In E. A. Silinsh (Ed.), *Organic Molecular Crystals: Their Electronic States* (pp. 1–46). Springer. doi:10.1007/978-3-642-81464-8_1

Singh, V., Kumar, A., Alam, M., Kumar, A., Kumar, P., & Goyat, V. (2022). A study of morphology, UV measurements and zeta potential of Zinc Ferrite and Al2O3 nanofluids. *Materials Today: Proceedings*, *59*, 1034–1039. doi:10.1016/j.matpr.2022.02.371

Su, J., & Tajitsu, Y. (2016). Piezoelectric and Electrostrictive Polymers as EAPs: Materials. In F. Carpi (Ed.), *Electromechanically Active Polymers: A Concise Reference* (pp. 509–531). Springer International Publishing. doi:10.1007/978-3-319-31530-0_22

Sunderland, A., Ciegis, R., Henty, D., Kågström, B., & Žilinskas, J. (2009). *Parallel Diagonalization Performance on High-Performance Computers*., doi:10.1007/978-0-387-09707-7_5

Wang, X., Li, Y., Gao, Y., Yang, Z., Lu, C., & Zhu, T. (2018). A quantum mechanical computational method for modeling electrostatic and solvation effects of protein. *Scientific Reports*, *8*(1), 5475. doi:10.103841598-018-23783-8 PMID:29615707

Wilson, E. B. Jr. (1939). A Method of Obtaining the Expanded Secular Equation for the Vibration Frequencies of a Molecule. *The Journal of Chemical Physics*, *7*(11), 1047–1052. doi:10.1063/1.1750363

Wilson, E. B., Decius, J. C., & Cross, P. C. (1955). *Molecular Vibrations: The Theory of Infrared and Raman Vibrational Spectra*. McGraw-Hill.

Wilson, E. B., Decius, J. C., & Cross, P. C. (1980). *Molecular Vibrations: The Theory of Infrared and Raman Vibrational Spectra*. Courier Corporation.

Chapter 11
Industry Requirement and Future Prospects of Lightweight AlMg2Si Functionally Graded Materials for Automotive Engine Components: Review

Subhash Chandra Ram
Tula's Institute, Dehradun, India

Awani Bhushan
Vellore Institute of Technology, Chennai, India

Sunkulp Goel
University of Science and Technology, Nanjing, China

ABSTRACT

The present study provides a comprehensive state-of-the-art on currently available information about the advancement of Al-Mg2Si in-situ FG-composites for various automotive application domains, as well as their inherent benefits and drawbacks. Additionally, the chapter explored how a functionally graded material (FGMs) might serve as a suitable replacement material for automotive components. FGMs can be practically modified through the modification of their constituents to meet

DOI: 10.4018/978-1-6684-9224-6.ch011

particular functional needs. The production of FGMs employs an extensive range of approaches. Taking into consideration Al-Si-Mg alloy as the starting materials, the fabrication methods are categorized as follows: liquid processes, gaseous processing, and solid particle methods. This chapter describes the typical centrifugal casting technique for Al-Mg2Si in-situ functionally graded lightweight composites. Optical and SEM analysis of reinforced phase Mg2Si particles, Al-Si eutectic phases, and α-Al grains size were presented in detail.

1. INTRODUCTION

Particularly in the automotive and aerospace industries, the production and demand for lightweight materials have recently expanded. Conventional components such as steel as well as iron casting have higher tensile strength but are heavier in weight than aluminum alloy. The majority of researchers and educators now believe that cast iron and steel can be replaced by alloys based on aluminium and silicon in the production of engine parts. A356 and A319, the two most widely used aluminium alloys, are employed in automobile parts but do not have the necessary tribological qualities (Watanabe et al., 2011; Sobczak et al., 2009; Zhang et al., 1999). The manufacturing of functionally graded materials (FGM) and their design made it possible to produce composite materials with different microstructures and proportions.Moreover, the wear resistance will be offered if the degree of distribution of hard reinforcing particles in the surface layers is altered. The flexibility and hardness of the middle and opposing side layers will eventually grow as a result of their gradual depletion throughout the reinforcing phase. Potential candidate materials include functionally graded aluminum-based composites reinforced with Mg2Si for use in aerospace, automotive, and other applications.

Functionally graded composites (FG-Composites) are made of heterogeneous materials that have been purposefully adjusted to have a variety of conflicting properties. FGMs are employed in a variety of industries, including automobiles, chemicals and petrol, mitigation, and space exploration (Kumar et al., 2021; Ram et al., 2017; Ram et al., 2023). The production of functional qualities of any component is only achievable with a gradient distribution of dispersed phase particles (Xu, F. M., et al. 2004). Aluminium has several applications, including the fabrication of lightweight structural components with varying qualities using various alloys. Aluminium and its alloys are commonly used in manufacturing businesses as well as research endeavours due to their outstanding castability and desirable features. Many desired qualities of aluminium and its alloys include high stiffness, ductility, high strength-to-weight ratio, good thermal stability, good conductivity and durability.

They are also less expensive when compared to other readily available and commonly used low-density alloys (Melgarejo et al., 2008). Because of their great flowability and castability, Al-Si-based alloys are ideal for use in the aerospace and automotive industries. Aluminium, various aluminium alloys, and ceramic composite materials are the most often used materials in composite production (Ram et al., 2016). The bulk of automobile components are frequently stressed either mechanically or thermally. Nowadays, there is a growing market for lightweight materials, particularly in the automotive sector to reduce fuel consumption. To attain the required functional performance, however, specialised advanced composites must be designed, such as low-density replacement materials that have proven structurally stable. Due to a variety of properties, functionally graded materials based on AlSi alloys may be one of the best solutions to achieve functional performance.The enormous potential for the development of other aluminum-based FGMs and FG-Composites with considerably more desired properties exists in future study. Zhang et al. investigated the use of centrifugal casting to create an Al-Mg_2Si FGM tubular product with Mg_2Si reinforcements(Vieira et al., 2009).The reinforced particle Mg_2Si has a remarkable strength and a lower in density (1.99x 10^3 kg m^{-3}), melting temperature (1085 °C), thermal expansion coefficient is low (7.5x 10^{-6} K^{-1}), and reasonably higher hardness (4.5x10^9 Nm^{-2}). Due to these characteristics associated with the reinforcements, Al-Mg_2Si FG-composites containing Mg_2Si reinforcement have become a desired alternative material for aerospace components, vehicle engine components, and other uses (Shi, J. L., et al., 2011; Deuis et al., 1997; Prabhu et al., 2017). However, the tough and brittle primary Mg_2Si particles will reduce the product's ultimate strength and flexibility compared to ordinary composites.For the development of Al-Mg_2Si in-situ FG composites, this problem might be overcome by using a gradient structure and reinforcement from Mg_2Si particles (Rao et al., 2010; Loh et al., 2018; Ram et al., 2019; Nai et al., 2003).

2. AUTOMOTIVE DEMANDS FOR LIGHTWEIGHT STRUCTURAL MATERIALS

Innovative components are critical for increasing modern automotive fuel economy while retaining performance and security. By using less weight parts while decreasing overall vehicle load, automakers can accept more complicated exhaust control mechanisms, protection characteristics, as well as integrated electrical installations. Since it takes less energy to accelerate a lighter object than a heavier one, lightweight components offer a high potential for increasing vehicle economy. Fuel efficiency may increase by 6% to 8% with a 10% reduction in automotive weight. The weight of a vehicle's bodies could be reduced by replacing cast iron and traditional steel components with lighter alternatives including exceptionally strong steel. The alloys based on magnesium, alloys made of aluminium (Al), carbon fibre, fibreglass, and polymer-based composites can be utilized. Using lightweight alloys may reduce the weight of the chassis by up to 50%, and reducing fuel consumption. Using lightweight products with high-efficiency engines provided by current materials in almost twenty percent of the US fleet could save more than five trillion gallons of gasoline annually by 2030. Lighter materials study and creation has become necessary for cost reduction, enhancing their recycling capability, facilitating them to be incorporated into automobiles, and maximising their fuel efficiency benefits.

Table 1. Lightweight metal and non-metal alloys with % Mass reduction

LIGHT WEIGHT ALLOYS/MATERIALS	REDUCTION IN %MASS
Magnesium Alloys	30-75%
Carbon fiber based composites	50-80%
Aluminum Alloys matrix composites	30-65%
Titanium Alloys	40-60%
Glass fiber based omposites	25-40%
High strength alloys steel	15-30%

Table 1 depicts the many alloys with adequate qualities for the fabrication of metal or ceramic matrix-based functional materials, as well as the equivalent amount of mass reduction in the vehicle.Maximum mass reduction may be feasible with carbon fiber-based composites, but one drawback of carbon fibre is an increase in vehicle cost. Al-Si Alloys-based matrix materials may additionally be employed to make FG-materials with excellent strength and wear resistance. The commercial aluminium alloys AA356 or A319 utilised as matrix provided a significant weight decrease while increasing mechanical strength.

3. Al-Mg_2Si LIGHT WEIGHT FUNCTIONAL MATERIALS

Aluminium alloys Al-Mg-Si are increasingly employed as automobile bodies in place of heavier metals due to economic and political demand to reduce fuel consumption and CO2 emissions. Over the previous few decades, The average amount of aluminium alloys used in passenger automobiles has increased significantly in the European automotive sector. The design of the thermo-mechanical treatment has a considerable impact on the mechanical properties of the alloys. For instance, the rolling pattern and annealing temperatures have an impact on the ability to form, whereas the annealing and quenched conditions have an impact on ultimate strength. Commercial aluminium casting alloys containing silicon as the principal alloying essential elements as a result of their exceptional casting abilities. Si additions to pure aluminium result in great fluidity, good feeding qualities, low shrinkage, and strong hot cracking resistance.Magnesium as well as silicon may be combined to generate the intermetallic hardening phase Mg2Si, which precipitates in the -aluminium matrix thus boosts yield strength. Magnesium that cannot be dissolved in the matrix is free to generate undesirable intermetallic complexes. Iron, the most detrimental contaminant for AlSiMg alloys, forms several intermetallic compounds with Si and Mg, including (Fe2SiAl8), (FeSiAl5), and (FeMg3Si6Al8). Cast Al-Si-Mg alloys are potential candidate materials for a variety of automotive applications, including engine blocks or cylinder heads, due to their excellent castability, resistance to corrosion, and excellent strength-to-weight proportion, resulting in improved performances and fuel efficiency.

The primary limitations of Al-Si-Mg component casting are caused by the significant impact of its solidification processes onto the final morphology. The finished cast component will certainly have a number of flaws, such as oxide coatings, shrinkage, and gas porosity, which will have a substantial impact on the mechanical properties, particularly the fatigue behavior.

4. Al-Mg_2Si IN-SITU FG-COMPOSITES AND ITS ANALYSIS OF GRADIENTS CHARACTERISTICS

The highly lightweight material includes aluminum matrix-Mg_2Si reinforcement composite materials. The reinforcement Mg_2Si have high hardness (4.50 GPa), reduced thermal expansion coefficient (7.5x 10^{-6} K^{-1}), greater melting temperature (108 C), low in density (1.99 gcm^3), and high elastic modulus (120x 10^3MPa) are just a few of the excellent features of the Mg_2Si reinforcement (Khorshidi et al., 2018; Bai et al., 2016; Zhu et al., 2018; Ram et al., 2017). Various composites are attracting the attention among material scientists and are currently being investigated to be potential replacements for the traditional SiC and Al2O3 reinforcements in aluminum matrix. Furthermore, structural applications and the automotive industry both frequently use the Mg_2Si reinforced Al matrix composite (Farahany et al., 2017; Shafieizad et al., 2015). The manufacturing process has a considerable impact on the microstructure and mechanical behavior of Al-Mg2Si FG-composites.In-situ and ex-situ are the two most used ways for synthesising Particulates-based composite materials. The advantages of in-situ synthesised include an excellent particle-wetting process, ease of manufacture, homogeneous reinforcement distribution, thermal stability, and robust interfacial bonding. The Al-Mg2Si and Al-Si binary eutectic systems, on the other hand, exhibit exactly the same solidification behavior (Soltani et al., 2014; Farahany et al., 2016).

Figure 1. Al–Mg2Si phase diagram
(Biswas et al., 2020)

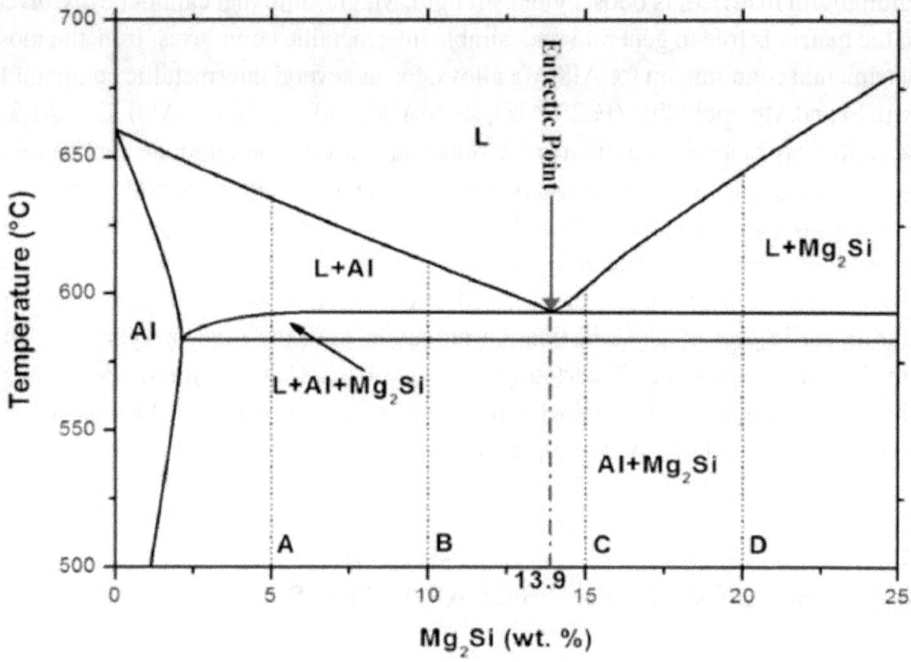

Similar to the Al-Si system, the Al-Mg_2Si phase schematic (Figure 1) shows that the system includes three compositional zones (hyper, eutectic, and hypo). Additionally, a system of Al-Mg_2Si showed eutectic composition for Mg_2Si content upto13.9%. It has become known that getting good mechanical properties in particulate-reinforced composites necessitates a detailed investigation of particle dispersion. In the present study, Al-Si-Mg FGMs cylinders made by centrifugal casing were fabricated together with solid aluminum cylinder liners. The fragmentation ratios, volume fractions, and major Si/Mg2Si particle size distributions in the FGMs tubes made using different procedure parameters (Watanabe et al., 2002; Watanabe et al., 2004).

Figure 2 depicts the casting of Al-based functional gradient-reinforced composites with the addition of Si and Mg_2Si particles utilizing centrifugal forming principles (Zhang et al., 2006). The primary Si as well as Mg_2Si particles separated due to a drop in melt temperatures when a rotational mould was filled with the hypereutectic Al-19Si-5Mg alloy after it had melted. As the primary Si & Mg_2Si particulates have a lower mass per unit of volume than the Al melt, the centrifugal force leads them to flow towards the axis, as seen in Figure 2(a). As a result, the particles congregate in the inner zone. Finally, gradients in-situ composites have been produced in the centrifugal casting mold, as can be seen in Figure 2(b).

Figure 2. Graphical representation of particles movement due to centifugal force (Zhang et al., 2006)

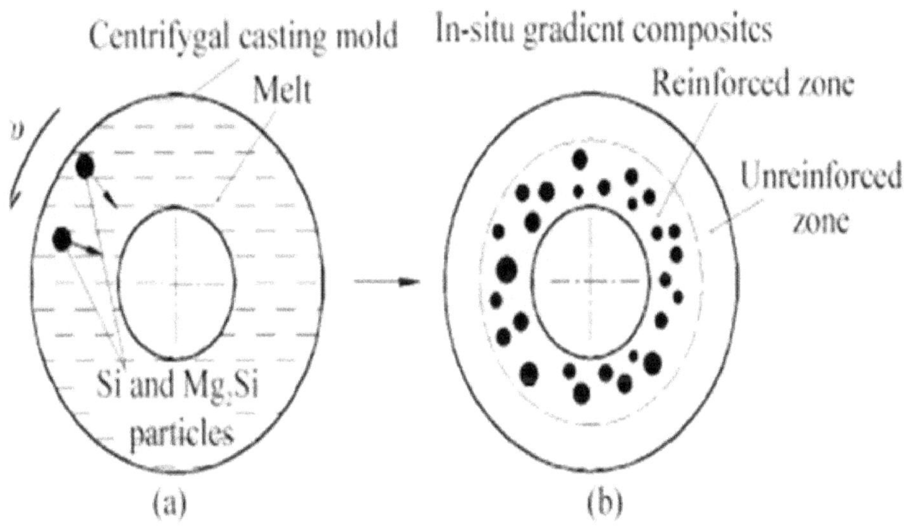

Figure 3 depicts the cross-section of the cylindrical sample that was centrifugally cast, and Figure 4 depicts the corresponding zone wise microscopic structures. The outer zone's microstructures are shown followed by in Figures 4(d, e and f). Figure 4(f) demonstrates that there are very few primary particles in the outermost zone. Other than that, no additional particles in the outer zone have been observed, although Figures 4(d) as well as 4(e) clearly indicates there is some dendritic primary Al present. The matrix that provides the combined materials together has a tertiary eutectic composition for Al-Si-Mg. The innermost zone's eutectic Si and Mg2Si structures are significantly thicker than those in the outermost zone.

Figure 3. Cross-section of tubular product cast by centrifugal method
(Zhang et al., 2006)

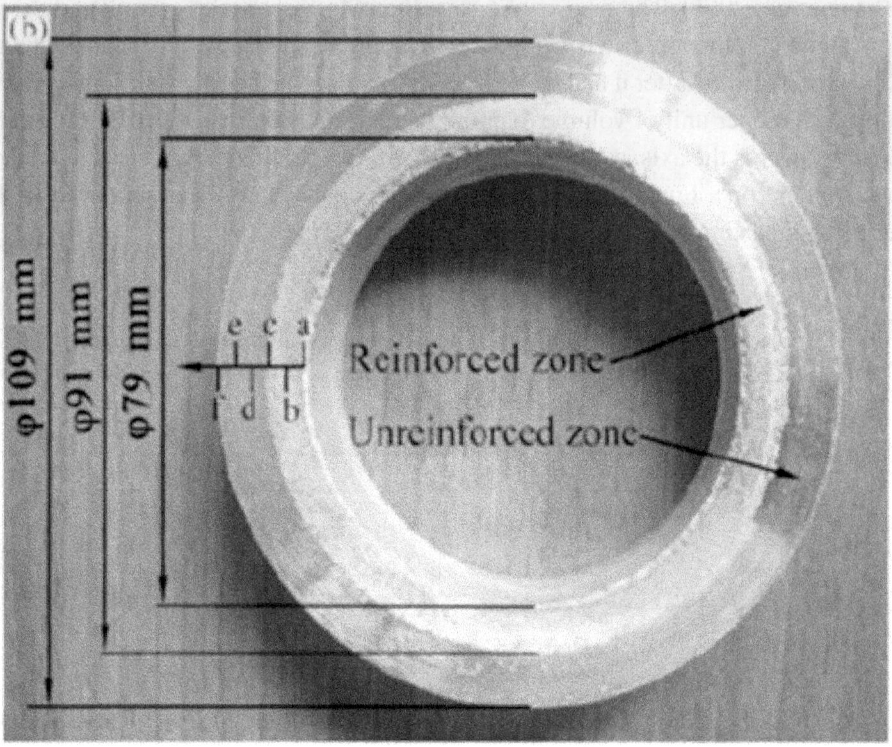

(Ram et al., 2018) used centrifugal casting with varying magenesium concentrations to fabricate the tubular components. They observed that the amount of magnesium was in charge of the creation of in-situ Mg_2Si particles.These particles serve as reinforcement, and the number of Mg_2Si particles formed was slightly higher than in cases when less Mg was added. The results of vickerhardness change in cast tubular components are shown in Figure5. Due to greater Mg_2Si reinforcement, the hardness of the inner surfaces increased.Due to increased production of Mg_2Si particles, the A356+10Mg cast component had the highest value of hardness.Additionally, as shown in Figure 5, the absence of Mg_2Si reinforcement caused the value of hardness decreased in the tube's perimeter.

Figure 4. Corresponding microstucture of tubular product cast by centrifugal method *(Zhang et al., 2006)*

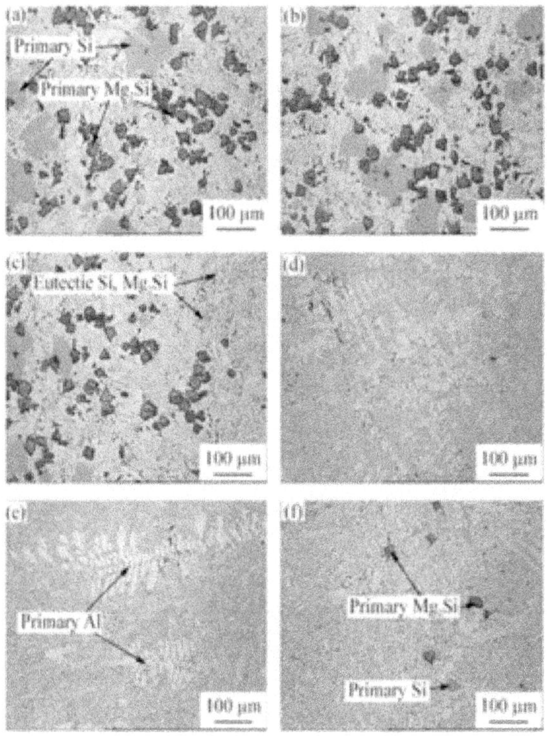

5. INDUSTRIAL SCOPE OF FABRICATED Al-Mg_2Si FG-COMPOSITES

Currently, the casting process, forming, extrusion process, and powder metal fabrication are the primary conventional techniques used to create aluminum alloy components for structures. Fortunately, the aforementioned conventional processes have produced an abundance of aluminum alloy merchandise that has been widely used, but the manufacturing and utilisation processes still have a lot of issues. Initially, the poor freezing rate for casting processes leads the microstructure to be coarser, and there are numerous defects. In cast aluminium metal alloys, frequent flaws include offset irregularities, shrinkage pores, dross incorporation, and constituent dispersion. These flaws reduce the physical and mechanical properties of the components. Furthermore, high-performance aluminum alloy products' processing and shaping processes are usually segregated, leading to a lengthy manufacturing path with restricted flexibility. The structural and functional performance requirements for aluminium alloy components are also consistently rising due to modern technological advances. In particular, the thermally protected system of airplanes or engines in cars frequently incorporates a layered or cell structure in order to fulfil the engineering requirements of high thermal conductivity, light weight, and great load carrying capacity. Integrated forming for complicated structural elements minimizes the amount of weight and concentrations of stress commonly linked with welding and various other joining techniques, as well as the time and tools required for the manufacturing and putting together of small to medium-sized parts (Kumar et al., 2022; Kumar et al., 2023; Singh et al., 2022). In the future, one of the main goals of study and development will involve the manufacturing of aluminium alloy components having a variety of structural configurations, dimensional precision, and close-to-net form.The lightweight composites especially fabrictaed with Al-Mg-Si light alloys are extensibly used for making the varieties of structural components as shown in Figure6.

Figure 5. shows the vaariation of hardness in cast tubular product
(Ram et al., 2018)

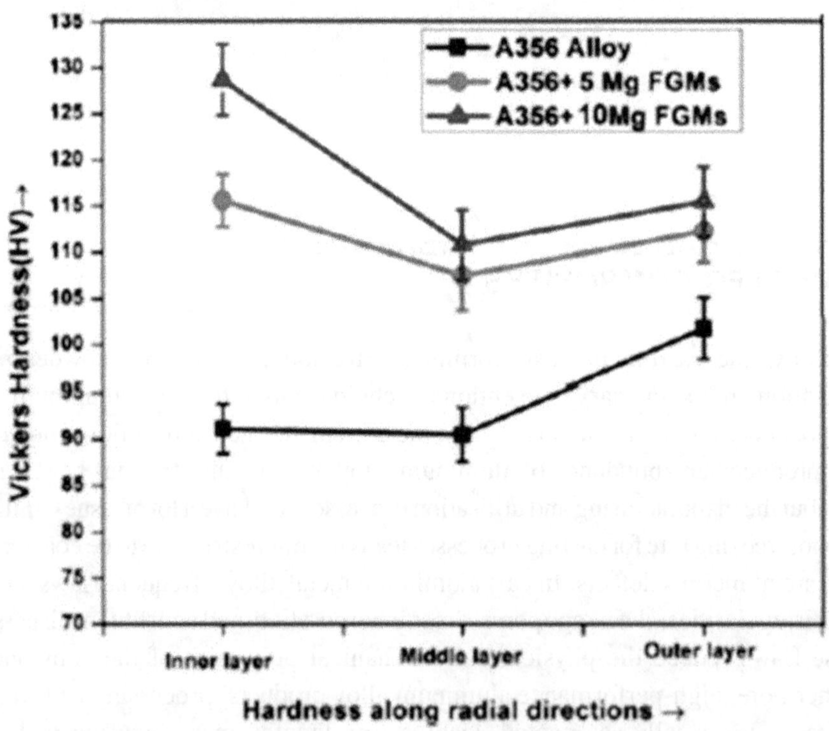

Figure 6. Scope and industrial applications

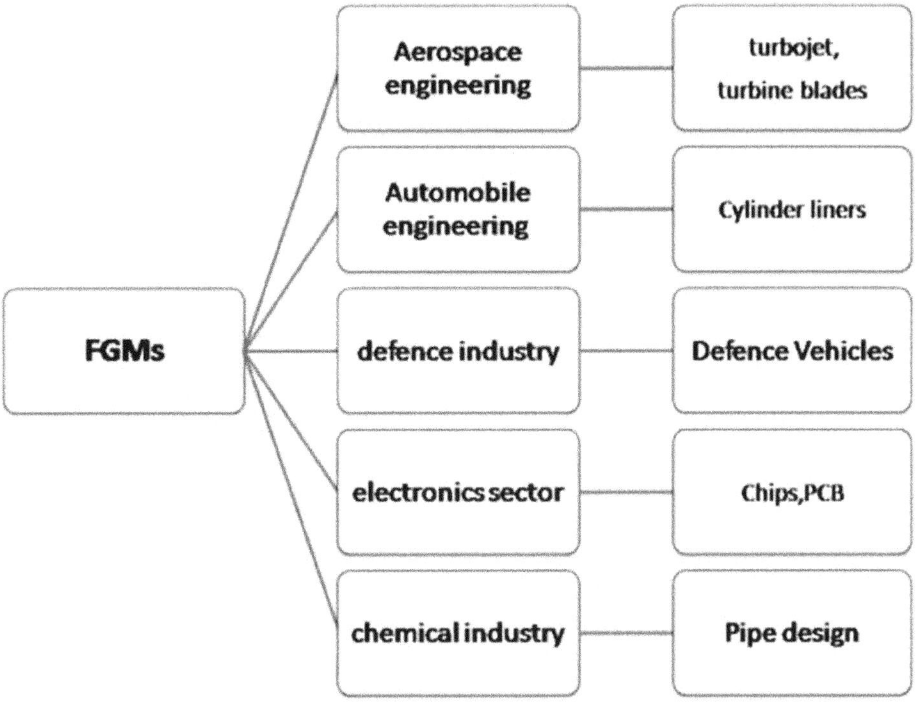

Nowdays, the automotive engine part like cylinder liner, disc brakes and gears are required high hardness, wear properties along with light weight. Interestingly, Al-Mg_2Si FG composites remain the ideal materials for fabricating the aforementioned automotive components.

Automotive internal combustion engine parts such as the cylinder block and liner, which are presented in Figure 7(a), (b), and (c). Theses engine parts play as an important role while rotating in respect to each other parts of the engine despite having subjected to thermally and mechanically repeated stresses. High degree of hardness great endurance against fatigue, and excellent resistance to wear are necessary for an engine block to last a long time.Some of the researcher worked to fabricate the light weight materials using additive fabrication technologies (Kumar et al., 2022; Kumar et al., 2023; Singh et al., 2022; Ye et al., 2003). Additive manufacturing have recent advanced techniques and can be used to produce 3D objects. Al-Si alloys are a great substitute for iron casting when making components for engines since they have a higher resistance to heat and a lower engine density. Thermal characteristics of Al-Si alloys Because of the increase in the highest possible operating temperature and engine pressure, mechanical efficiency must be increased (Javidani et al., 2014;

Ajay et al., 2023; Singh et al., 2019; Joshi et al.,2021; Singh et al.,2023). Al-Si-Mg alloys are progressively replacing metals such as steel and iron in engine components as well as engine components for vehicles.

Figure 7. Scope and applications of FGMMCs (a) Cross–section of a cylinder in an i.c. engine, (b) Air cooled 4-Stroke Aluminum motorcycle engine Cylinder Block, (c) Aluminum engine blocks for car

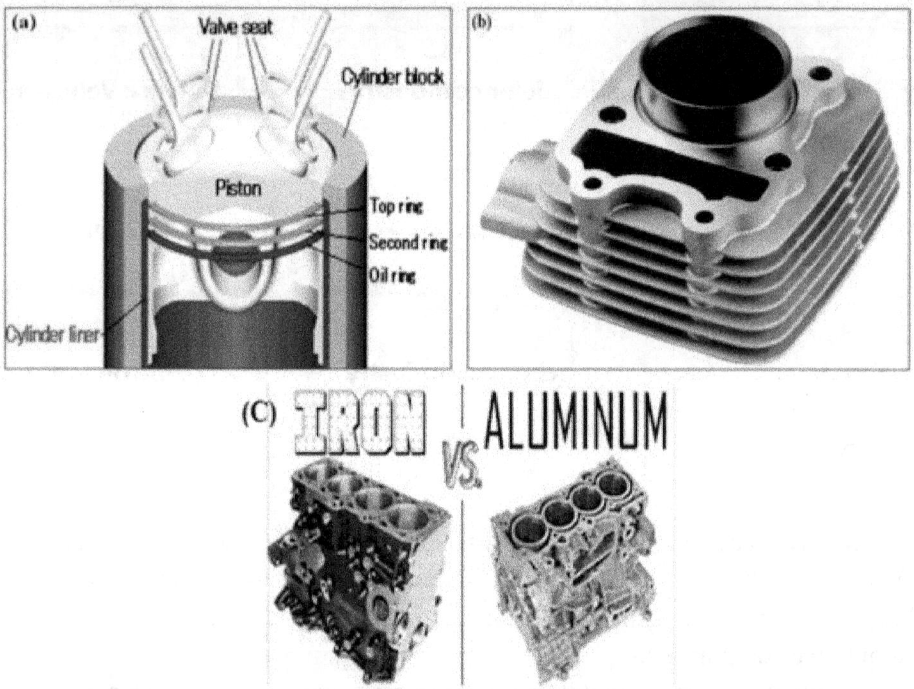

6. CONCLUSION

The current review emphasized mostly lighter materials with functional characteristics. In the literature, it was found that most of the researchers working on Al-Si-Mg alloys. Centrifugally cast Al-Mg_2Si insitu FG-composites are a novel type of advanced material with gradient properties. Mechanical; chemical, physical, and microscopic properties of materials differ with a given orientation from one surface to another. Al-Mg_2Si FG-composites provides a distinct advantage over traditional materials as a result of their significant strength-to-weight proportion. In situ Al-Si alloy matrix functionally graded Al-Mg_2Si composites are tremendously useful in structural component design, particularly for automobile engine parts, aerospace industry, as well as a variety of structural elements. Engine parts having gradient properties, including cylinder blocks as well as liners, can be designed to increase the service life as well as the performance of automobile engine components. Al-Mg_2Si insitu produced Mg_2Si particles have a low weight, strong hardness, low thermal contraction coefficient, a substantial modulus of elasticity, exceptional toughness, and outstanding wear resistance. The most typical practical applications for Al-Mg_2Si functionally graded composites are cylinder liners, brakes with discs and gears.

REFERENCES

Ajay, P. Ahmad, S., Sharma, J., & Gambhir, V. (Eds.). (2023). Handbook of Sustainable Materials: Modelling, Characterization, and Optimization (1st ed.). CRC Press. https://doi.org/ doi:10.1201/9781003297772

Bai, G., Liu, Z., Lin, J., Yu, Z., Hu, Y., & Wen, C. (2016). Effects of the addition of lanthanum and ultrasonic stirring on the microstructure and mechanical properties of the in situ Mg2Si/Al composites. *Materials & Design*, *90*, 424–432. doi:10.1016/j.matdes.2015.10.159

Biswas, P., Mandal, D., & Mondal, M. K. (2020). Failures analysis of in-situ Al–Mg_2Si composites using actual microstructure based model. *Materials Science and Engineering A*, *797*, 140155. doi:10.1016/j.msea.2020.140155

Deuis, R. L., Subramanian, C., & Yellup, J. M. (1997). Dry sliding wear of aluminium composites—A review. *Composites Science and Technology*, *57*(4), 415–435. doi:10.1016/S0266-3538(96)00167-4

Farahany, S., Ghandvar, H., Nordin, N. A., & Ourdjini, A. (2017). Microstructure Characterization, Mechanical, and Tribological Properties of Slow-Cooled Sb-Treated Al-20Mg 2 Si-Cu In Situ Composites. *Journal of Materials Engineering and Performance*, *26*(4), 1685–1700. doi:10.100711665-017-2624-8

Farahany, S., Ghandvar, H., Nordin, N. A., Ourdjini, A., & Idris, M. H. (2016). Effect of primary and eutectic Mg2Si crystal modifications on the mechanical properties and sliding wear behaviour of an Al–20Mg2Si–2Cu–xBi composite. *Journal of Materials Science and Technology*, *32*(11), 1083–1097. doi:10.1016/j.jmst.2016.01.014

Javidani, M., & Larouche, D. (2014). Application of cast Al-Si alloys in internal combustion engine components. *International Materials Reviews*, *59*(3), 132–158. doi:10.1179/1743280413Y.0000000027

Joshi, G. P., Tyagi, A., Meena, S. L., Chourasia, S., Saxena, N. S., & Kumar, A. (2021, August). Optical Band Gap Determination of Ni-Al Doped Polyaniline at Room Temperature and Different Annealing Temperatures. [). IOP Publishing.]. *Journal of Physics: Conference Series, 1950*(1), 012091. doi:10.1088/1742-6596/1950/1/012091

Khorshidi, R., Honarbakhsh-Raouf, A., & Mahmudi, R. (2018). Effect of minor Gd addition on the microstructure and creep behavior of a cast Al–15Mg2Si in situ composite. *Materials Science and Engineering A, 718*, 9–18. doi:10.1016/j.msea.2018.01.081

Kumar, A., Kumar, P., Mittal, R. K., & Gambhir, V. (2023). Materials processed by additive manufacturing techniques. In Advances in Additive Manufacturing Artificial Intelligence, Nature-Inspired, and Biomanufacturing (pp. 217-233). Elsevier. doi:10.1016/B978-0-323-91834-3.00014-4

Kumar, A., Mittal, R. K., & Haleem, A. (Eds.). (2022). *Advances in Additive Manufacturing: Artificial Intelligence, Nature-Inspired, and Biomanufacturing*. Elsevier. doi:10.1016/C2020-0-03877-6

Kumar, S. N., Devarajaiah, R. M., & Prabhu, T. R. (2021). Review on aluminium based functionally graded composites. *Materials Today: Proceedings, 39*, 1743–1749. doi:10.1016/j.matpr.2020.06.307

Loh, G. H., Pei, E., Harrison, D., & Monzón, M. D. (2018). An overview of functionally graded additive manufacturing. *Additive Manufacturing, 23*, 34–44. doi:10.1016/j.addma.2018.06.023

Melgarejo, Z. H., Suárez, O. M., & Sridharan, K. (2008). Microstructure and properties of functionally graded Al–Mg–B composites fabricated by centrifugal casting. *Composites. Part A, Applied Science and Manufacturing, 39*(7), 1150–1158. doi:10.1016/j.compositesa.2008.04.002

Nai, S. M. L., Gupta, M., & Lim, C. Y. H. (2003). Synthesis and wear characterization of Al based, free standing functionally graded materials: Effects of different matrix compositions. *Composites Science and Technology, 63*(13), 1895–1909. doi:10.1016/S0266-3538(03)00158-1

Prabhu, T. R. (2017). Processing and properties evaluation of functionally continuous graded 7075 Al alloy/SiC composites. *Archives of Civil and Mechanical Engineering, 17*(1), 20–31. doi:10.1016/j.acme.2016.08.004

Ram, S. C., Chattopadhyay, K., & Bhushan, A. (2023). A literature review on Al-Si alloy matrix based in situ Al-Mg_2Si FG-composites: Synthesis, microstructure features, and mechanical characteristics. *Proceedings of the Institution of Mechanical Engineers. Part C, Journal of Mechanical Engineering Science*, *237*(4), 919–940. doi:10.1177/09544062221124064

Ram, S. C., Chattopadhyay, K., & Chakrabarty, I. (2016). Dry Sliding Wear Behavior of A356 Alloy/Mg 2 Sip Functionally Graded in-situ Composites: Effect of Processing Conditions. *Tribology in Industry, 38*(3).

Ram, S. C., Chattopadhyay, K., & Chakrabarty, I. (2017). High temperature tensile properties of centrifugally cast in-situ Al-Mg_2Si functionally graded composites for automotive cylinder block liners. *Journal of Alloys and Compounds*, *724*, 84–97. doi:10.1016/j.jallcom.2017.06.306

Ram, S. C., Chattopadhyay, K., & Chakrabarty, I. (2018, July). Effect on microstructures, hardness and SDAS of primary Mg2Si/Al-Si eutectic phases of centrifugally cast functionally graded Al-(Mg2Si) p in-situ composites. [). IOP Publishing.]. *IOP Conference Series. Materials Science and Engineering*, *390*(1), 012012. doi:10.1088/1757-899X/390/1/012012

Ram, S. C., Chattopadhyay, K., & Chakrabarty, I. (2019). Microstructures and high temperature mechanical properties of A356-Mg2Si functionally graded composites in as-cast and artificially aged (T6) conditions. *Journal of Alloys and Compounds*, *805*, 454–470. doi:10.1016/j.jallcom.2019.07.075

Shafieizad, A. H., Zarei-Hanzaki, A., Abedi, H. R., & Al-Fadhalah, K. J. (2015). The Mg2Si phase evolution during thermomechanical processing of in-situ aluminum matrix macro-composite. *Materials Science and Engineering A*, *644*, 310–317. doi:10.1016/j.msea.2015.07.060

Shi, J. L., Yan, H. G., Su, B., Chen, J. H., Zhu, S. Q., & Chen, G. (2011). Preparation of a functionally gradient aluminum alloy metal matrix composite using the technique of spray deposition. *Materials and Manufacturing Processes*, *26*(10), 1236–1241. doi:10.1080/10426914.2010.544814

Singh, V., Gupta, M., & Kumar, A. (2019, November). Investigating the morphology and thermophysical properties of Al2O3 and CuO nanofluids. [). IOP Publishing]. *IOP Conference Series. Materials Science and Engineering*, *577*(1), 012052. doi:10.1088/1757-899X/577/1/012052

Singh, V., Gupta, M., Kumar, A., Luthra, S., & Kumar, A. (2023). Experimental investigations of thermo physical properties and convective heat transfer of Al2O3 and CuO nanofluids in a copper tube: Proposing New Correlations. *Biointerface Research in Applied Chemistry*, *13*(3), 229.

Singh, V., Kumar, A., Alam, M., Kumar, A., Kumar, P., & Goyat, V. (2022). A study of morphology, UV measurements and zeta potential of zinc ferrite and Al2O3 nanofluids. *Materials Today: Proceedings*, *59*, 1034–1039. doi:10.1016/j.matpr.2022.02.371

Sobczak, J. J., & Drenchev, L. B. (2009). *Metal Based Functionally Graded Materials*. Bentham Science Publishers. doi:10.2174/97816080503831090101

Soltani, N., Nodooshan, H. J., Bahrami, A., Pech-Canul, M. I., Liu, W., & Wu, G. (2014). Effect of hot extrusion on wear properties of Al–15 wt.% Mg2Si in situ metal matrix composites. *Materials & Design*, *53*, 774–781. doi:10.1016/j.matdes.2013.07.084

Vieira, A. C., Sequeira, P. D., Gomes, J. R., & Rocha, L. A. (2009). Dry sliding wear of Al alloy/SiCp functionally graded composites: Influence of processing conditions. *Wear*, *267*(1-4), 585–592. doi:10.1016/j.wear.2009.01.041

Watanabe, Y., Kawamoto, A., & Matsuda, K. (2002). Particle size distributions in functionally graded materials fabricated by the centrifugal solid-particle method. *Composites Science and Technology*, *62*(6), 881–888. doi:10.1016/S0266-3538(02)00023-4

Watanabe, Y., & Sato, H. (2011). Review fabrication of functionally graded materials under a centrifugal force. *Nanocomposites with unique properties and applications in medicine and industry*, *133*, 150.

Watanabe, Y., Sato, R., Matsuda, K., & Fukui, Y. (2004). Evaluation of particle size and particle shape distributions in Al-Al3Ni FGMs fabricated by a centrifugal in-situ method. *Science and Engineering of Composite Materials*, *11*(2-3), 185–200. doi:10.1515/SECM.2004.11.2-3.185

Xu, F. M., Zhu, S. J., Zhao, J., Qi, M., Wang, F. G., Li, S. X., & Wang, Z. G. (2004). Effect of stress ratio on fatigue crack propagation in a functionally graded metal matrix composite. *Composites Science and Technology*, *64*(12), 1795–1803. doi:10.1016/j.compscitech.2004.01.009

Ye, H. (2003). An overview of the development of Al-Si-alloy based material for engine applications. *Journal of Materials Engineering and Performance*, *12*(3), 288–297. doi:10.1361/105994903770343132

Zhang, B. M. (2006). *Centrifugal Casting* [M]. Machinery Industry Press.

Zhang, J., Wang, Y., Yang, B., & Zhou, B. (1999). Effects of Si content on the microstructure and tensile strength of an in situ Al/Mg_2Si composite. *Journal of Materials Research, 14*(01), 68–74. doi:10.1557/JMR.1999.0012

Zhu, J. N., Zhou, T. T., Zha, M., Li, C., Li, J. H., Wang, C., Gao, C.-L., Wang, H.-Y., & Jiang, Q. C. (2018). Microstructure and wear behaviour of Al-20Mg2Si alloy with combined Zr and Sb additions. *Journal of Alloys and Compounds, 767*, 1109–1116. doi:10.1016/j.jallcom.2018.07.032

Chapter 12
Nanowires Synthesis and Application as Smart Biochemical Sensors and Energy Storage Materials

Felix Yao Huemabu Kutsanedzie
Accra Technical University, Ghana

Richard Barnes Asamoah
https://orcid.org/0000-0003-2508-5533
Accra Technical University, Ghana

Peter Kwesi Nyanor
https://orcid.org/0000-0003-4388-2771
Accra Technical University, Ghana

Isaac Asempah
Accra Technical University, Ghana

ABSTRACT

Nanowires are a class of smart materials with unique physicochemical properties owing to their electron confinement along their thickness which differs them from their bulk counterparts. Their enhanced physical, chemical, optical, and mechanical properties have heightened their suitability for a host of sensing applications. The profound sensitivity underlines their preferential applications in lighting, computing, communication, energy harvesting, and medicine for detecting and analyzing signals and molecules with a relative greater efficiency. Advanced synthesis methods have been developed to engineer the properties of nanowires for tailored sensing applications. In this chapter, captured various synthesis techniques and potential sensing applications of nanowires underpinned by their unique properties. The chapter output will enlighten recent advances in nanowires and its subsequent sensing technologies utilized in healthcare, environment, and electronic devices. Consequently, the concepts are expected to open vistas in allied nanomaterials and sensing technology development.

DOI: 10.4018/978-1-6684-9224-6.ch012

1. INTRODUCTION

Requirements for contemporary and advancement of devices for applications in electronics, and biosensing among others need enhanced material development contrary to their bulk properties. Nanomaterials are a class of nano-scale dimension materials possessing unique properties from their bulk counterparts (Marrese, Guarino, and Ambrosio, 2017). The nanoscale dimension produces material properties for the development of a new era of devices for quality functionalities and enriched performance whiles allowing a relatively lower cost of production. Nano is a prefix denoting one billionth of a meter. In a more common comparison, one nanometer is one-thousandth (1/1000) of the diameter of a hair. A nanometer is however about 1000 times the size of an atom. The properties of materials at this dimension are greatly altered to have properties, unlike their bulk counterparts. At the nanoscale, materials approach their atomic or molecular level, they are a few atoms in dimension. The phenomenon greatly influences their chemical, mechanical, thermal, electrical, and magnetic properties. Changes in the physicochemical properties at the nanoscale are applied for the development of a new generation of enhanced devices. Nanomaterials are materials possessing novel properties with at least one of their size dimensions within the 1-100nm range. Therefore, nanomaterials are classified as zero-dimensional (0-D), one-dimensional (1-D), and two-dimensional (2-D). Zero-dimensional nanomaterials are also called quantum dots or nanoparticles (Lue, 2007). These are materials that have all their dimensions within the nano scale range with zero or no degree of freedom; none of the dimensions is within the micro or macroscopic scale. Zero-dimension materials have all size dimensions within the 1-100nm range. However, quantum dots are typically 1-10nm range semi-conductive characteristic materials possessing optical and or electronic properties that are size dependent. Quantum dots including fullerenes, metallic, inorganic, and magnetic nanoparticles exhibit quantum confinement properties (Sudha *et al.*, 2018). They display a coloration property dependent on the particle size. Quantum dots have a valence band, a band gap and a conduction band. The absorbance of adequate energy photons allows the excitation of electron from the valence band to the conduction band. The deexcitation of the electron causing a drop to the valence band from the conduction band emits (releases) an energy difference between the conduction band and the valence band which produces an equivalent fluorescent wavelength hence the color emitted. The phenomenon is called quantum or electron confinement. The color emitted by the quantum dot is dependent on its size (Chen and Jia, 2014). According to the Planck-Einstein relation ($E=hv=hc/\lambda$; E is the energy of the bandgap, h is the Planck's constant, v is the frequency in hertz, c is the speed of light and λ is the wavelength), the wavelength of emitted light is dependent on the bandgap. The particle size determines the bandgap. The bandgap is inversely proportional to the particle size (Ravindra *et al.*, 2019). The bandgap increases for decreasing particle size (radius) hence the fluorescence produced. The smaller the quantum dot nanoparticle, the higher the energy emitted. 2nm nanoparticles produces blue color, 3nm produces green color, 6nm generates red color, etc (Goh, Xu and McCormick, 2014). Thus, as the particle size increases, the wavelength increases from the ultraviolet to the infrared spectrum in the order of decreasing energy hence the color change.

Electron confinement leads to a restriction in the electronic wave function which leads to alteration in the physicochemical properties such as optical properties, phase changes, magnetic properties, etc. For example, the optical properties are altered by enhancement in the electronic band gap. It affects the band gap by influencing the potential electron mobility hence changes in the optical properties (Ravindra *et al.*, 2019). The levels of state of energy of an electron are no longer continuous but discrete in a confined region (nanoscale) whiles they are continuous in a region of degree of freedom (macroscopic scale or bulk region) (Weinberger, 2014). The creation of discrete levels of energies is a signature feature in nanostructure responsible for their unique properties form their bulk counterparts.

Size reduction in nanomaterials is also responsible for creating high surface area. This leads to high surface to volume ratio. Enhanced surface to volume ratio allows more atoms to be present on the surface of the nanowire. Increasing the surface to volume ratio increases the number of atoms present on the material surface. For instance, a 100nm particle has about 2% of its atoms on its surface whiles a 10nm particle has about 10-25% of its atomic constituents on its surface (Koshy *et al.*, 2011). Decreasing the particle size further to say 3nm will enhance the surface to volume ratio. Thus, it will increase the number of atoms present on the particle surface to a significant 45-60%. These atoms which are present on the surface possess free bonds for chemical reactivity. The unsaturated atomic bonds present on the particle surface influences the properties of materials at the nano scale.

One dimensional material has two of its dimensions in the nano scale whiles the third is microscopic or macroscopic. Electrons in a one-dimensional material have electrons confined in two directions whiles they have a degree of freedom in one direction. Thus, one dimension of the material is in the micro or macroscopic scale. One dimensional material includes nanorods, nanowires, nanofibers, nanofilaments and nanotubes. Likewise, as zero dimensional materials, one dimensional material can be amorphous or crystalline; single or poly crystalline, and metallic or ceramic (Lin, Szwacki and Yakobson, no date). Two-dimensional materials are materials having a thickness within 1-100nm or less. The electrons in the material are confined in one direction whiles they have a degree of freedom in two dimensions. Whiles one dimension of the material is within the nano scale (1-100nm) the other two dimensions are within the micro or macroscopic scale. Thus, it has two degrees of freedom.

Bulk materials have all their three dimensions above 100nm scale. None of the dimensions is within the nano scale hence the electrons exhibit no quantum confinement effects (Mukherjee *et al.*, 2012). Their macroscopic nature also produces a lower surface to volume ratio hence reduced or insignificant surface atoms with unsaturated bonds for chemical reactivity. Bulk materials have a relative lower bandgap than their nanomaterial counterparts since bandgap increases with decreasing size. For instance, bulk copper oxide has a bandgap of 1.2eV whiles copper oxide nanoparticles have a band of 1.7eV (Wang *et al.*, 2016). Changes in the bandgap affects the optical properties.

Quantum confinement coupled with high surface to volume ratio effects of nanomaterials over their bulk counterparts is principally responsible for enhanced properties suitable for applications in the growth of a new age group cohort of devices. Nanowires as an example of nanomaterial possesses the like features of quantum confinement and high surface to volume ratio for enhanced properties. The characteristic features of nanowires for an appropriate application are dependent on the development of appropriate synthesis procedures. This chapter enumerates the various synthesis procedures and applications underpinned by suitable properties of nanowires.

Figure 1. Classification of nanomaterials
(Morber, 2008)

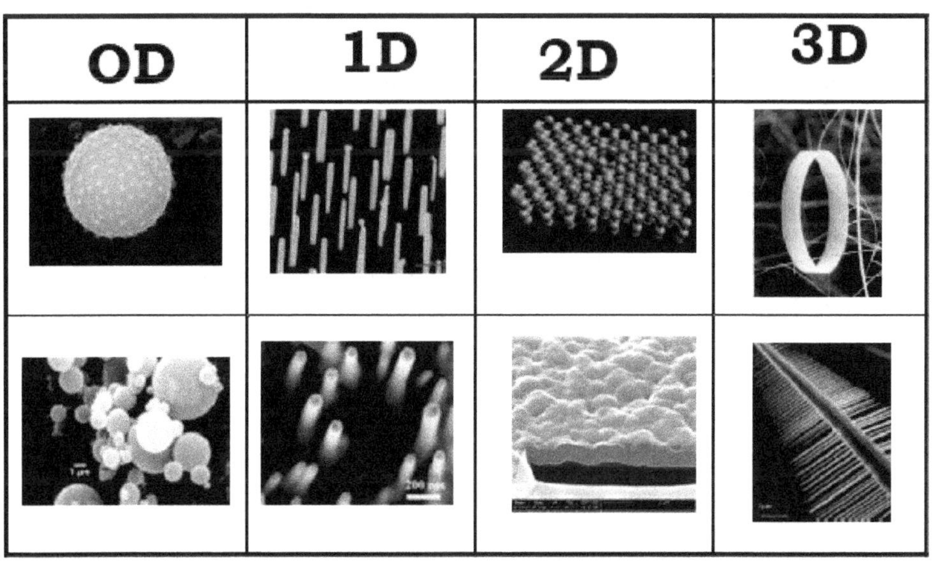

Figure 2. A comparison of the size scale of objects in the nanoscale and macroscale

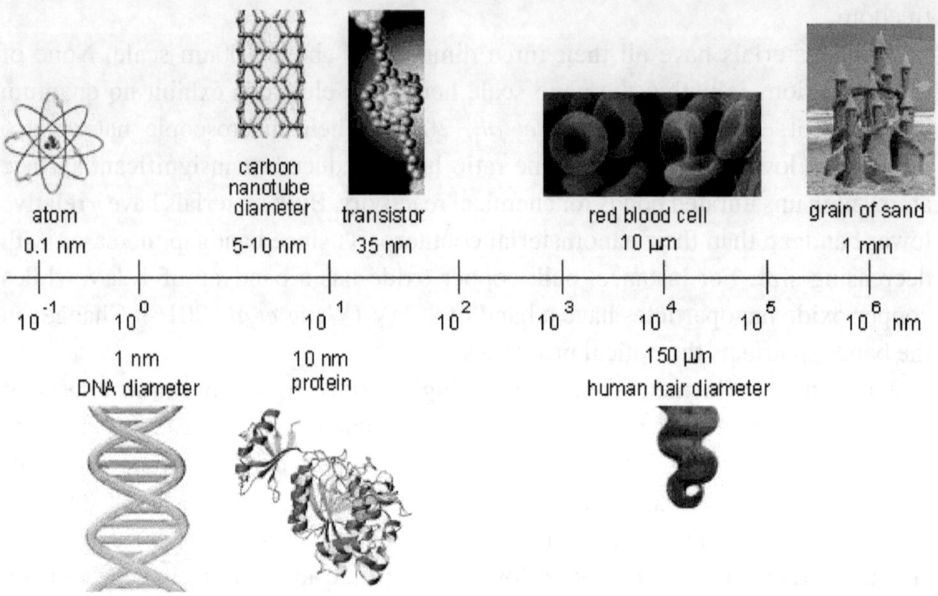

2. WHAT ARE NANOWIRES?

Nanowire is a cylindrically shaped nano structural material. A nanowire has at least one of its dimensions in the nano scale range. This means a nanowire can be best described as a one-dimensional (1-D) material (Chester, 1980). At least one dimension of the nanowire is expected to be below 100nm. The diameter may be a few ones to a few tens of nanometers thick whiles its length may extend into several tens of hundreds of nanometers or microns to centimeters or even longer. Hence, whiles the diameter of a nanowire is in the nano scale range, the length is in micro or macroscopic (bulk) range. Thus, nanowires have high aspect ratio; the ratio of the diameter (thickness or width) to the length. The aspect ratio (length-to-width ratio) may be up to a 1000 or more (Higgins and Stevens, 2021). The electrons in a nanowire are confined along its diameter whiles they have degree of freedom along the length. The nano scale dimension experienced along the diameter of a nanowire has discrete levels of energy states unlike the continuous levels of energy states in bulk or three-dimensional construct counterparts (Liu *et al.*, 2015). Thus, introducing in the material, quantum level properties unique at the nano scale such as the bandgap which affects the optical properties. The bandgap of a nanowire depends on its diameter. The bandgap can be tuned by tuning the diameter of the nanowire. The electron confinement along the diameter are significant to the nanoscale properties of nanowires which vary from its bulk counterparts. In accordance with the quantum confinement effects in nanowires, they are also called quantum wires. The size reduction effect in nanowires as well allows high surface to volume ratio (Jeong *et al.*, 2022). The physicochemical properties of the nanowire are affected by the electron confinement along its diameter which influences the bandgap, high surface to volume ratio which affects the surface chemistry for a range of applications, carrier density of the material and its morphology. Different varieties of nanowires have been developed, including; metallic (e.g., Nickel, Platinum and gold), semiconducting (e.g., Indium Phosphide (InP) and Gallium Oxide and Nitrogen (GaN)), insulating (e.g., Silicon oxide (SiO_2) and Titania (TiO_2)) conductive polymers (Polypyrrole, Polyaniline (PANI)), and molecular nanowires made up of organic DNA molecule or inorganic molecules. A range of synthesis procedures have been developed to influence the material properties for an application.

Figure 3. A Transmission electron microscope images of nanowires with a diameter of 15nm and a high aspect ratio
(Salhi, Hossain and Mukhaimer, 2016)

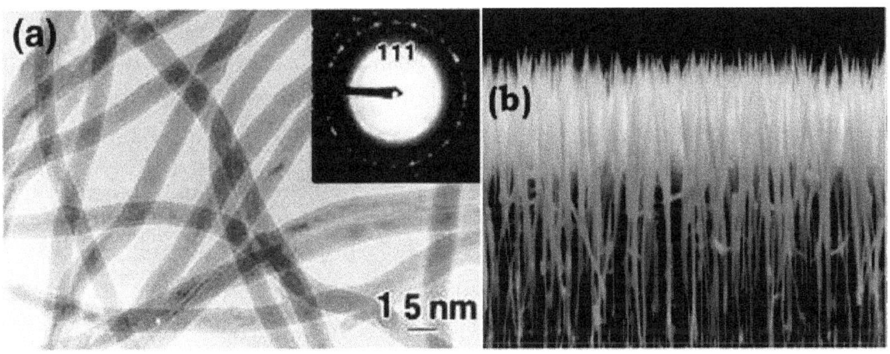

3. SYNTHESIS OF NANOWIRES

The synthesis procedures determine the growth of the nanowire and its consequent associated dimensions. There are two (2) major types of synthesis of nanomaterials which apply to nanowires. The top-down and the bottom-up synthesis are the two main types of the synthesis of nanomaterials including nanowires. Top-down synthesis requires the size reduction of bulk materials into the required nanostructure such as a nanowire. Centrally, bottom-up synthesis is the growth of nanostructures including nanowires from the atomic or molecular level. Milling and lithography are examples of methods of size reduction in the top-down synthesis of bulk materials to form nanostructures as nanowires. Nanowires can also be formed by using chemical synthesis approaches as examples in the bottom-up method to form nanostructures.

Figure 4. Nanomaterials synthesis; top-down and bottom-up synthesis

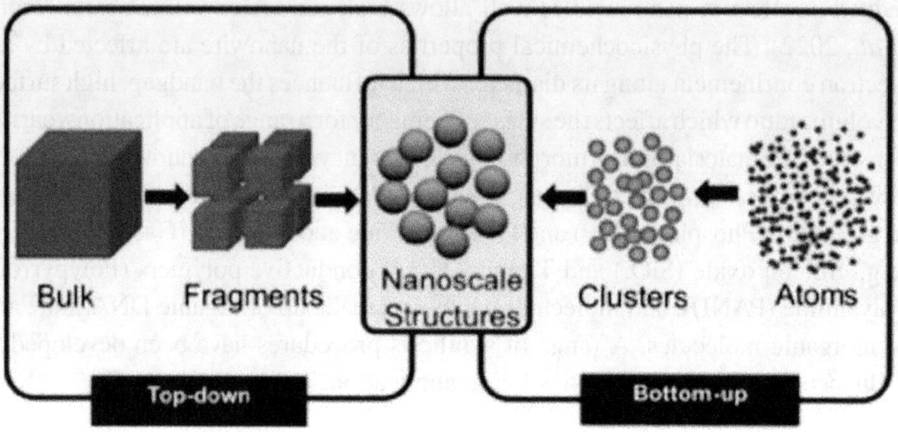

3.1 Top-Down Synthesis of Nanowires

The top-down synthesis can be described as the sculpting of a statue from marble. It relies on precise instrumentation in cleanroom facilities to create nanowire structures from bulk wafers. The advantage of top-down fabrication is that the nanowire structures produced have less defect and have a more homogeneous morphology. However, the machinery involved could be sophisticated and costly hence practiced by established nanofabrication firms. Primarily among the top-down synthesis of nanowires are lithography and chemical etching procedures.

3.1.1 Lithography

This method of nanowire fabrication follows a procedural approach in the microelectronic and semiconductors industry. In this approach, a focused electron beam is used to create patterns for structures including nanowires from a bulk wafer. It involves the deposition of a resist material on the bulk wafer. The resist material acts as a photographic film for focusing high intensity electron beam > 30kV (Tiginyanu, Ursaki and Popa, 2011). The electron beam is scanned across the required surface to trace out the required pattern in a cleanroom (a cleanroom has limited number of airborne particles, contaminants and pollutants). Polymethyl methacrylate is an example of a resist material for electron bean lithography. Electron beam lithography has a considerably high resolution. It allows the fabrication of patterned structures down to the nano scale at 10nm and even lower. Electron beam lithography has a longer depth penetration as compared to other lithographic methods. Longer depth penetration allows the creation of nanostructures especially nanowires with high aspect ratios. Photolithography is however limited by the wavelength of the light used. It is thus difficult to attain resolutions beyond certain dimensions or size.

Scanning probe lithography (SPL) is a modified lithographic procedure which is mask-less, requires no cleanroom, patterning of non-polar surfaces and the ability to pattern proteins without denaturing. SPL uses a probe similar to scanning electron microscopy and atomic force microscopy to fabricate a structure from the bulk wafer or substrate. SPL creates patterns on surfaces through chemical adsorption and physical mechanical scratching. SPL techniques are able to produce nanowire arrays in nano scale up to the resolution of 5nm from different types of materials including polymers and oligonucleotides.

Figure 5. Lithography approach to the synthesis of nanowires. (A) Masking followed by anisotropic etching. (B) Anisotropic etching without masking (exposed region)

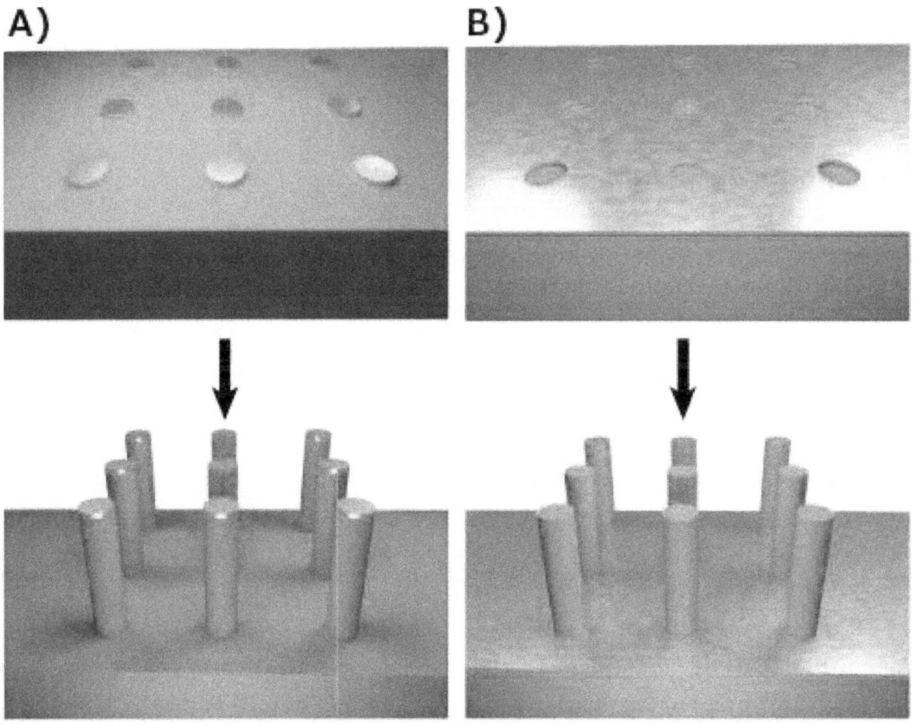

3.1.2 Chemical Etching

Chemical etching is the use of etchants to create components from a bulk wafer or substrate which is typically a metal sheet (Kawamura *et al.*, 2020). Chemical etching is used for the fabrication of nanowires by the removal of layers from the metal sheet. The chemical etchant removes layers that are not with photoresist. This controls the morphology (size and shape) of the resulting particle (nanowire) produced. Chemical etching is a least invasive method of producing nano scale structures by top-down synthesis as compared to lithography. The chemical etching process uses etchants as ferric chloride to remove designs already printed on a metal sheet. The printed designs are exposed to the acidic chemical etchant. Unexposed parts are not removed because they are covered with photoresist. The process is effective for producing highly homogeneous nanostructures as nanowires. For example, Si nanowires are produced by chemical etching technique. A clean Si wafer or substrate is electroplated with silver (Ag) by immersing in a solution of $AgNO_{3.}$ This is followed by etching with hydrofluoric acid and hydrogen peroxide. The Ag catalyst is removed by immersing in $HNO_{3.}$ Size controlled nanowire arrays are produced by scratching them from the metal surface.

3.2 Bottom-Up Synthesis

The bottom-up synthesis allows the growth of nanostructures from the self-assembly of atomic or molecular units. The self-assembly into nanostructures depends on physical and or chemical forces operating at the atomic or molecular levels. The inspiration for the bottom-up synthesis is drawn from natural biological systems where structures are built from the basic elemental units by making use of their chemical forces. A variety of methods have been developed for the synthesis of nanowires form the atomic unit. They typically range from the condensation of atomic vapor to the coalescence of atoms in solution.

3.2.1 Vapor-Liquid-Solid Synthesis

Vapour-Liquid-Solid Synthesis (VLS) is popular for the growth of nanowire structures. The method uses a precursor which is in the gaseous state, a nanocrystal liquid as a catalyst and a substrate which is a solid (Weng *et al.*, 2020). Whiles the liquid droplet catalyst is deposited on the solid substrate before deposition into a chemical vapor deposition tube furnace. The gaseous source precursor is produced from the substrate through heating. As the gaseous precursor begins to supersaturate the liquid catalyst, the source gas precursors solidify between the catalyst-substrate interface. In this case a one-dimensional nanostructure is formed which is a nanowire. The diameter of the nanowire depends on the size of the nanocrystal catalyst. Si nanowires are commonly grown from silane substrate and a gold (Au) nanocrystal catalyst.

Similar to the VLS approach is the vapor-solid-solid synthesis. It uses a lower temperature and a solid catalyst. The source precursor diffuses through the catalyst for the growth of a nanowire between the substrate and catalyst.

Non-catalytic approaches to growing nanowires have been developed. The techniques is to grow more chemically pure nanowires free from contaminants. Researchers have demonstrated the growth of copper oxide be heating up copper foils in air. Nanowire morphologies were determined by the process parameters including the annealing temperature and the time for the growth of the nanowire.

Figure 6. A schematic representation of the VLS synthesis mechanism

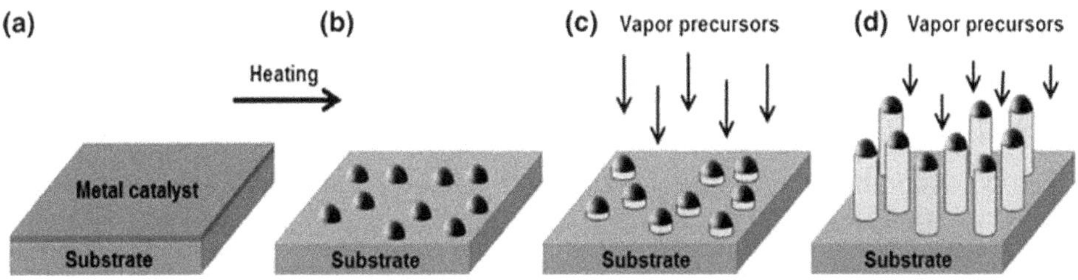

3.2.2 Solution Phase Synthesis

The solution phase synthesis of nanowires is a facile and a cost-effective means of producing nanowires from different kinds of materials (Cargnello, Gordon and Murray, 2014). Contrary to the vapor phase, the precursor is dissolved in a liquid to form a wet chemical solution. The seed particles are grown by reducing the dissolved precursor with a reducing agent such as sodium borohydride. In the case of metal oxides as copper oxide, the dissolved precursors are caused to precipitate by using a precipitating agent as sodium hydroxide. Silver, gold, platinum, copper oxide, zinc oxide, etc. nanowires have been generated by the solution phase synthesis. Copper oxide nanowires have been produced at ambient conditions using facile single pot wet chemical precipitation method. The rods measure a diameter of 4-10nm and a length of 100-400nm. Zinc oxide nanowire of similar morphology have also been synthesized using the wet chemical precipitation method. The resulting products after precipitation in this case were however calcined at 400°C for 4 hours. Nanowire sizes were dependent on synthesis parameters. Table 1 summarizes different synthesis methods and the associated characteristic dimensions of the resultant nanowire produced.

Table 1. The effect of different synthesis method on nanowire dimensions

Type of Nanowire	Type of Synthesis	Diameter of Nanowire (nm)	Length of Nanowire (μm)	Reference
Ag	Hard template	7	.>1	(Braun et al., 1998)
Ag	Soft template	~1	~1	(Hong et al., 2001)
Ag	Hydrothermal	45-65	.200-500	(Xu et at., 2006)
Ag	Solvothermal	50	.250	(Li et al., 2019)
CuO	Wet chemical	15-30	0.15-0.4	(Li et al., 2019)
Silicon	Lithography	100	1.6	(Wu et al. 2022)
GaAs	Dielectrophoresis	30-50	3-5	(Garciá Núez et al. 2020a)
InAs	Sulphur passivation	200	10	(Xiong et al 2017)

4. APPLICATIONS OF NANOWIRES

Nanowires continue to gain interest among researchers and the fabrication industrial sector due to its potential for applications in the development of new and advanced devices. The applications of nanowires are buttressed by its one-dimensional nanoscale nature which have divergent properties from their parent counterpart materials owing to its size effect. They possess unique electronic, magnetic, mechanical, chemical among other properties. For instance, the optical electronic properties of nanowires are dependent on its diameter, they also have a high aspect ratio contributing to high surface to volume ratio and a consequent surface chemical reactivity which influence its physicochemical properties. The properties of nanowires underpin its applications for a range of devices and systems ranging from diodes, transistors, logic gates, and sensor applications because of its chemically sensitive surfaces. Below, we consider some of the critical uses of nanowires for contemporary devices and future developments.

4.1 Electronics

Nanowires have many applications in the electronics industry. The electronic applications of nanowires are typically dependent on its enhanced bandgap and electron density. The bandgap of a nanowire is enhanced by the size of its diameter which is in the nanoscale whiles the electron density is low relative to its dimensional bulk material counterpart. The electron applications of nanowires include; p-n junction diodes and transistors.

Field effect is an essential principle for the operation of nanowire sensors (Hobson, 2019). The field effect phenomenon allows an external electric field to alter the conductivity of a materials which is this case is a nanowire. Bulk material has comparatively high electron density as compared nanoscale materials as nanowire. When an external field is applied to a bulk material it has a lower effect since it can penetrate to a shorter distance because of the higher electron density. In a nanowire, the applied external field has a higher effect hence can penetrate deeper due to the lower electron density. The field effect affects the conductivity of the nanowire which serves as the gate to the channel of the transistor. The gate modifies the carrier charges between the channel of the transistor to control its "on" and "off" states. Transistor channels are n-type (n channel) and p-type (p channel). The electric field of the gate may either cause an enhancement or depletion of the majority charge carriers of the channel. Whiles enhancement is an increase in the majority charge carriers, depletion is a decrease in the majority charge carrier. Examples of transistor are the metal oxide semiconductor field effect transistor (MOSFET), metal semiconductor field effect transistor (MESFET) and junction field effect transistor (JFET).

Figure 7. (a)Nanowire field effect transistor, where Vg is the field effect potential or the gate voltage which is externally applied, I_{sd} is the source to drain voltage, V_{sd} is the source to drain voltage, μ is the electron mobility, C is the gate capacitance and L is the length of the channel. (b) and (c) dI / dVg=μ (C/L²)Vsd

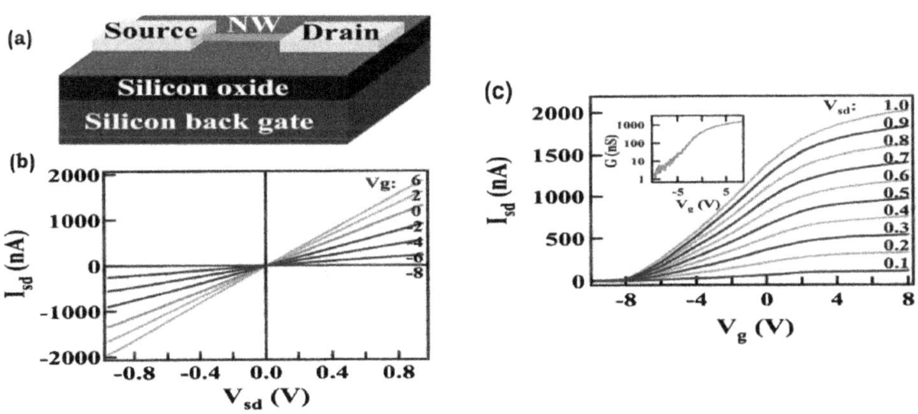

4.2 Biochemical Sensing

Prominent among the applications of nanowires is biological and chemical detection. Nanowires are a class of sensing materials for detecting biological and chemical species (Vicente *et al.*, 2020). Nano-biosensors are central to the detection of pathogens among other biomolecules at the molecular level for disease diagnosis, monitoring and drug discovery. The nanoscale dimension of nanowire is comparable to the size of biomolecules hence presenting a good interaction to detect microscopic signals. Nano have large surface to volume ration coupled with a size effect. They have free atoms with unsaturated bonds. The unsaturated bonds serve a property for biological and chemical sensing through their reactivity. Inorganic nanowires also have optical properties for the development of sensitive probes.

A pH sensor was designed by the functionalization of amine (-NH_2) and Si-OH on the surfaces of Si nanowire. -NH_2 and Si-OH covalently grafted on the surfaces of Si nanowire were responsive to pH changes. They could detect the pH of a solution by the corresponding changes in their charge density (Yong *et al.*, 2019). The charge density was affected by the protonation and deprotonation of the probe was caused by the -NH_2 and Si-OH groups. The changes in the electrical charge on the surfaces of the probe influenced its conductance. The pH was a linear function dependent on the electrical conductance of the modified Si nanowire probe.

Biomolecules are generally electrically charged in aqueous solution. Hence, they are easily detected by nanowires due to the changes in electrical charge they present. Nanowire field effect transistors can act as sensor probes to measure the changes in the electrical conductance. For example, biotin conjugated Si nanowire is highly sensitive to streptavidin at even very low concentrations as 10^{-12} moles (10 pico molar) (Biswas *et al.*, 2017). The biotin grafted Si nanowire shows a rise in conductance in a solution of streptavidin.

The influenza virus A was detected by a nanowire field effect transistor (Felix, Kollu and Grace, 2019). The unfunctionalized probe receptor showed a rise in electrical conductance upon a single viral attachment. The conductance is normalized upon detachment of the virus.

Similarly, the SARS virus N protein is detected by In_2O_3 nanowire. Fibronectin based modified In_2O_3 nanowire was found to selectively detect the biomarker for the virus by changes by electrical conductance (Cao *et al.*, 2019).

Single stranded DNA molecules have also been detected by Si nanowires (Kornilova *et al.*, 2019). The electrical conductance of protein nucleic acid modified Si nanowire is found to increase in conductance at the presence of single stranded DNA biomolecule.

Figure 8. Schematics of various biochemical detections using nanowires; (a) pH detection using nanowire probe (b) detection of biotin with streptavidin (c) detection of DNA protein nucleic acid (d) Viral detection

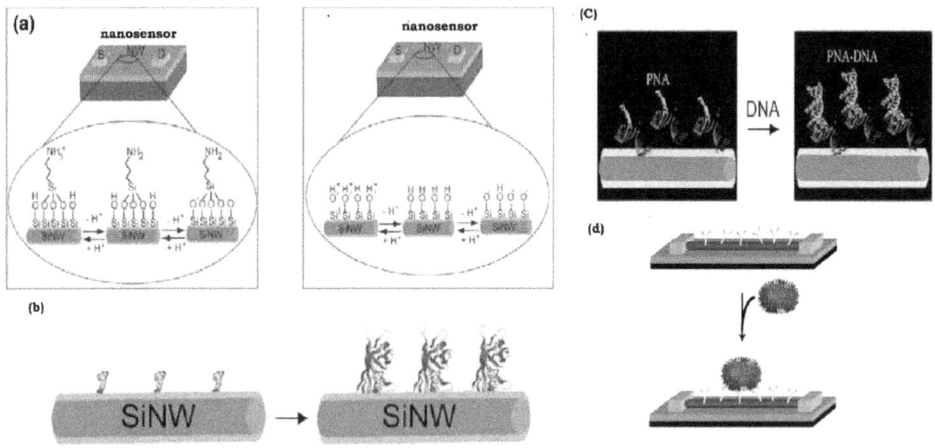

4.3 Energy

Nanowires has gained prime interest among researchers to revolutionize the energy sector. They are potential candidates for new generation of solar cells and batteries for energy storage.

4.3.1 Solar Cells

Solar cells utilize the sunlight to produce a renewable energy source through an incident light (photon) transfer into electric current mechanism. When light is incident on a surface it creates an electron excitation which generates an electron hole pair. The availability of a junction allows a charge separation for the electrons and holes. The harvesting of the electrons through a device as a conductive wire produces electric current. A junction allows the creation of a depletion layer where there is current production. A solar cell device is therefore produced by the appropriate extrinsic or intrinsic doping of a semiconductor material into regions of majority electron (negative charge) carriers (N-type region) and a majority hole (positive charge) carriers (P-type region). A depletion is created between the N-type and P-type regions. Typically, in solar cell devices, the N-type region is constructed thin enough to allow light passage for electron hole pair generation preceding current production. In response to incident light in the N-type region, an electron is excited to create an electron hole pair. The excited electron is energized to escape from the depletion region. If a wire is connected to the N-type region (cathode), the excited electron would flow in the wire to create electric current. The electric circuit is created by created by connecting an electric wire from the N-type region (cathode) to the P-type region (Anode) where the current generated in the wire passes through the load, crating an electric current flow as seen in figure 9 below. The electrons flowing through the wire recombine at the P-type region to restore neutrality. The holes created by electron excitation escape through the electrical contact.

Figure 9. A solar cell consisting of an N-type and P-type regions with a separating depletion zone. Electron flow causes current to flow through the load. Complete electric circuit is realized to restore electrical neutrality achieved by electron-hole recombination at the P-type region

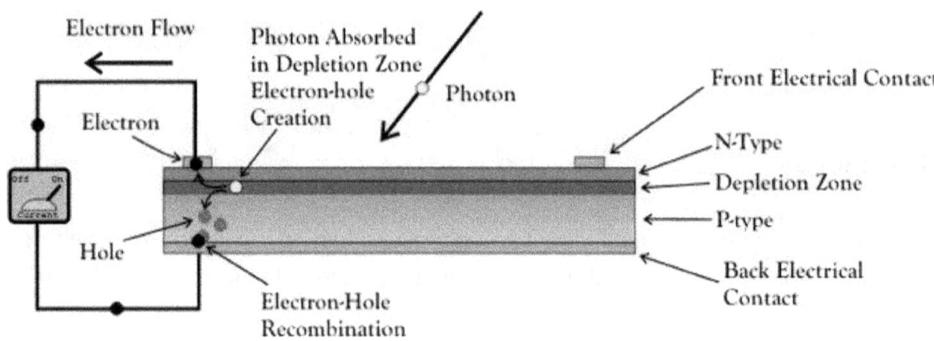

Nanowires are promising for fabricating solar cell systems due to their one-dimensional nanostructure which permits efficient thermoelectric harvesting. Nanowire geometry has the advantage of limited light reflection, improved light absorption and a fine-tuned bandgap. The geometric configurations have different impact on properties such as absorption, charge transfer and charge carrier collection. Junctions are important parts of a nanowire solar cell which account for its geometry. There are the redial junction and the axial junction (Garnett *et al.*, 2011). In the radial junction a doped core nanowire is surrounded by an oppositely doped shell. Axial junction is a vertical stalk of layers oppositely doped. Radial junction permits the optimization of light absorption and charge separation for a highly efficient nanowire solar cells. There are two main junctions of nanowire solar cells which are defined by the nanowire geometry. In the axial junction, charge separation occurs along nanowire length. Charge separation in radial junction happens along the radial axis of the nanowire. In axial junction, a charge carrier must move far along the axial direction of the nanowire length to cross the junction to reach a contact and be extracted or separated. This travel distance by charge carriers is more efficient in radial nanowires. In radial junctions, the charge carrier only hops from the outer shell to the core shell to be extracted or separated. A drawback in the operation of radial junction however lies in their saturation current. The saturation current is dependent on the surface area. The open circuit voltage decreases with increasing surface area. The outer or external shell introduced in radial junction leads to an increase in the surface area hence a negative impact on its saturation current. The nanowire geometry impacts its absorption (Magnusson, Paul and Nys, 2012). For instance, in a bulk planar silicon solar cell the high refraction index of 3.9 (n=3.9) than air causes a reflection of incident light leading to about 30% loss in absorption. This can however be reduced by introducing a layer on intermediate wavelength but this efficient for only a single wavelength. A continuous gradient of refractive index between air and the solar cell is required for an ideal antireflection. This achieved in nanowires. A nanocone with a tapering cone-like nanostructure has a refractive index at the tip similar to the refractive index of air. The refractive index of the nanocone gradually increases downwards. Hence, the reflection decreases significantly reduces in a nanocone solar cell. Thus, nanocone permits efficient absorption of incident light.

4.3.2 Energy Storage

Portable devices such as mobile phones and laptops are enabled by the use of energy storage devices such as a battery. Batteries are storage devices for electrical power. It converts chemical energy directly into electrical energy through electrochemical process of oxidation and reduction reaction. They have external connections for delivering the energies stored in them to power electrical devices. The batteries have a capacity in the order of discharge per hour. For example, one-watt hour (1Wh) is equivalent to a discharge of 1 watt per hour or 2 watts per 0.5 hours or o.5 watts per 2 hours. Unlike fossil fuels, batteries produce free carbon dioxide and other carbon related waste. Thus, they are environmentally friendly hence the battery technology is highly pursued to produce decarbonized transportation system. The battery technology can be classified by their respective energy densities (Wh/Kg) indicating the amount of energy a battery can contain in respect to its weight. The first kind of battery is the lead-acid battery which is still used in vehicles. It has an energy density of about 100Wh/Kg. This was improved by the nickel-cadmium battery and subsequently the nickel-metal hydride. The major breakthrough in battery technology came by the advent of the lithium-ion battery which has an energy density of approximately 372Wh/Kg. This is the battery common in most portable electrical devices as laptops to contemporary electrical vehicles. Current lithium-ion battery is made up of an anode, a cathode, a microporous separator and

Figure 10. Respective schematics of (a) axial junction solar cell (b) radial junction solar cell. α denotes absorption coefficient, Ln and Lp respectively denote diffusion length of holes and electrons

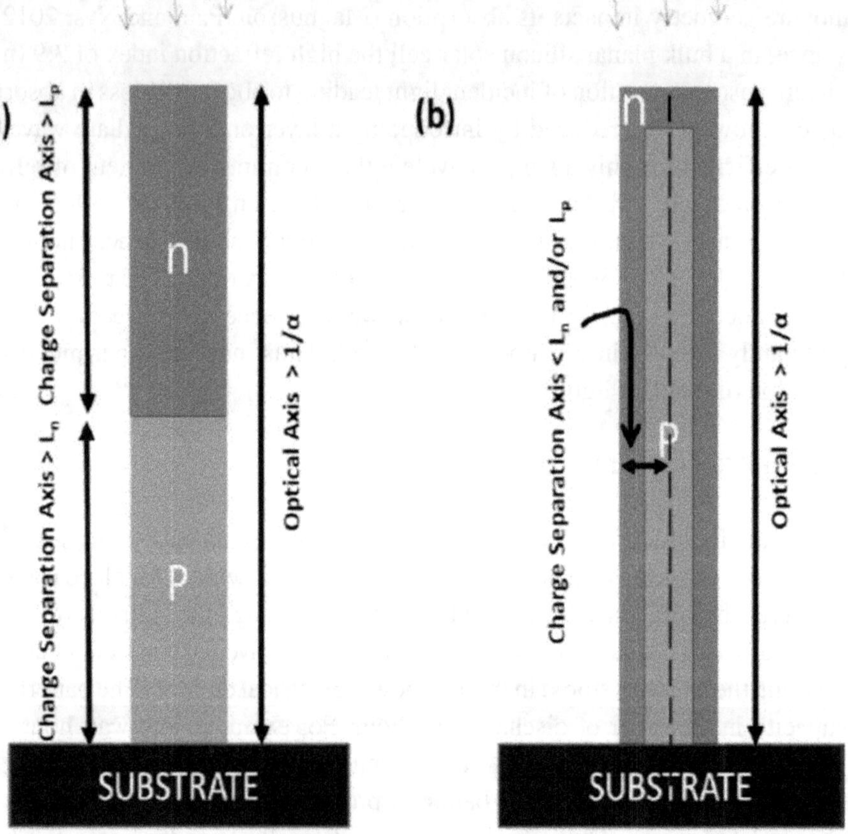

an electrolyte. The anode consist of a porous graphite whiles the cathode is a porous transmission metal oxide. The electrolyte is a liquid to allow flow of lithium ions from the anode to the cathode and vice versa. When the battery is fully charged, lithium ions exist between the graphite layers in the anode. Upon discharge, the lithium ions migrate to the transition metal oxide. External connectors are respectively connected for current transfer and supply to a load. Future batteries are expected to achieve high energy density of about 700Wh/Kg, longer life time from the current approximate 8 years to about 15 years, lower cost and recyclable. So far, silicon has shown great potential of increasing the battery capacity to about 4200Wh/Kg when its used as the anode in modern lithium-ion batteries to replace the current graphite used. Silicon is however susceptible to swelling deficiencies during lithium ion removal and insertion during battery charge and discharge. This leads to about 400 times increase in volume and a fall of the electrode. Nanowires holds the potential to improve battery technology (Salhi, Hossain and Mukhaimer, 2016). Nanowires have large surface area to volume ratio. Nanowires increase the surface area at the anode or cathode or both. They possess a significant larger surface area for storing lithium ions. The enhancement would increase the energy density of the battery. For example, the swelling anomaly in silicon is better amortized when silicon nanowires are used. The large surface area in nanowires improves the mechanical stresses to accommodate lithium ion removal and insertion. In addition, silicon nanowires in combination with other materials such as tin oxide are also being investigated to achieve a stability of the electrode from swelling effect to enhance the battery capacity (Salhi, Hossain and Mukhaimer, 2016). Research continues to advance in several methods to enhance battery capacity for higher energy storage.

Figure 11. Relative incident light absorption performance of thin films < nanowires < nanocones
(Garnett et al., 2011)

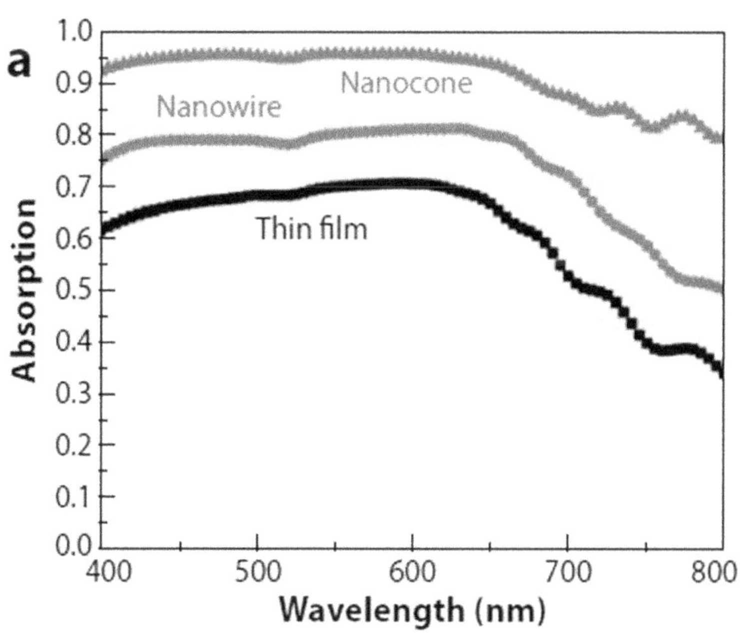

Figure 12. A cross sectional illustration of the battery showing its cells and operational mechanism

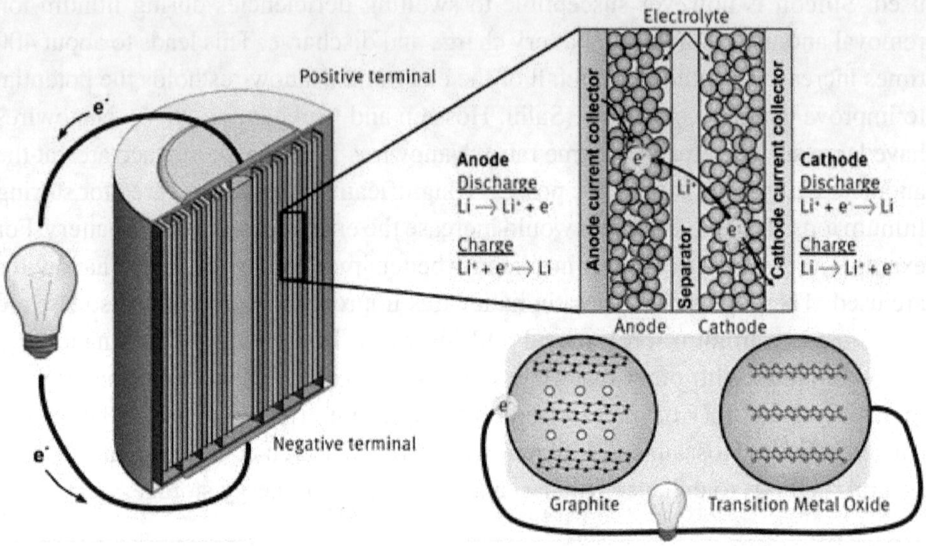

Table 2 contains an abridged application of nanowires. It summarizes the varying nanowire applications as a biosensor, sensors in electronic devices for different physical quantities and also its use as energy storage devices. It is realized that different types of nanowires materials do not always share identical applications. They exhibit different applications in most cases. The different types of nanowires typically stemmed from their different material types and also the method of synthesis of the nanowire produce divergent range of physicochemical properties unique to the nanowire type. As such, among others, they have varying range of optical or bandgap property, electron confinement, surface to volume ratio and chemical reactivity. These properties of the nanowire determine its peculiar use for a specific application. Hence the different usages observed.

5. CONCLUSION AND FUTURE DIRECTIONS

The research trends in nanowire fabrication and associated application creates a knowledge blend bridging the physical sciences and engineering. Principles in chemistry and physics are harnessed for the various synthesis paradigms of nanowires. Engineering is relied to realize devices from the as-synthesized nanomaterials whose applications spans through biological procedures to energy and consumable electronics. The broad requisite knowledge for the field necessitates a multidisciplinary

Table 2. Types of Nanowire and associated application

S.No.	Type of Nanowire	Application of Nanowire	Reference
1	Silicon	Electronics, biosensors, battery, electroluminescence	(Adu, Gutierrez, and Eklund 2007; Zhong, Yang, and Lieber 2007; Vlad et al., n.d.)
2	Germanium	Electronics, IR electrodes	(Zhao et al. 2020)
3	GaAs	High speed electronics, photoconductive roles	(Garciá Núez et al. 2020b)
4	InAs	Photodetector solar cell NWFET, tunnel diode.	(Alexander-Webber et al., 2005)
5	InP	NWFET, Single-photon detectors, tunnel diode	(Zafar and Iqbal 2016)
6	Ag nanowires,	Wearable devicesumma	(Kumar et al. 2022)
7	CuO	Electronic conductor	(Hoa et al. 2010)
8	Zn nanowires	Optoelectronics, light emitters and lasers	(Ravindra et al. 2019)
9	Cadmium-telluride	IR detectors foor various wavelengths	(Zweibel, Moskowitz, and Fthenakis 1998)

skill comprehending the analytical techniques from synthesis to application. Future scientist and researchers in nanowire technology would be required to improve the synthesis methods, understand the basic science principles for altering nanowire properties from the synthesis paradigm resulting in enhanced efficiency of the material. Improved material properties through synthesis methods are expected to result in efficient application. Future works in nanowire would be more significant as they focus on tailoring the material properties by altering the synthesis paradigm for a specified application. This chapter has introduced the concepts of nanomaterials as a basics to update on the broader picture of nanowires. The chapter considered the main types of synthesis for nanowires and the various significant applications of nanowires. Different techniques including; lithography, etching, VLS and solution phase were examined as the most common fabrication techniques of nanowires. Silicon nanowire among others are identified as a common nanowire of great research interest. They are found to be adaptable to different synthesis techniques with vast potential applications. The applications of nanowires were found useful in field effect transistors (FET), sensor applications, photovoltaic cells and batteries. The sensing applications together with the energy storage devices are primarily founded on the principles of nanowire properties which remains to be continued to be exploited for improved device fabrication. The chapter contributes to knowledge for further advancement of nanowire technologies.

REFERENCE

Adu, K. W., Humberto, R. G., & Eklund, P. (2007). *Phonons in Silicon Nanowires. Nanosilicon: new photonic and electronic material at the transition between solid and molecule*. Elsevier Ltd. https://doi.org/. doi:10.1016/B978-0-08-044528-1.50008-7

Alexander-Webber, J. A., Groschner, C. K., Sagade, A. A., Tainter, G., Gonzalez-Zalba, M. F., Di Pietro, R., Wong-Leung, J., Tan, H. H., Jagadish, C., Hofmann, S., & Joyce, H. J. (2017). 'Supporting Information for Engineering the Photoresponse of InAs Nanowires'. *ACS Applied Materials & Interfaces*, *9*(50), 43993–44000. doi:10.1021/acsami.7b14415 PMID:29171260

Biswas, P., Karn, K., Balasubramanian, P., & Kale, P. (2017). 'Biosensors and Bioelectronics Biosensor for Detection of Dissolved Chromium in Potable Water : A Review'. *Biosensors and Bioelectronic*. . doi:10.1016/j.bios.2017.03.043

Cao, L., Kiely, J., Piano, M., & Luxton, R. (2019). *A Copper Oxide / Zinc Oxide Composite Nano-Surface for Use in a Biosensor*. MDPI. . doi:10.3390/ma12071126

Cargnello, M., & Thomas, R. (2014, October 08). Gordon, and Christopher B Murray. 2014. 'Solution-Phase Synthesis of Titanium Dioxide Nanoparticles and Nanocrystals'. *Chemical Reviews*, *114*(19), 9319–9345. doi:10.1021/cr500170p PMID:25004056

Chen, Y. M., & Hong, W. J. (2014). *Synthesis and Characterization of ZnO Nanoparticles Templated by SBS-OH. Advanced Materials Research* (Vol. 864–867). Elsevier Ltd. doi:10.4028/www.scientific.net/AMR.864-867.694

Felix, S., Kollu, P., & Grace, A. N. (2019). Electrochemical Performance of Ag–CuO Nanocomposites towards Glucose Sensing. *Materials Research Innovations*, *23*(1), 27–32. doi:10.1080/14328917.2017.1358507

Garnett, E., Brongersma, M., Cui, Y., & Mcgehee, M. (2011). 'Nanowire Solar Cells'. . doi:10.1146/annurev-matsci-062910-100434

Goh, E. G., Xu, X., & McCormick, P. G. (2014). 'Effect of Particle Size on the UV Absorbance of Zinc Oxide Nanoparticles'. *Scripta Materialia*, *78–79*, 49–52. doi:10.1016/j.scriptamat.2014.01.033

Higgins, S. G & Stevens, M. (2021). 'Europe PMC Funders Group High-Aspect-Ratio Nanostructured Surfaces as Biological Metamaterials' 32 (9): 1–100. https://doi.org/. doi:10.1002/adma.201903862.High-Aspect-Ratio

Hoa, N. D., Van Quy, N., Jung, H., Kim, D., Kim, H., & Hong, S. K. (2010). 'Synthesis of Porous CuO Nanowires and Its Application to Hydrogen Detection'. *Sensors and Actuators. B, Chemical*, *146*(1), 266–272. doi:10.1016/j.snb.2010.02.058

Jeong, H., Noh, Y., Young, G., Lee, H., & Lee, D. (2022). 'Roll-to-Roll Processed Silver Nanowire / Silicon Dioxide Microsphere Composite for High-Accuracy Flexible Touch Sensing Application'. *Surfaces and Interfaces*, *30*(April), 101976. doi:10.1016/j.surfin.2022.101976

Kawamura, H., Matsusaka, S., Nomoto, K., Kodaka, H., Hidai, H., Chiba, A., & Morita, N. (2020). 'Improvement in Etching Efficiency of Borosilicate Glass by Dissolving Internal Silver Precipitates'. *Precision Engineering*, *64*(April), 108–112. doi:10.1016/j.precisioneng.2020.03.016

Kornilova, A. V., Gorbachevskii, M. V., Kuralbayeva, G. A., Jana, S., Novikov, A. A., Eliseev, A. A., Vasiliev, A. N., & Timoshenko, V. Y. (2019). 'Plasmonic Properties of Halloysite Nanotubes with Immobilized Silver Nanoparticles for Applications in Surface-Enhanced Raman Scattering'. *Physica Status Solidi (A). Physica Status Solidi. A, Applications and Materials Science*, *1800886*(14), 1–5. doi:10.1002/pssa.201800886

Koshy, J., Samuel, M. S., Chandran, A., George, K. C., Predeep, P., Thakur, M., & Varma, M. K. R. (2011). 'Optical Properties of CuO Nanoparticles'. *AIP Conference Proceedings*, *1391*(1), 576–578. doi:10.1063/1.3643615

Kumar, A., Kumar, M., Goyat, M. S., & Avasthi, D. K. (2022). A Review of the Latest Developments in the Production and Applications of Ag-Nanowires as Transparent Electrodes. *Materials Today. Communications*, *33*(September), 104433. doi:10.1016/j.mtcomm.2022.104433

Liu, H., Duan, C., Yang, C., Chen, X., Shen, W., & Zhu, Z. (2015). 'A Novel Nitrite Biosensor Based on the Direct Electron Transfer Hemoglobin Immobilized in the WO 3 Nanowires with High Length – Diameter Ratio'. *Materials Science and Engineering C*, *53*, 43–49. doi:10.1016/j.msec.2015.04.016 PMID:26042689

Magnusson, M., Paul, N., & De Nys R. (2012). *For Oil-Based Biofuel and Bioproducts Total Lipid and Fatty Acid Composition of Seaweeds for the Selection of Species for Oil-Based Biofuel and Bioproducts*. Wiley. . doi:10.1111/j.1757-1707.2012.01175.x

Marrese, M., Guarino, V., & Ambrosio, L. (2017). 'Atomic Force Microscopy: A Powerful Tool to Address Scaffold Design in Tissue Engineering'. *Journal of Functional Biomaterials*, *8*(1), 7. doi:10.3390/jfb8010007 PMID:28208801

Mukherjee, B., Varghese, B., Zheng, M., Karthik, K. R. G., Mathews, N., Mhaisalkar, S. G., Tok, E. S., & Sow, C. H. (2012). 'Synthesis, Characterization and Electrical Properties of Hybrid Zn 2GeO 4-ZnO Beaded Nanowire Arrays'. *Journal of Crystal Growth, 346*(1), 32–39. doi:10.1016/j.jcrysgro.2012.02.008

Núez, G., & Carlos, A. F. (2020a). 'Single GaAs Nanowire Based Photodetector Fabricated by Dielectrophoresis'. *Nanotechnology, 31*(22), 225604. doi:10.1088/1361-6528/ab76ee PMID:32187022

Ravindra, P., Baral, M., Biswas, T., Nand, M., Jha, S. N., Athresh, E., Ranjan, R., Jain, M., Ganguli, T., & Avasthi, S. (2019). 'Electrical and Optical Properties of Low-Bandgap Oxide Zn 2 Mo 3 O 8 for Optoelectronic Applications'. *Thin Solid Films, 677*(March), 95–102. doi:10.1016/j.tsf.2019.03.023

Salhi, B., Hossain, M. K., Mukhaimer, A. W., & Al-Sulaiman, F. A. (2016). 'Nanowires : A New Pathway to Nanotechnology-Based Applications'. *Journal of Electroceramics, 37*(April), 34–49. doi:10.100710832-016-0037-y PMID:32214875

Sudha, P. N., Sangeetha, K., Vijayalakshmi, K., & Barhoum, A. (2018). *Nanomaterials History, Classification, Unique Properties, Production and Market. Emerging Applications of Nanoparticles and Architecture Nanostructures*. Elsevier Inc. doi:10.1016/B978-0-323-51254-1/00012-9

Tiginyanu, I., Ursaki, V., & Popa, V. (2011). *Nanoimprint Lithography (NIL) and Related Techniques for Electronics Applications. Nanocoatings and Ultra-Thin Films*. Woodhead Publishing Limited. doi:10.1533/9780857094902.2.280

Vicente, A., Santano, D., Zubiate, P., Urrutia, A., Del, I., & Ruiz, C. (2019, April). 2020. 'Sensors and Actuators B : Chemical Lossy Mode Resonance Sensors Based on Nanocoated Multimode-Coreless- Multimode Fi Bre'. *Sensors and Actuators. B, Chemical, 304*, 126955. doi:10.1016/j.snb.2019.126955

Wang, Y., Lany, S., Ghanbaja, J., Fagot-Revurat, Y., Chen, Y. P., Soldera, F., Horwat, D., Mücklich, F., & Pierson, J. F. (2016). 'Electronic Structures of C U2 O, C U4 O3, and CuO: A Joint Experimental and Theoretical Study'. *Physical Review. B, 94*(24), 245418. Advance online publication. doi:10.1103/PhysRevB.94.245418

Weinberger, P. (2014). 'Niels Bohr and the Dawn of Quantum Theory'. *Philosophical Magazine, 94*(27), 3072–3087. doi:10.1080/14786435.2014.951710

Xiong, Y., Tang, H., Wang, X., Zhao, Y., Fu, Q., Yang, J., & Xu, D. (2017). Significantly enhanced thermal conductivity of indium arsenide nanowires via sulfur passivation. *Scientific Reports, 7*(1), 13252. doi:10.103841598-017-13792-4 PMID:29038573

Yong, Y., Cui, H., Zhou, Q., Su, X., Kuang, Y., & Li, X. (2018, June). 2019. 'Journal of Physics and Chemistry of Solids Ultrathin Nanowire Based on Icosahedral W @ Au 12 and Application as NO Gas Sensor'. *Journal of Physics and Chemistry of Solids*, *127*, 68–75. doi:10.1016/j.jpcs.2018.12.008

Zafar, F., & Iqbal, A. (2016). Indium Phosphide Nanowires and Their Applications in Optoelectronic Devices. *Proceedings of the Royal Society A: Mathematical, Physical and Engineering Sciences*. Royal Society of London. 10.1098/rspa.2015.0804

Zhao, Y., Li, L., Liu, S., Wang, J., Xu, J., Shi, Y., Chen, K., Cabarrocas, P. R., & Yu, L. (2020). 'Germanium Quantum Dot Infrared Photodetectors Addressed by Self-Aligned Silicon Nanowire Electrodes'. *Nanotechnology*, *31*(14), 145602. doi:10.1088/1361-6528/ab647e PMID:31860876

Zhong, Z., Yang, C., & Lieber, C. (2007). *Silicon Nanowires and Nanowire Heterostructures. Nanosilicon: new photonic and electronic material at the transition between solid and molecule*. Elsevier Ltd. doi:. doi:10.1016/B978-0-08-044528-1.50006-3

Zweibel, K., Moskowitz, P., & Fthenakis, V. (1998). *Thin-Film Cadmium Telluride Photovoltaics: ES&H Issues, Solutions, and Perspectives*. OSTI. https://www.osti.gov/bridge

Chapter 13
Applications and Types of Smart Materials

Rituparna Roy
University of Calcutta, India

Debjani Bhakta
University of Calcutta, India

Subhasis Roy
https://orcid.org/0000-0003-4197-535X
University of Calcutta, India

ABSTRACT

Different materials have been used for various purposes for the advancement of humanity ever since the beginning of human civilization. The term "smart material" encapsulates the change taking place in the field of material science. Smart materials are the key that can be used to unlock our impossible desires, such as the ability of a mobile phone screen to repair itself, the ability of some buildings to react to weather, the ability of some medical fields to release drugs into the bloodstream at the very first stage of a disease, etc. Because of its appealing and intelligent qualities, smart materials are known by other names such as responsive and intelligent materials. Overall, it can be concluded that it demonstrates a vivacious spirit and wise behavior. Temperature, pressure, tension, an electric or magnetic field, moisture, or pH can all cause these materials to react. This chapter discusses recent reviews of smart materials based on their state of the art. The authors also explore different types of smart materials and their various ways and sectors for their applications.

DOI: 10.4018/978-1-6684-9224-6.ch013

1. INTRODUCTION

Smart materials possess properties that can be altered in a controlled manner in response to external stimuli, such as temperature, light, electricity, pressure, or magnetic fields. These materials are designed to exhibit unique characteristics that go beyond the traditional properties of conventional materials. Based on sensing and recovering ability against a huge range of external stimuli, it is getting a huge demand and application in various fields. Its attractive properties like self-sensing, self-recovery, huge sensitive adaptability, cleverness, smartness, thinking ability, and multiple functionalities dominate conventional materials. It is also known for intelligent, active, and smart materials. Its general performance, with its inherent intelligence characteristics in some unpredictable places or environments, snatches our interest to reach our desired conditions. The presence of characteristics of integrational functionalities in their structure can feel and accepts the external or inner stimuli and react against them. It can be used as a diagnostic tool for their changing properties, which helps identify defects, cracks, and faults. For this, it can properly activate itself and compensate for the fault when embedded into a host material, this phenomenon is also known as self-repairing property (Shivaji S. Aher et al., 2015). Smart materials attempt a big role in terms of a sensor that can sense the disturbance that comes along with the input signal and actuators that respond against the disturbance by changing their position, shape, and frequency to make the system stable or pursue the steady state condition in which the system runs. The difference between the traditional structure and the smart material structure those are (i) traditional structures can perform in some general and certain environments or conditions like speed, life span, load, etc. In contrast, the smart material can work in some unpredictable environment. (ii) Another important difference is that traditional material is unable to change or modify itself under the introduction of any change into the system. In contrast, smart material is famous for its self-repairing property and self-sensing properties. This smart material can be called a self-efficient material. Smart materials have their thinking ability, and we can say they have mental alertness, quick perceptions, a huge activity in speed, high effectiveness, and all show spirited liveliness in nature. The smart structure has five components: (i) Data Acquisition, which means collecting raw data required for appropriate sensing and monitoring the structure. (ii) Data transmission, also known as sensory nerves of the system. Raw data is forwarded to the local or central command and control system by this part. (iii) command and control unit which is known as the brain of the smart material, to manage or control the total system by analyzing the data and reach to the appropriate conclusion. (iv) Data instructions, which are known as the motor nerves of the smart material which transmit the decision the member of the smart material structure. (v) Action devices, which can be said as muscles of the smart material as it triggers the controlling unit or devices. An attractive application of smart materials is repairing mobile phone screens on their own, drugs released on their own on sensing the sign of infection in the bloodstream, and some sportswear having some ventilation valves sensitive to temperature and humidity. It is up when the wearer breaks out in a sweat and closes when the body becomes cooled, etc (Jaishree Damodharan et al., 2018). Achieving those conditions shows the condition of dreams coming true.

2. A BRIEF HISTORY OF THE DEVELOPMENT OF SMART MATERIALS IN CONCERN WITH INDUSTRIAL IMPORTANCE

In the condition of trending multifunctionality property, researchers also want to find this property in the materials. The greed for finding multifunctional property everywhere and every field starts by noticing nature's behavior. Full, fill this property in the field of material is termed as "smart", so we can describe the material as having this multifunctionality property as a smart material. From the very initial stage of any research, it is required that the research material can pass through some unpredictable situations and can exist easily. Rogers, in 1988, conclude that smart material can exist easily by changing its physical properties in a well-maintained manner as a response to the introduction of any external stimuli (Rogers et al., 1988). Later, Takagi (in 1990) gave a conclusion by calling smart material intelligent material, as it can give response against external stimuli not only in a specific manner but also in an optimized way and an opium condition and deform its physical properties along with the demand of the environment created due to the introduction of the external stimuli (Takagi et al., 1990). Dagani 1995, was able to conclude the similarity between biopolymers and the smart polymer (Dagani et al., 1995). The first article on smart polymer was published in the early nineties when there was a collaboration of 6000 scientific works on smart polymer along with 500 and more numbers of review articles. Fairweather, in 1998 defined smart materials as those with the property to change their physical and chemical properties by changing their geometry and materialistic property when they are introduced to some change of electric field, magnetic field, thermal field, and the ability to acquire transducer energy inherently.

This history proves that the research craze on smart polymer and smart material is increasing daily, which indicates a huge jump in the material science research field that will lead us to a highly advanced and to our imaginary world.

3. SMART POLYMERS

The scenario of increasing demand for new materials and new technologies in the field of polymer is very warming. Polymers are versatile, and nowadays, they are available to us in different forms intended to achieve our desired situation by improving their characteristics and properties. Polymers in various types like natural polymers, synthetic polymers, and also in the form of semisynthetic polymers. In the field of smart materials, polymers also play a vital role. Smart materials are also available in polymer composites and polymeric nanocomposites. Defining a smart polymer is difficult, but those polymers can respond against external stimuli, are sensitive to environmental changes, and play a sensitive and active role in soluble and insolubility perspectives. Smart polymers already have huge applications in the fields of biotechnology, medicinal purposes, and also in engineering aspects. According to Galaev and Mattiasson (1999), life is a strong representation of polymer and its ability and polymers can explore its ability, properties, and magic in the name of life, as living organisms build of different natural polymers in the form of proteins, carbohydrates, and nucleic acids (Galaev et al., 1999). According to Kumar et al., (2007) for mimicking biopolymers and their properties, synthetic polymers are good replacement options for biopolymers to achieve our desired requirements in industrial and scientific applications (Kumar et al., 2007). Smart polymer is also known as "intelligent polymers" according to Kikuchi and Okano, (2002), "stimuli-responsive smart polymers" according to Hoffman et al., (2000), and also "environmental-sensitive smart polymer" according to Qiu and Park, (2001) (Kikuchi et al., 2002; Hoffman et al., 2000; Qiu et al., 2001). According to Aguilar et al., 2007, when the surrounding atmosphere changes slightly, smart polymers or stimuli-responsive polymers experience reversible, significant physical or chemical changes in their characteristics (Aguilar et al., 2019). Near the transition conditions, linear and solubilized smart macromolecules will transition from a mono-phasic to biphasic rise to reversible sol-gel states. When a smart cross-linked network transitions from a collapsed to an expanded state, chain restructuring takes place.

Smart surfaces with responsive interfaces alter their hydrophilicity in response to an external stimulus. All of these modifications may be used in the union of smart devices for a variety of uses, such as the development of low-invasive injection systems (Nguyen et al., 2010), pulsatile drug transportation mechanisms (Duarte et al., 2011), and novel substrates for cell culture and tissue engineering (Sato, Katayama, Suzuki, & Shiomi, 1998). Smart polymer behavior was studied using a different medium like water by Galaev and Mattiasson, in 1999, in polymer blends studied by Sato et al., in 1998, and in the organic solvent d by Osada and Gong, in 1998 (Galaev et al., 1999; Ougizawa et al., 1986; Osada et al., 1998; Vanderroost et al., 2014). In the water medium, the balance between hydrophobicity and hydrophilicity is the main reason behind the phase transition, which can be considered a response against external stimuli, using this property, researchers have become able to use smart polymers in the biotechnology and medicine application. Parallel with the application of smart material in biotechnology and medicine it also has application in the food packaging section. These active and advanced packaging forms a transfer of material relation between the packaging material and the food, which provides the extended life of the food material by releasing antimicrobials as an active material (Vanderroost et al., 2014). Many polymeric smart materials can effortlessly be functionalized by pre-polymerization or post-polymerization by incorporating receptors and other functional molecules into the structure (Arnold et al., 2012). All these achievements create a burning research thirst on this topic. Table 1 shows a compact overview of different smart polymers' classification, principles, synthesis, properties, and applications.

4. SMART CERAMICS

At a nearly exponential speed, silicon-based circuits and silica-based optical assemblies are being merged with electro-ceramic parts and gadgets on a smaller scale. The materials we use are a product of an inorganic evolution that will soon be more complicated and on a larger scale than biological evolution. One of the thrilling technical adventures of the next century will be how these smart ceramics will mix with human culture, potentially leading to a hybridized kind of longevity (Newnham R. E., 1999).

Smart ceramic components are mainly utilized as actuators and sensors and in dental applications, sometimes bone replacement applications. Actuators transform the system's electrical information into various physical traits appropriate for their surroundings. Sensors transform the physical properties of their surroundings into electrical signals that the system can use (Newnham et al., 1990) (Figure 1). The smart ceramic material can be in the form of smart bio-ceramic material and piezo-ceramic material. Piezoceramics are always polycrystalline. According to model-based control law, the patches in the piezoceramic components can sense the strain and, as a response, can also show bending results. In the field of transducers, piezoceramics play some role. To use material above the curie point, the point above which a material loses its ferroelectric and piezoelectric property, its piezoceramic plays an attractive role. Ceramic material shows a suitable polarization value, which is why it best fits sensing devices. Lead zirconate titanate (PZT), and lead metabolite are ceramic materialized in the production of the ring, disc, and plate form. Ceramics has huge applications due to their easy availability and low manufacturing cost. Some of the recent uses of smart ceramics have been briefed in Table 2:

Table 1. Classification, principle, synthesis, properties, and application of smart polymers

Name	Classification	Principle	Synthesis	Application
Temperature-responsive Polymers (Hoogenboom, 2019)	Shape memory materials	On being heated above the greatest glass transition temperature, shape-memory materials may be distorted in any shape, creating a permanent shape. Formation of a transient shape when deformed amid the two different transition temperatures, which can then be fixed by freezing the distorted state under the switching temperature.	Free radical polymerization, Cationic, and anionic polymerization, Copolymerization, Controlled radical polymerization, Reversible Addition-Fragmentation Chain-Transfer (RAFT) polymerization, Atom Transfer Radical Polymerization (ATRP), Ring-opening polymerization	Polymeric thermometers, Controlling and modifying the side-chain functionalities (ex., Binding properties) in different molecules, Controlling and examining the osmotic strength and enabling the reversible transport of water through a permeable membrane in polymeric solutions, Preparing thermos-responsive hydrogels, Controlling the absorption and release of water Collecting water from fog Designing multi-responsive polymers
	Liquid crystalline materials	Liquid crystalline phase along with glassy or rubbery state Anisotropic order of mesogens comprising the polymer Up to 40% compression can be achieved upon heating.		
	Responsive polymer solutions	The transition from a homogeneous solution into a concentrated and dilute phase is accompanied by a transition from a clear to the cloudy solution at the cloud point temperature. Phase transition upon heating indicates lower critical solution temperature (LCST) Phase transition upon cooling indicates higher critical solution temperature (LCST)		
pH-responsive polymers (Fernández et al., 2019)	Basic	Comprises of primary groups (ex., Imidazole, and pyridine compounds) Can alter the number of groups that are positively charged as a response to changes in the external pH	Emulsion Polymerization (Micro-emulsion, Mini-emulsion), Reversible addition-fragmentation chain-transfer (RAFT) polymerization, Atom transfer radical polymerization (ATRP)	Drug delivery systems Gene carriers Biosensors Chromatography Controlling surface wettability of various compounds after modifying pH-responsive polymers Nano bio-fouling membranes after modifying pH-responsive polymers
	Acidic	Comprises of acidic groups (ex. Carboxylic and sulfonic acids, sulfonamides)		
	Natural	They are desirable due to their prevalence in nature, strong biocompatibility, non-toxic property, and ease with which they may be altered by basic chemistry. (ex. Chitosan, alginate, hyaluronic acid) To external stimuli, they respond linearly in an all-or-nothing fashion. Small changes are made in response to changing parameters up to a critical point, after which there is no noteworthy response from the system because the transition occurs within a restricted range of the parameter variance.		

continues on following page

Table 1. Continued

Name	Classification	Principle	Synthesis	Application
Photo-responsive polymers (Xiong et al., 2019)	Main chain conjugated polymer (photochromic)	Photosensitive groups may be added to the framework of a conjugated chain of polymers to change the material's optoelectronic capabilities between a non-conjugated and a conjugated structure. (ex. Diarylethene, spiropyran)	ATRP, Copolymerization, click chemistry, UV irradiation, photo-treatment, conjugation techniques, Dimerization	Controlled drug delivery Functional micropattern on a substrate Polymeric hydrogels responsive to stimulus Photodegradable substances Incorporation of photo-switchable units into liquid crystalline polymers for remote actuation
	The photo-responsive terminal group in polymers	A polymer chain's terminal photochromic groups have been employed to modify the polymer chains' physical characteristics. The end-group effects of this approach are only present in polymer chains with low molecular weights, and they disappear as molecular weight increases. (ex. 2-diazo-1,2-naphthoquinone) Shows fluorescence resonance energy transfer (FRET) characteristics		
	Side chain photochromic polymers	Chain loops that develop in response to light have less interchain entanglement and more cloud points. (ex. Copolymerization of di-methyl-aminoethyl methacrylate with a methacrylic monomer based on coumarin)		
	Photochromic side chain block copolymers	High self-assembling characteristics The majority of these systems were created to control micelle production by modifying the chain's hydrophilicity and hydrophobicity and balancing them. (ex. a block copolymer made of a poly(t-butyl acrylate-co-acrylic acid) sequence that is randomly distributed		
	Dendritic polymers	Low solution or melt viscosity levels, strong solubility, a huge population of terminal functional groups, heavily branched polymers with controlled topologies, and dendritic polymers are all characteristics. Through the use of artificial procedures, the size, extent of branching, as well as functionality can all be modified and controlled. (ex. Azobenzene core-based dendrimers)		
	Supramolecular polymers	These are created when telechelic polymers with 2-ureido-4-pyrimidone (UPy) units as their terminals self-assemble. UPy may dimerize through highly affine quadruple H-bonding, creating lengthy chains. The proportion of monofunctional UPy introduced to a solution, which regulates the extent of the polymerization and, consequently, the viscosity, shearing impacts, and viscoelastic qualities.		

continues on following page

Table 1. Continued

Name	Classification	Principle	Synthesis	Application
Enzyme-responsive polymers (Asha et al., 2019)	Supramolecular assembly Polymer hydrogels Polymer nanoparticles Self-immolation polymers	Enzymatic processes can influence the way that stimuli-responsive polymers change structurally. One of the most effective techniques is the construction of micelles and other supramolecular structures from amphiphilic block copolymers. On incorporating an external stimulus, the functional groups composing the responsive block shift from double-hydrophilic towards amphiphilic. By changing the solubility of the nanostructures, it is possible to modify their size and form, which gives the enzymatic process substantial influence over the triggering process with an elevated level of selectivity.	Living Anionic or cationic Polymerization, Living Free Radical Polymerization, ATRP, RAFT polymerization, Nitroxide mediated polymerization	Drug delivery Tissue Engineering Injectable scaffolds
Inflammation-responsive polymers (Cano et al., 2019)	Passively targeted polymer	To encourage aggregation at the illness site, the polymeric system time for circulation needs to be increased. Inflammatory cells' sequestration and consumption of the polymer-drug complex play a critical role in retention rates. Simple internalization may be accomplished by customizing the polymeric carriers' shape, dimension, and surface.	ATRP, Copolymerization, RAFT, conjugation techniques, Redox reactions	Drug delivery Treatment against inflammatory diseases Bioactive films as implant coating for reducing the foreign body responses to implanted biomaterials
	Cell-mediated targeted polymer	Prolonged cell life exposure Spherical Nanoparticles are more suited for the technique requiring endocytosis, while non-spherical counterparts are favored for the leukocytes' interface attachments. Rounded forms promote cell internalization of nanocarriers.		
	Redox responsive polymers	Inflammation-sensitive output is facilitated by sulfur-based polymeric materials' sudden shift in polarity while sulfur (II) is oxidized to sulfoxide (IV) or sulfone (IV), and a new implantation with embedded nanoparticles was developed to lessen the post-implantation external body response to microfabricated cortical neuroprostheses. Fast tissue growth rate and biodegradation (ex. Poly-oxalates)		

Figure 1. Working process of an actuator and a sensor

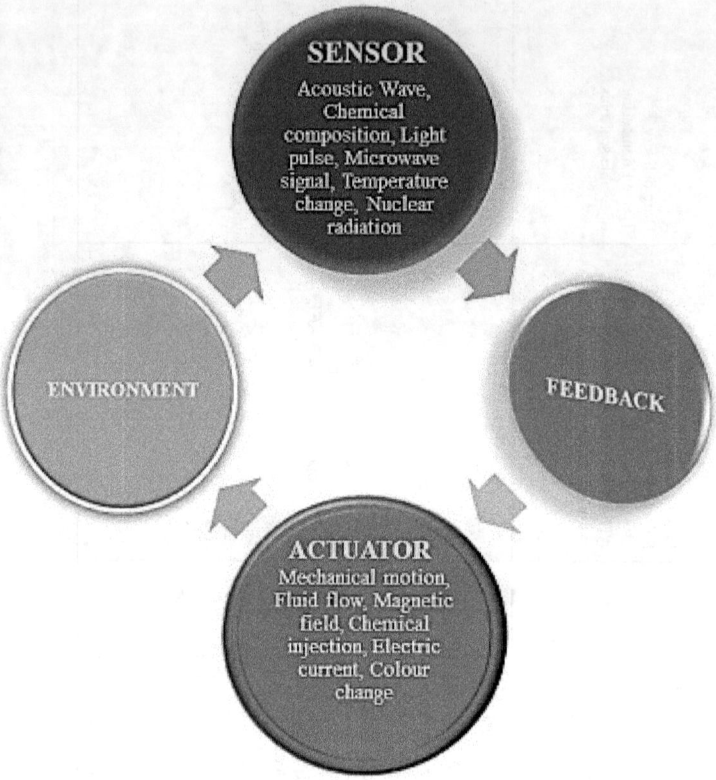

Table 2. Recent uses of smart ceramics

Device	Uses and principle
Hydrophone materials (Newnham et al., 1990)	Piezoelectric components create hydrophones, underwater listening instruments that react to hydrostatic pressure waves. Sonar devices for submarines, geophysical prospecting gear, fish finders, offshore oil platforms, and seismic sensors are just a few uses for hydrophones. Effective hydrophone transceivers will be used to record and broadcast fish communication and track the development of underwater ecosystems.
Modulated suspension systems (Newnham et al., 1990)	Toyota's piezo-TEMS (i.e., Toyota Electronic Modulated Suspension) system, intended to increase the car's drivability and stability while enhancing travelers' comfort, uses regulated conformity to piezoelectric ceramics. Essentially being a highway stability sensor and a shock adjustor, the TEMS quickly modifies the shock absorbers to deliver a lighter or more firm damping force, based on what is required to reduce distress while retaining control over the vehicle. As driving situations change, shock absorbers can be repeatedly readjusted to prevent rocking and trembling on soft jolts.

continues on following page

Table 2. Continued

Device	Uses and principle
Electro-rheological fluids (Newnham et al., 1990)	Electrorheological fluids are solutions of small elements in a fluid media; applying an electrically charged field may significantly alter the suspension's viscosity. Particles are aligned in fibril-like divisions in the path of the supplied electric field. When the electrical field is eliminated, the alignment vanishes, resulting in the required attribute of full cyclic repeatability. Electrorheological fluids are a sophisticated category of self-tuning composites that will be extensively used in vibration control applications (Kumar, Bhalla, & Cross, 1994). This sort of device is also appealing for incorporation into multi-function, self-sufficient smart materials due to its easy integration with contemporary solid-state electronics.
Actuators (Newnham et al., 1990)	Materials with piezoelectric properties, electrostrictive materials, magnetostrictive materials, shape-memory metal alloys, and thermally adjustable materials can all be utilized to regulate actuation strain. Creating structures with built-in vibration and form control capabilities is feasible using a system with distributed actuators. Shape memory metals and ceramics are two of the most significant actuator materials. Alloys that go through thermoelastic martensite transitions show the shape memory effect. A body-centered cubic metal undergoes a first-order displacive transition in which, upon cooling, it shears into a martensitic phase. Shape memory alloys can recover deformation during the martensitic low-temperature phase and return to their original shape once heated to a temperature when martensite returns to its initial body-centered cubic form.
Aerospace applications (Hero1dschmidt', et al.)	The specific need for aerospace constructions is to achieve great rigidity and tensile strength at minimal weight. Thus, carbon fiber reinforced plastics (CFRP) are used to make a variety of present-day components, including helicopter blades, aircraft rudders and tails, spacecraft antennas, and various payload-carrying components. Many of those functions involve damping vibrations which are applied to structures. Active approaches must be used since passive damping frequently has low efficiency. Structures that might profit from active damping include sensitive space payloads that experience high acoustic loads throughout the launch, vertical aircraft tails that are subject to buffet vibrations, solar arrays on satellites that have very low intrinsic damping rates and are susceptible to thermal shocks, and aircraft panels that flutter when operating at supersonic speeds.

5. SMART CARBON MATERIALS

Due to their relatively low density, excellent conductive properties, mechanical adaptability, and durability in harsh conditions, 3-dimensional and porous materials made of carbon, like graphene as well as aerogels based on carbon nanotubes, are currently showing great potential for a variety of purposes, including actuators, sensors, insulation from heat, and transforming energy. Mechanical versatility is crucial in determining their robustness and dependability in real-world applications. As a result, compressive fragility has been effectively addressed, and several sophisticated carbon blocks have been created with amazing super-compressibility and super-elasticity. Some of the recent innovations in smart technologies using carbon materials have been elaborated in Table 3 below.

Table 3. Advantages of recently innovated smart carbon materials

Material	Advantages
Carbon-nanotube-reinforced Cementous compounds (Han, et al., 2016)	Carbon nanotubes (CNTs) are carbon graphite tubes made at the molecular level and have exceptional qualities. They exhibit amazing electrical properties, and are one of the stiffest and strongest fibers ever discovered, amongst numerous other distinctive qualities. These characteristics, together with the CNTs' prominent aspect ratio, short diameter, lightweight, and exceptional chemical along with thermal properties, make them suitable for augmentation in the creation of multipurpose and intelligent materials based on cement. Exceptional mechanical, piezoresistive, electrical, heating, and damping capabilities may be found in CNT-reinforced cement-based products.
Smart vibration/ magnetic sensors (Huai-Ling Gao, et al., 2021)	The compression along with the torsional fragility of a pore-filled carbon-based bulk may be successfully averted by creating a novel prolonged lamellar multi-arch microscopic structure, resulting in a true spring-like carbon-based substance with dependable strength and longevity. The carbon spring is a useful foundation for building powerful contact vibration and noncontact magnetic sensors due to its exceptional mechanical efficiency, superior conductivity, and ease of modification. Additionally, both sensor types can function in conditions with severe temperatures, which can suit the needs of specialized missions like space missions.
Smart textiles (Zhai, et al., 2015)	Pitch or polyacrylonitrile-based carbon fibres provide superior physical and electrical qualities. Active carbon fibres may be employed to create all-carbon solid-body yarn supercapacitors. The mix of carbon compounds in the fabric electrode was tuned to make yarn supercapacitors with a unique longitudinal capacitance in addition to gravimetric capacitance. The constructed yarn supercapacitors' electrolytic and physiochemical characteristics were assessed, and the results showed some of the greatest specific longitudinal capacitance along with total capacitance values as compared to any fiber supercapacitors that have ever been documented in the available literature. Three lengthy fiber supercapacitors have been weaved into wearable bracelets that can illuminate an LED indication because of their exceptional mechanical durability, revealing their versatility in smart textiles.
Healthcare materials/ Medical research (Oka, Chang, Gao, & Chung, 2015)	The idea of using biochar in medical facilities is supported by the extensive tradition of charcoal usage in medicines and its powerful capacity to absorb pollutants. The current use of charcoal-based composites in the form of biomaterial to encourage differentiated growth of human stem cells from embryos significantly suggests the effectiveness of biochar for medicinal uses. Active carbon is employed for the normal detoxifying or cleansing of poison or for a drug overdose. Deoxynivalenol, a frequent crop contaminant that can arise in the growing process or when in preservation, can cause cytotoxicity and DNA damage. However, the active carbon itself is also efficient in protecting against these effects. In medical research, the concept of a smart biochar platform may be used for things like filler media as well as in mediums for delivering drugs.
Semiconductor photocatalysis via carbon dots (Yu, et al., 2016)	Due to their simplicity in formation, distinctive framework, adaptable composition, simplified surface modification, exceptional electron-transfer effectiveness, as well as an adjustable light-harvesting spectrum, carbon dots have recently become recognized as an emerging category of metal-free photocatalysts. They exhibit semiconductor-like photoelectric characteristics and exhibit outstanding activity in an extensive number of photoelectrochemical as well as photocatalytic fields.

continues on following page

Table 3. Continued

Material	Advantages
Energy storage (Zhang, Lv, Tao, & Yang, 2013)	The electrochemical efficiency of supercapacitors depends upon carbon materials, and improving performance calls for carbon electrodes featuring an optimal structure. Three of the most crucial characteristics of carbon-based materials are their surface chemistry, the amount of surface area, along with pore geometry. The appropriate structure of the pores determines the ion propagation and rate effectiveness of the supercapacitor. In contrast, a large surface area leads to greater capacitance due to its numerous readily available active sites. Henceforth, the superior conductivity, large area of contact, and porosity of nanomaterials composed of carbon make them ideal for preserving energy.

6. SMART NANOMATERIALS

As there was a reference to carbon nanotubes in cementitious compounds in the previous section for smart carbon materials, it is evident that nanomaterials play a great role in the innovation of smart technologies. Due to the extraordinary thermal, electrical, optical, and physical capabilities, smart nanostructures have emerged in the 21st century as one of the outstanding substances. Additionally, nanotechnology is important for innovation in sustainability and environmentally friendly development. It is considered a way to produce non-nano substances and goods that are more ecological or to assist in resolving financial, ecological, and societal problems like the lack of energy supply or freshwater contaminant problems (Gottardo, et al., 2021). Recently nanomaterials are also being used in smart food packaging all throughout the world. Silver nanoparticles of different sizes and forms have drawn interest in developing hydrogel-forming polysaccharides. Hydrogels made of polysaccharides are intelligent substances often employed in everyday life, medicine, and cosmetics. Nanoparticles are also increasingly used in biochemical, spectroscopic, and clinical analysis in medicinal research (Sharma & Hussain, 2018). Recent advances in nanomaterials have made new cancer treatment and diagnostic technologies possible. Nanoscale instruments easily engage with biomolecules inside as well as outside of cells due to their tiny size. Fullerenes and its derivatives are examples of nanoparticles that interact well with the body's antibodies and work as chemotherapy drugs. On the other hand, some nanomaterials may specifically use the particular tumor environment, including their improved permeability, retention impact, and the particular acidic milieu, to deliver active medicines to cancerous cells. The upcoming types of nanoparticles have become the subject of intense research and hold great promise for developing individualized and customized cancer therapies (XiaoHui, Cheng, Laurent, & ChunYing, 2010).

7. SMART BIOMATERIALS

On innovating anything technology or other things, a general environmental concern challenges the researcher about the environmental friendliness of this outcome product. Considering the environmental concern, researchers tried to use natural resources to make the product as a raw material. Inorganic materials can be used along with this application for improving different physical properties like electrical and chemical properties and sometimes for the mechanical properties of those materials. Cellulose is the most available material in the environment. It can be available from plants, and it has some attractive characteristics due to being renewable, biologically compatible, and as it is a biopolymer it is biodegradable also, the uses of the derivatives of this include protein immobilization, immobilization of antibodies, formation of the optical film, coating and in various application, make it suitable for using it as a biodegradable raw material (Klemm et al., 2005; Robert J. Moon et al., 2011). Those applications make cellulose a competitor to synthetic polymers. Bacterial cellulose is one such smart biomaterial made out of cellulose. Smart biomaterials have a wide range of applications (Figure 2). Cellulose parallel to the synthetic polymer can also show active characteristics that make it useable as a sensor and actuator (Kim J. Y. et al., 2006).

Cellulose is used as active material in applying electroactive paper (EAPap). This EAPap has easy modification characteristics along with low production cost. It shows bending displacement when a voltage is introduced to the electrode. This bending displacement also depends on the frequency and adhesive nature. This EAPap works based on different interactions between cellulose and paper materials. Interactions like the dipolar moment, piezoelectric effect, and ion migrations. Cellulose shows the piezoelectricity equivalent to quartz crystal (Fukada, 1955; Bazhenov, 1961). It is also able to create self-powered vibration and can able to act as a transducer having self-energy scavenging properties. Based on those characteristics, it can be used in flexible speakers, electronics, etc (Kim J. et al., 2007; Suresha et al., 2008; Kim J et al., 2006). The coating of multi-walled carbon nanotubes improves the outputs of the EAPap (Yun S. et al., 2006; Yun S. et al., 2009; Yun S. et al., 2011). Sometimes the composites of zinc oxide with EAPap or the coating of zinc oxide on EAPap can increase the performance and show new features and properties (A. John et al., 2015; A. John et al., 2011; Mun et al., 2015). Bioceramics plays a vital role in bone tissue regeneration. Bioceramics and bioactive glasses known as BAG are combinations of active inorganic materials and biodegradable polymers (Cao et al., 1996; Zheng Y et al., 2017; Saveleva et al., 2019). Examples of bioceramics are calcium phosphates and hydroxyapatite (HA), etc. BAG has had a clinical application since 1985 and Larry Hench first invented it in 1969 (Hench L. L et al., 1971; Hench L. L et al., 2006). Undergoing some surface reactions by developing adhesive bonds between the hard and soft tissue, BAG can form an apatite-like layer in the living organism. BAG also has applications in the dental field for implant coating and reinforcing agents and the bone replacement field, periodontal surgery, and the treatment of intervertebral disc disease, etc (Popa et al., 2017; De Caluwé et al., 2017; Kim D.-A. et al., 2017; Najeeb et al., 2016; Kumar V. P. et al., 2017; Rosenqvist et al., 2017; Ke et al., 2018; Malat et al., 2018). Biomaterials are also going parallelly in various ways with synthetic polymers or other materials, inspiring researchers to research more intensely in this field.

Figure 2. Applications of smart biomaterials

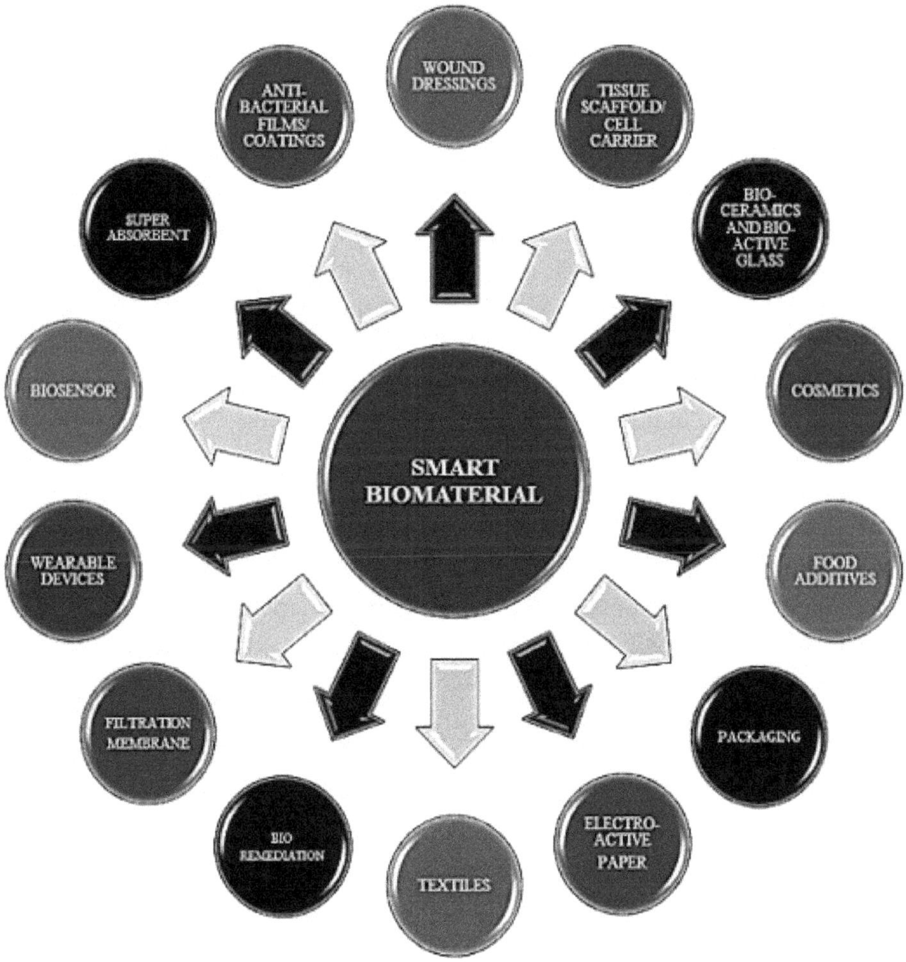

8. SMART BIODEGRADABLE AND BIO-BASED MATERIALS

In response to the rising demand for readily available and better-quality foods, several novel concepts concerning food packaging, such as smart features, have been created recently. The market for certain bio-based and biodegradable plastics is anticipated to develop dramatically over the coming decades (e.g., Bio-polyethylene), while others are anticipated to converge (e.g., polyamides) due to low manufacturing capacity. The ability to produce bioplastics worldwide is anticipated to rise, from 2.11 million tons in 2019 to 2.43 million tons by 2024. Over the past ten years, various techniques have been proposed to make bio-based materials, which may either be made from bio-based sources, be biodegradable, or possess both traits. These materials can have a wide range of properties and uses. The detailed classification of biodegradable and bio-based plastics can be seen in Figure 3. Proactive coats and physical/chemical sensors paired with the packaging materials are examples of smart functions of food packaging. While sensors play a part in tracking physical and chemical factors that impact or reflect the overall quality of food items, smart antimicrobial coats extend the longevity of products (Halonen, et al., 2020). These modifications inevitably have favorable health, ecological, and socioeconomic consequences, which can be strengthened even more by carrying out clever functions utilizing resilient technology and sustainable natural resources (Ajay, Parveen, Ahmad, Sharma, & Gambhir, 2023). Dodecanol was shown to work well as a nucleating substance for both the polycaprolactone and polylactide structures, enabling these structures to store a significant quantity of heat with no supercooling. These novel ultra-thin-structured biomaterials hold promise as energy-storing systems to effectively encapsulate and wrap temperature-sensitive products within refrigeration systems and serve as intelligent food, medical as well as pharmaceutical packages (Perez-Masia, Lopez-Rubio, Fabra, & Lagaron, 2013).

A biologically compatible and biodegradable polymeric stent that was 3D printed was recently created with a wireless pressure sensor embedded into it. Because the Poly-D-lactide polymer can disintegrate and get absorbed by the body gradually, it was often utilized as an ecologically friendly drug covering material. The Poly-D-lactide powders were dissolved into tetrahydrofuran solution to create a thin coating, which was then applied to the sacrificial layer to create the pressure sensor. A complicated mesh framework was created using a biocompatible substance (Polycaprolactone) and the method of 3D printing. The wireless pressure sensor demonstrated excellent linearity within the pressure gap of 0-120 mm Hg, a sensitivity of 50 kHz/mmHg, as well as an oscillatory frequency in the region of 137-150 MHz. Due to its great flexibility, the manufactured wireless sensor may be intimately connected to the innermost layers of the stent and eventually become thoroughly absorbed by the tissues of the body (Park, Kim, Park, & Lee, 2018).

Figure 3. Categorisation of bio-based plastics for food packaging

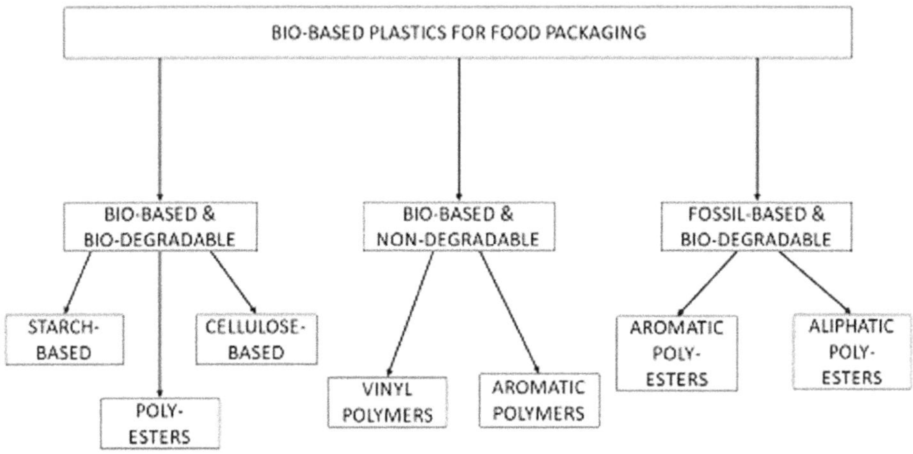

9. SMART METALS AND METAL OXIDES

Due to their exceptional behavioral advantages, metals and metal oxides have already gone a long way in contributing to smart technologies. In terms of adsorption, as well as absorption traits, magnetic operations, chemical redox reactions, ligand sequestration, and other characteristics, nanoparticles made of metal and metal oxide display special qualities (Sarkar, Sengupta, Guibal, & Quignard, 2012). To do this, efforts have been initiated to utilize them for a variety of purposes namely as catalysts for the manufacture of biodiesel, alkaline water electrolysis, dye-sensitized solar cells, screens, and smart windows in addition to lithium-ion batteries, super-capacitors, as well as biosensors (Vayssieres, Hagfeldt, & Lindquist, 2000). The varied applications of smart metals and metal oxides have been elaborated in Figure 4.

It has been suggested that alkaline water electrolysis, utilizing power from renewable resources, is a green approach for producing the significant amounts of hydrogen gas needed by a potential hydrogen economy. It has been investigated how oxidized Ni, Fe, or Co oxides behave in an alkaline environment in terms of redox, charge retention, as well as electrocatalytic behavior. First, hydrous micro-dispersed oxides are easily formed using the repeating cyclic potential sweep approach with an initial metal (Fe, Co, or Ni) in an aqueous alkaline solution. An inner, mainly anhydrous compact oxide and an outer, hydrated micro-dispersed oxide layer that demonstrates significant electrocatalytic activity about the anodic evolution of oxygen make up the dual oxides/solution interface. Secondly, electrochemical and ecological variables like lower level as well as higher potential sweep limits, the achievable sweep percentage, the baseline concentration used, the mixture's temperature, & the pH of the solution affect the charge penetration capabilities exhibited by the hydrous oxide. Thirdly, the acidic and basic responses of anodically produced transition metal oxides are crucial when addressing redox cycling and oxygen evolution. Hydrous oxides are much harder to reduce than less-hydrated and compact materials, which can greatly impactmpacthe cathodic ORR's catalysis (Rebouillat, Lyons, Brandon, & Doyle, 2011).

Roughly 15% of the global energy use goes to cooling and heating buildings. Improved adaptive and switchable windows can assist buildings in using energy as efficiently as possible for their temperature control and lighting functions by enabling the autonomous management of solar energy. Because of their effectiveness as well as adjustable control over the near-infrared portion of the sun's spectrum while not impacting the dynamic visible transmission of the smart windows, energy-effective plasmonic electrochromic smart window systems (PESWs) constructed from metal oxide nanocrystals are currently attracting lot of interest. The PESW technology centered around metal oxides has progressed quickly during the past ten years. Recently, scientists have created several PESWs from metal oxides such as ITO, titanium dioxide, tungsten oxide, cadmium oxide, and niobium oxide (Wang, et al., 2021).

Using Sr-Ti combined metal oxide as a heterogeneous base catalyst in the transesterification process was effective to generate biodiesel from used cooking oil. To identify the physical and chemical features of the catalyst, powder XRD, XPS, SEM, EDX, FT-IR, and BET surface area analyses were carried out. The polymer precursor process was alsoptimized to get the maximum fatty acid methyl ether conversion (Sahani, Roy, & Sharma, 2020).

Solid-binding peptides (SBPs) have become increasingly popular as building blocks of molecules in nanobiotechnology during the past ten years. The exteriors of a wide variety of solid substances, comprising metals (Au, Pd, Pt, Ag), metal oxides, metal complexes, magnetic compounds, semiconductors, carbon-based polymers, as well as minerals, are highly receptive to such peptides and exhibit selectivity. They are endowed with the power to coordinate the synthesis and building of nanoparticles and intricate nanostructures, and they can regulate the assemblage and functioning of materials. The possible uses for solid-binding peptides increase as quickly as the supply of freshly created nanomaterials (Care, Bergquist, & Sunna, 2015).

Figure 4. Prime applications of smart materials and metal oxides

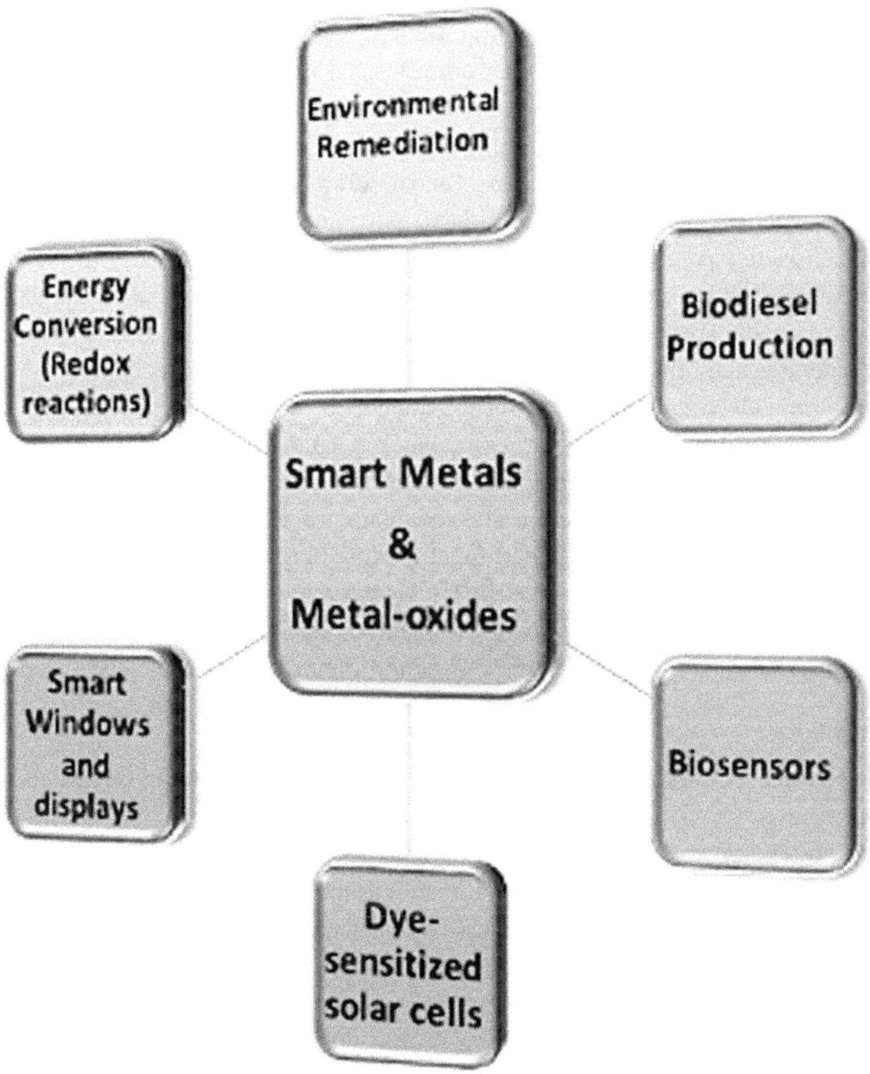

10. SMART COMPOSITES

Smart composite is a mixture of materials such as metals, and polymers, reinforced with fibers and particles. It includes some smart components, including sensors, actuators, and microelectronics. These components play the main role on behalf of the composites to sense changes happening in their environment, take actions against those changes, and adapt their properties accordingly. Intelligent composites or multifunctional composites are the other names of smart composites. It is an advanced material that combines traditional composites with integrated sensing, actuating, and adaptive capabilities. These composites provide some improved performance and functionality compared to other conventional materials. One key application of smart composites is to monitor structural health. The sensors embedded in the composites can detect stress, strain, or damage, allowing the smart composites for real-time structural integrity assessment and timely maintenance or repair if required. Smart composites enable shape control and morphing through shape memory alloys or electroactive polymers. These properties allow the smart materials for adaptive structures in aerospace and other industries. Energy harvesting and storage is another application of smart composites. When piezoelectric or photovoltaic elements incorporate each other, the resulting composites can convert mechanical or light energy into electrical energy, powering embedded sensors and reducing the need for external power sources. Smart composites are another engineering solution for actively controlling temperature gradients in structures like electronics or aerospace systems for thermal management. The advantages of smart composites in terms of development and use are enhanced functionality, play a role in weight reduction, improved efficiency, and increased structural integrity. These materials can potentially revolutionize industries such as aerospace, automotive, energy, civil infrastructure, and consumer electronics, enabling the creation of innovative and adaptive products.

11. FUNCTIONALLY GRADED PIEZOELECTRIC MATERIALS

Piezoelectricity was first discovered by the brothers Jaques and Pierre Curie in 1880 on the Rochelle salt, and they defined it as the change of electric polarization influencing the change of applied stress later, it became very use full for playing the role of a smart character. Piezoelectric material is the most common type of smart material that can respond by producing voltage against the introduction of external stimuli in terms of change in the stress or the application of stress, and vice versa is also true. It is mainly used as the sensor and actuator. Lead zirconate titanate ($PbZr_{1-x}Ti_xO_3$) is a famous type of piezoelectric material. It plays an important role in forming the n-type and p-type material, which leads to a huge application of the piezoelectric material with a better dielectric value and the compressive forces that are relatively attached to the crystal structure.

The device becomes advanced in terms of its optical adaptivity, hydrophones, sonobuoys, fuse device, sounders having the depth-sensing ability, thickness measuring properties, flaw detection, level indicating property, alarm systems, strain gauge, fatal heart detector, pressure indicator, etc. piezoelectric material is of two types they are piezoceramics and the piezo-electrostriction. The difference between them is the representation of reverse characteristics of the electric field. Microscopically we can say that a piezoelectric effect can be observed in the crystal in the form of displacement of ionic charge. The crystal's charge distribution is symmetric until external stimuli are introduced as strain. Otherwise, it will also show a dipole moment value of zero. If there is an influence of external stress, the charge will be displaced, the charge distribution also becomes unsymmetrical, and the dipole moment value will be hampered. All the piezoelectric crystal materials have a perovskite structure. So, on the other hand, perovskite has a smart structure and it can also be a smart material. Perovskite can be prepared in open-air conditions (Moyez S. A. et al., 2018).

Figure 5. Sensory applications of perovskites (Reprinted with permission from (Shellaiah et al., 2020))

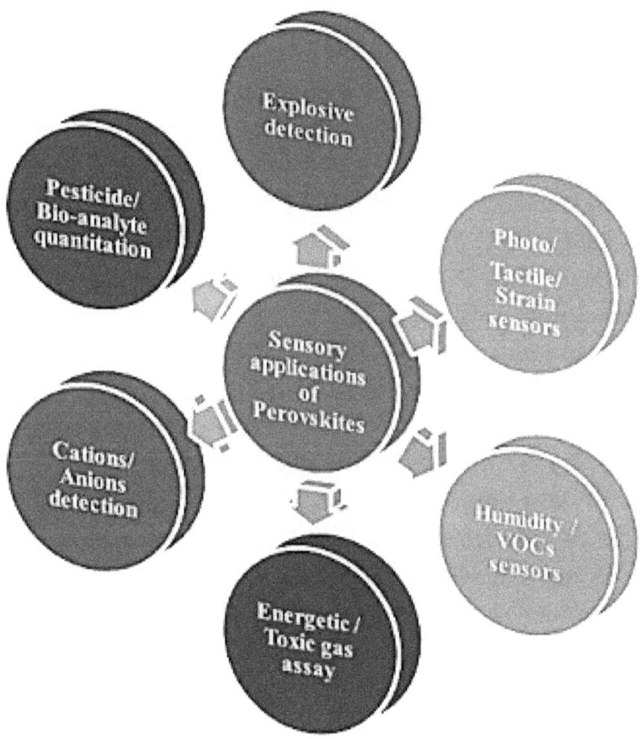

Perovskite has huge advanced applications. The perovskite material $CH_3NH_3Pb_{(1-x)}Sn_xCl_3$ has some attractive optical properties in thin film perovskite for photovoltaic applications (Moyez S. A. et al., 2020). Perovskites also play a role in photocatalytic water-splitting with the property of forming an electron transport bilayer in the case of hematite having TiO_2/CO dropped (Roy et al., 2018). Perovskite also plays an essential role in defect engineering, having a photovoltaic application (Bera et al., 2022). In demand for thin film, perovskite lead-free perovskites can equally contribute to their role (Moyez S. A. et al., 2017). Perovskites also have various sensory applications (Figure 5) (Shellaiah et al., 2020). Overall researchers get a huge scope to research in this field and can use this material broadly.

12. ADVANCES IN 4D PRINTING USING SMART MATERIALS

Scientists have produced complicated, bio-inspired, multi-material designs, remotely controlled robots, layouts of drug delivery mechanisms derived from algorithmic optimization, and microenvironments for bio tissues via 3D printing (Kumar, Mittal, & Haleem, 2022). A varied range of biomaterials can be used in additive manufacturing techniques (Kumar, Kumar, Mittal, & Gambhir, 2023)Traditional 3D printing techniques concentrate on creating static structures using a single material, which is insufficient to fulfill the requirements of dynamic functionality required for products like soft surgical grippers, self-folding packaging, and adaptable wind turbines. Momeni et al.'s research indicates that printing wind turbine blades using shape memory polymers (SMP) can significantly enhance blades' structural and mechanical characteristics. A blade may change shape whenever heated to adjust with the wind and it keeps its original form on being reheated. A smart blade is created in a single print cycle without electromechanical components or post-assembly. This type of functionality gains from the addition of time, the fourth printing dimension, that permits change. The concept of an additional dimension gave rise to a new field of research in additive manufacturing (AM) known as four-dimensional(4D)-printing. As a result, 4D printing produces metamaterial frameworks with various overlaid structural reactions brought on by modifications in their operating environment (Zhang, Demir, et al., 2019). The material that is selected for the printer should be compatible. This is evident in most situations where the structure's dynamism is produced by incorporating a stimulus which will cause an alteration in form or functioning, thanks to its shape memory behavior. In certain other situations, the deformation or mismatch brought on by a variance in the material's characteristics, for example, its coefficient of expansion under heat, is the basis for creating the dynamism of the component. The success of 4D printing depends on a certain stimulus. Numerous stimuli, including light, heat, microwaves, water/moisture, magnetic fields, etc., are suitable for smart materials. Since most stimulus reactions are not inherent to the material, fillers like carbon nanotubes (CNTs), magnetite, silver nanowires (AgNWs), gold nanoparticles (AuNPs), etc. are used to tune the responses. By using 4D printing, objects that are 3D-printed may change under the influence of a stimulus, such as their form or functionality. The majority of technical and medical uses in which dynamic instead of static elements are needed benefit from this (Melly et al., 2020). Applications of 4D printing have been elaborated in Figure 6.

Figure 6. Applications of 4D Printing
(Reprinted with permission from (Falahati et al., 2020))

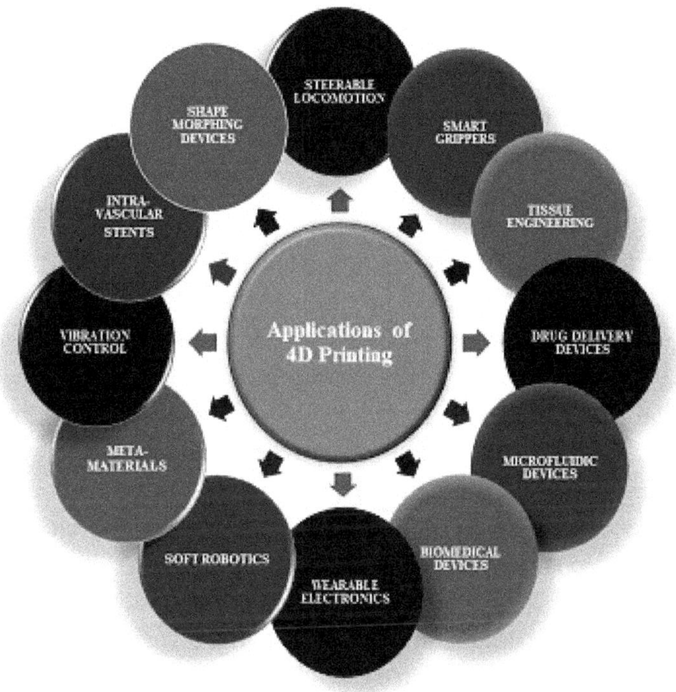

13. REGULATIONS FOR SMART MATERIALS THROUGH ISO AND ASTM STANDARDS

ASTM standards stand for the American standards test method. Every time researchers need a standard method for analyzing any property of the material or anything.

For example, nanotechnologically updated sportswear specifications that can be measured are described here. A spray test having ISO 4920:2012 in which the spraying of a specific amount of distilled water, the fabric resistance to surface wetting is measured by comparing the sample with the standard pictures. This test is followed for the material water-proof breathable fabrics or the sportswear included nanotechnology. The resistance created by the water penetration along with hydrostatic pressure is measured through a methodological way that has ISO 811: 1989, BS EN 20811: 1992, this method works by steadily increasing the water pressure on one face until penetration occurs. The resistance faced by the air-water vapor on the permeation is also measured through the ISO 15492: 2004 test where the change of fabric weight is measured that will result in the breathability of the sportswear. Bundesmann rain-shower test ISO 9856:1991 and BS EN 29865: 1993, is performed to measure the ability of sportswear in terms of water-repulsive properties, along with this test AATCC 35-2006 and BS 5066: 1974 rain test is also performed associating other standards with water-proof fabrics having ASTM,1988; ASTM D737-04:2012 and BS EN ISO 9237:1995 (Kanjana et al., 2018).

There are many other smart objects where the necessary ASTM standards can be found in the standard books for ASTM standards.

14. CONCLUSION AND FUTURE SCOPE

For new material that has to satisfy different properties to achieve our desired function or objective. Smart material can fill our objectives in all possible ways. As well as smart materials can fill all demands of all engineering sectors and the medical field. Smart materials open ups the door for researchers to apply this in every field with huge possible ways having a minimum failure.

The primary objectives of the study in this area are to comprehend and regulate the chemical composition and microstructure of all novel materials, which is essential for creating high-quality smart materials. The creation of novel substances for implementation in various applications shall be sped up by the knowledge acquired by studying how a material's crystallographic core structure behaves under varied thermal, mechanical, and chemical conditions. With time, more and more innovations will be made in the smart technology field, making our daily personal and professional chores much easier.

REFERENCES

Aguilar, M. R. J. S. (2019). Introduction to Smart Polymers and Their Applications. Smart Polymers and their Applications. Elsevier.

Aguilar, M. R., & Román, J. S. (2019). Introduction to Smart Polymers and Their Applications. Smart Polymers and their Applications. Elsevier. doi:10.1016/B978-0-08-102416-4.00001-6

Ajay, P. Ahmad, S., Sharma, J., & Gambhir, V. (Eds.). (2023). Handbook of Sustainable Materials: Modelling, Characterization, and Optimization (1st ed.). CRC Press. https://doi.org/ doi:10.1201/9781003297772

Arnold, R. M., Huddleston, N. E., & Locklin, J. (2012). Utilizing click chemistry to design functional interfaces through post-polymerization modification. *Journal of Materials Chemistry*, *22*(37), 19357–19365. doi:10.1039/c2jm31708g

Asha, A. B., Srinivas, S., Hao, X., & Narain, R. (2019). Enzyme-Responsive Polymers: Classifications, Properties, Synthesis Strategies, and Applications. Smart Polymers and Their Applications. Elsevier.

Bera, S. S. (2022). Review of defect engineering in perovskites for photovoltaic application. *Materials Advances* (*Vol.* 3, Issue 13). Royal Society of Chemistry (RSC). doi:10.1039/D2MA00194B

Cano, E. E., Aguilar, M. R., Vázquez, B., & Román, J. S. (2019). Inflammation-Responsive Polymers. Smart Polymers and Their Applications, Elsevier, 219-254.

Cao, W., & Hench, L. L. (1996, January). Bioactive materials. *Ceramics International*, *22*(6), 493–507. doi:10.1016/0272-8842(95)00126-3

Care, A., Bergquist, P. L., & Sunna, A. (2015, May). Solid-binding peptides: Smart tools for nanobiotechnology. *Trends in Biotechnology*, *33*(5), 259–268. doi:10.1016/j.tibtech.2015.02.005 PMID:25796487

Dagani, R. (1995). Polymeric smart materials respond to changes in their environment. *Chemical & Engineering News*, (*Vol. 73*, Issue 38). American Chemical Society (ACS). . doi:10.1021/cen-v073n038.p030

De Caluwé, T. V. (2017). Addition of bioactive glass to glass ionomer cements: Effect on the physico-chemical properties and biocompatibility. *Dental Materials* (*Vol. 33*, Issue 4). Elsevier BV.

Duarte, A. R. C., Mano, J. F., & Reis, R. L. (2011, February). Thermosensitive polymeric matrices for three-dimensional cell culture strategies. *Acta Biomaterialia*, *7*(2), 526–529. doi:10.1016/j.actbio.2010.09.032 PMID:20883833

Falahati, M., Ahmadvand, P., Safaee, S., Chang, Y.-C., Lyu, Z., Chen, R., Li, L., & Lin, Y. (2020). Smart polymers and nanocomposites for 3D and 4D printing. *Materials Today*, *40*, 215–245. doi:10.1016/j.mattod.2020.06.001

Fernández, L. G., Boza, A. M., & Ortega, F. R. (2019). pH-Responsive Polymers: Properties, Synthesis, and Applications. *Smart Polymers and Their Applications*. Elsevier. . doi:10.1016/B978-0-08-102416-4.00003-X

Fukada, E. (1955, February 15). Piezoelectricity of Wood. *Journal of the Physical Society of Japan*, *10*(2), 149–154. doi:10.1143/JPSJ.10.149

Galaev, I. (1999, August 1). "Smart" polymers and what they could do in biotechnology and medicine. *Trends in Biotechnology*, *17*(8), 335–340. doi:10.1016/S0167-7799(99)01345-1 PMID:10407406

Gottardo, S., Mech, A., Drbohlavová, J., Małyska, A., Bøwadt, S., Riego Sintes, J., & Rauscher, H. (2021, January). Towards safe and sustainable innovation in nanotechnology: State-of-play for smart nanomaterials. *NanoImpact*, *21*, 100297. doi:10.1016/j.impact.2021.100297 PMID:33738354

Halonen, N., Pálvölgyi, P. S., Bassani, A., Feorentini, C., Nair, R., Spigno, G., & Kordas, K. (2020). Bio-Based Smart Materials for Food Packaging and Sensors – A Review. *Frontiers in Materials*, *7*, 82. doi:10.3389/fmats.2020.00082

Han, B., Wang, Y., Ding, S., Yu, X., Zhang, L., Li, Z., & Ou, J. (2016, July 28). Self-sensing cementitious composites incorporated with botryoid hybrid nano-carbon materials for smart infrastructures. *Journal of Intelligent Material Systems and Structures*, *28*(6), 699–727. doi:10.1177/1045389X16657416

Hench, L. L. (1971). Bonding mechanisms at the interface of ceramic prosthetic materials. *Journal of Biomedical Materials Research* (*Vol. 5*, Issue 6). Wiley. doi:10.1002/jbm.820050611

Hench, L. L. (2006). The story of Bioglass®. In Journal of Materials Science: Materials in Medicine (Vol. 17, Issue 11). Springer Science and Business Media LLC. doi:10.100710856-006-0432-z

Herod-Schmidt, U., Floeth, E., Last, B., Schafer, W., Zaglauer Daimler-Benz, H., & Dornier, A. (1997), Qualification of smart composites for the use in aerospace applications. In J. M. Sater (Ed.), *SPIE Proceedings*. SPIE. 10.1117/12.274660

Hoffman, A. S., Stayton, P. S., Bulmus, V., Chen, G., Chen, J., Cheung, C., Chilkoti, A., Ding, Z., Dong, L., Fong, R., Lackey, C. A., Long, C. J., Miura, M., Morris, J. E., Murthy, N., Nabeshima, Y., Park, T. G., Press, O. W., Shimoboji, T., & Miyata, T. (2000). Really smart bioconjugates of smart polymers and receptor proteins. *Journal of Biomedical Materials Research*, *52*(4), 577–586. doi:10.1002/1097-4636(20001215)52:4<577::AID-JBM1>3.0.CO;2-5 PMID:11033539

Hoogenboom, R. (2019). Temperature-Responsive Polymers: Properties, Synthesis, and Applications. *Smart Polymers and Their Applications*. Elsevier. . doi:10.1016/B978-0-08-102416-4.00002-8

Huai-Ling Gao, W. Z.-Y., Cui, C., Bao, J.-Z., Zhu, Y.-B., Xia, J., Wen, S.-M., & Yu, S.-H. (2021). A Highly Compressible and Stretchable Carbon Spring for Smart Vibration and Magnetism Sensors. *Advanced Materials*, *33*(39), 1–9. doi:10.1002/adma.202102724 PMID:34387379

Jaishree Damodharan, A. S. (2018). A Review on Smart Materials, Types and Applications. *International Journal of Engineering Technology Science and Research, 5*(3).

John, A. (2015). Preparation and characterization of cellulose–ZnOhybrid film by blending method. *Macromolecular Research*, *23*(9), 814–818. doi:10.100713233-015-3111-4

John, A., Ko, H.-U., Kim, D.-G., & Kim, J. (2011). Preparation of cellulose–ZnOhybrid films by a wet chemical method and their characterization. *Cellulose (London, England)*, *18*(3), 675–680. doi:10.100710570-011-9523-1

Kanjana, S., & Nalankilli, G. (2018). Smart, Waterproof, Breathable Sportswear-AReview. *Journal of Textile and Apparel, Technology and Management*, *10*(3).

Ke, X., Zhang, L., Yang, X., Wang, J., Zhuang, C., Jin, Z., Liu, A., Zhao, T., Xu, S., Gao, C., Gou, Z., & Yang, G. (2018, January 8). Low-melt bioactive glass-reinforced 3D printing akermanite porous cages with highly improved mechanical properties for lumbar spinal fusion. *Journal of Tissue Engineering and Regenerative Medicine*, *12*(5), 1149–1162. doi:10.1002/term.2624 PMID:29222837

Kikuchi, A., & Okano, T. (2002, July). Intelligent thermoresponsive polymeric stationary phases for aqueous chromatography of biological compounds. *Progress in Polymer Science*, *27*(6), 1165–1193. doi:10.1016/S0079-6700(02)00013-8

Kim, D. A., Lee, J. H., Jun, S. K., Kim, H. W., Eltohamy, M., & Lee, H. H. (2017, July). Sol–gel-derived bioactive glass nanoparticle-incorporated glass ionomer cement with or without chitosan for enhanced mechanical and biomineralization properties. *Dental Materials*, *33*(7), 805–817. doi:10.1016/j.dental.2017.04.017 PMID:28535954

Kim, J., Bae, S. H., & Lim, H. G. (2006, May 17). Micro transfer printing on cellulose electro-active paper. *Smart Materials and Structures*, *15*(3), 889–892. doi:10.1088/0964-1726/15/3/026

Kim, J., Yun, S., & Lee, S. K. (2007, October 17). Cellulose Smart Material: Possibility and Challenges. *Journal of Intelligent Material Systems and Structures*, *19*(3), 417–422. doi:10.1177/1045389X07083140

Kim, J., Yun, S., & Ounaies, Z. (2006, May 19). Discovery of Cellulose as a Smart Material. *Macromolecules*, *39*(12), 4202–4206. doi:10.1021/ma060261e

Klemm, D., Heublein, B., Fink, H. P., & Bohn, A. (2005, May 30). Cellulose: Fascinating Biopolymer and Sustainable Raw Material. *Angewandte Chemie International Edition*, *44*(22), 3358–3393. doi:10.1002/anie.200460587 PMID:15861454

Ko, H. U., Kim, H. C., Kim, J. W., Zhai, L., & Kim, J. (2017, June 2). Fabrication Method Study of ZnO Nanocoated Cellulose Film and Its Piezoelectric Property. *Materials (Basel)*, *10*(6), 611. doi:10.3390/ma10060611 PMID:28772971

Kumar, A., Kumar, P., Mittal, R. K., & Gambhir, V. (2023). Materials processed by additive manufacturing techniques. In Advances in Additive Manufacturing Artificial Intelligence, Nature-Inspired, and Biomanufacturing (pp. 217-233). Elsevier. https:// doi.org/ doi:10.1016/B978-0-323-91834-3.00014-4

Kumar, A., Mittal, R. K., & Haleem, A. (Eds.). (2022). *Advances in Additive Manufacturing: Artificial Intelligence, Nature-Inspired, and Biomanufacturing*. Elsevier. doi:10.1016/C2020-0-03877-6

Kumar, A., Srivastava, A., Galaev, I. Y., & Mattiasson, B. (2007, October). Smart polymers: Physical forms and bioengineering applications. *Progress in Polymer Science, 32*(10), 1205–1237. doi:10.1016/j.progpolymsci.2007.05.003

Kumar, S., Bhalla, A., & Cross, L. (1994, September). Smart Ceramics for Broadband Vibration Control. *Journal of Intelligent Material Systems and Structures*, *5*(5), 673–677. doi:10.1177/1045389X9400500511

Kumar, V. P. (2017). *Future Perspectives of Bioactive Glasses for the Clinical Applications. Bioactive Glasses*. Springer International Publishing., doi:10.1007/978-3-319-45716-1_11

Malat, T. A., Glombitza, M., Dahmen, J., Hax, P. M., & Steinhausen, E. (2018, April). The Use of Bioactive Glass S53P4 as Bone Graft Substitute in the Treatment of Chronic Osteomyelitis and Infected Non-Unions – a Retrospective Study of 50 Patients. *Zeitschrift für Orthopädie und Unfallchirurgie*, *156*(02), 152–159. doi:10.1055-0043-124377 PMID:29665602

Melly, S. K., Liu, L., Liu, Y., & Leng, J. (2020, July 10). On 4D printing as a revolutionary fabrication technique for smart structures. *Smart Materials and Structures*, *29*(8), 083001. doi:10.1088/1361-665X/ab9989

Moon, R. J., Martini, A., Nairn, J., Simonsen, J., & Youngblood, J. (2011). Cellulose nanomaterials review: Structure, properties and nanocomposites. *Chemical Society Reviews*, *40*(7), 3941. doi:10.1039/c0cs00108b PMID:21566801

Moyez, S. A., Maitra, S., Mukherjee, K., Sengupta, A., & Roy, S. (2020, October 13). Structural Features and Optical Properties of CH3NH3Pb(1−x)SnxCl3 Thin-Film Perovskites for Photovoltaic Applications. *Journal of Electronic Materials*, *49*(12), 7133–7143. doi:10.100711664-020-08529-5

Moyez, S. A., & Roy, S. (2018, October). Dual-step thermal engineering technique: A new approach for fabrication of efficient CH3NH3PbI3-based perovskite solar cell in open air condition. *Solar Energy Materials and Solar Cells*, *185*, 145–152. doi:10.1016/j.solmat.2018.05.027

Mun, S., Maniruzzaman, M., Ko, H. U., Kafy, A., & Kim, J. (2015). Preparation and characterisation of cellulose ZnO hybrid film by blending method and its glucose biosensor application. *Materials Technology, 30*(sup7), B150–B154. doi:10.1179/1753555714Y.0000000197

Najeeb, S., Khurshid, Z., Zafar, M., Khan, A., Zohaib, S., Martí, J., Sauro, S., Matinlinna, J., & Rehman, I. (2016, July 14). Modifications in Glass Ionomer Cements: Nano-Sized Fillers and Bioactive Nanoceramics. *International Journal of Molecular Sciences*, *17*(7), 1134. doi:10.3390/ijms17071134 PMID:27428956

Newnham, R. (1999, May). Ceramics into next millennium. *British Ceramic Transactions*, *98*(5), 251–255. doi:10.1179/096797899680534

Newnham, R. E., Xu, Q. C., Kumar, S., & Cross, L. E. (1990). Smart ceramics. *Ferroelectrics*, *102*(1), 259–266. doi:10.1080/00150199008221486

Nguyen, M. K., & Lee, D. S. (2010, March 1). Injectable Biodegradable Hydrogels. *Macromolecular Bioscience*, *10*(6), 563–579. doi:10.1002/mabi.200900402 PMID:20196065

Ok, Y. S., Chang, S. X., Gao, B., & Chung, H. J. (2015, October). SMART biochar technology—A shifting paradigm towards advanced materials and healthcare research. *Environmental Technology & Innovation*, *4*, 206–209. doi:10.1016/j.eti.2015.08.003

Osada, Y., & Gong, J. P. (1998, August). Soft and Wet Materials: Polymer Gels. *Advanced Materials*, *10*(11), 827–837. doi:10.1002/(SICI)1521-4095(199808)10:11<827::AID-ADMA827>3.0.CO;2-L

Ougizawa, T., & Inoue, T. (1986, July). UCST and LCST Behavior in Polymer Blends and Its Thermodynamic Interpretation. *Polymer Journal*, *18*(7), 521–527. doi:10.1295/polymj.18.521

Park, J., Kim, J.-K., Park, S. A., & Lee, D.-W. (2018). Biodegradable polymer material based smart stent: Wireless Presor Sensor and 3D Printed stent. *Microelectronic Engineering*, *206*, 1–5. doi:10.1016/j.mee.2018.12.007

Perez-Masia, R., Lopez-Rubio, A., Fabra, M. J., & Lagaron, J. M. (2013, June 14). Biodegradable polyester-based heat management materials of interest in refrigeration and smart packaging coatings. *Journal of Applied Polymer Science*, *130*(5), 3251–3262. doi:10.1002/app.39555

Popa, A. C., Stan, G., Husanu, M. A., Mercioniu, I., Santos, L., Fernandes, H., & Ferreira, J. (2017, January). Bioglass implant-coating interactions in synthetic physiological fluids with varying degrees of biomimicry. *International Journal of Nanomedicine*, *12*, 683–707. doi:10.2147/IJN.S123236 PMID:28176941

Qiu, Y., & Park, K. (2001, December). Environment-sensitive hydrogels for drug delivery. *Advanced Drug Delivery Reviews*, *53*(3), 321–339. doi:10.1016/S0169-409X(01)00203-4 PMID:11744175

Rebouillat, S., Lyons, M. E., Brandon, M. P., & Doyle, R. L. (2011, November). Paving the Way to The Integration of Smart Nanostructures: Part II: Nanostructured Microdispersed Hydrated Metal Oxides for Electrochemical Energy Conversion and Storage Applications. *International Journal of Electrochemical Science*, *6*(11), 5830–5917. doi:10.1016/S1452-3981(23)18448-3

Rogers, C. A. (1988). U.S. Army Research Office Workshop on Smart Materials, Structures and Mathematical Issues Held in Blacksburg. *Defense Technical Information Center* (p. 5). Blacksburg: Technomic Publishing Co.

Rosenqvist, K., Gursoy, M., Kononen, E., Gursoy, U. K., & Juppo, A. M. (2017). A Pilot Study - Comparison between a Novel Combination of Bioactive Glass with Clodronate and Bioactive Glass Alone as a Treatment for Chronic Periodontitis. *Journal of Biotechnology & Biomaterials*, *07*(03). doi:10.4172/2155-952X.1000265

Roy, S., & Botte, G. G. (2018). Perovskite solar cell for photocatalytic water splitting with a TiO2/Co-doped hematite electron transport bilayer. *RSC Advances*, *8*(10), 5388–5394. doi:10.1039/C7RA11996H PMID:35542422

Sahani, S., Roy, T., & Sharma, Y. C. (2020, May). Smart waste management of waste cooking oil for large scale high quality biodiesel production using Sr-Ti mixed metal oxide as solid catalyst: Optimization and E-metrics studies. *Waste Management (New York, N.Y.)*, *108*, 189–201. doi:10.1016/j.wasman.2020.04.036 PMID:32360999

Sarkar, S., Guibal, E., Quignard, F., & SenGupta, A. K. (2012, January 25). Polymer-supported metals and metal oxide nanoparticles: Synthesis, characterization, and applications. *Journal of Nanoparticle Research*, *14*(2), 715. doi:10.100711051-011-0715-2

Sato, T., Katayama, K., Suzuki, T., & Shiomi, T. (1998, February). UCST and LCST behaviour in polymer blends containing poly(methyl methacrylate-statstyrene). *Polymer*, *39*(4), 773–780. doi:10.1016/S0032-3861(97)00339-X

Saveleva, M. S., Eftekhari, K., Abalymov, A., Douglas, T. E. L., Volodkin, D., Parakhonskiy, B. V., & Skirtach, A. G. (2019, April 4). Hierarchy of Hybrid Materials—The Place of Inorganics-in-Organics in it, Their Composition and Applications. *Frontiers in Chemistry*, *7*, 179. Advance online publication. doi:10.3389/fchem.2019.00179 PMID:31019908

Sharma, D., & Hussain, C. M. (2020, January). Smart nanomaterials in pharmaceutical analysis. *Arabian Journal of Chemistry*, *13*(1), 3319–3343. doi:10.1016/j.arabjc.2018.11.007

Shellaiah, M., & Sun, K. W. (2020, July 14). Review on Sensing Applications of Perovskite Nanomaterials. *Chemosensors (Basel, Switzerland)*, *8*(3), 55. doi:10.3390/chemosensors8030055

Shivaji S. & Aher, A. D. (2015). A Review On Smart Materials: Future Potentials In Engineering. *International journal of science technology and management, 10.*

Suresha, K. M., Yang, S. Y., Lee, M. H., Kim, J. H., & Kim, J. (2008, January). Actuator, sensor and MEMS devices based on regenerated cellulose. *Composite Interfaces*, *15*(7–9), 679–685. doi:10.1163/156855408786778393

Takagi, T. (1990). A Concept of Intelligent Materials. Journal of Intelligent Material Systems and Structures, 1(2), 149–156. doi:10.1177/1045389X9000100201

Vanderroost, M., Ragaert, P., Devlieghere, F., & De Meulenaer, B. (2014, September). Intelligent food packaging: The next generation. *Trends in Food Science & Technology*, *39*(1), 47–62. doi:10.1016/j.tifs.2014.06.009

Vayssieres, L., Hagfeldt, A., & Lindquist, S. E. (2000, January 1). Purpose-built metal oxide nanomaterials. The emergence of a new generation of smart materials. *Pure and Applied Chemistry*, *72*(1–2), 47–52. doi:10.1351/pac200072010047

Wang, K., Meng, Q., Wang, Q., Zhang, W., Guo, J., Cao, S., Elezzabi, A. Y., Yu, W. W., Liu, L., & Li, H. (2021, September 3). Advances in Energy-Efficient Plasmonic Electrochromic Smart Windows Based on Metal Oxide Nanocrystals. *Advanced Energy and Sustainability Research*, *2*(12), 2100117. doi:10.1002/aesr.202100117

XiaoHui, L., Cheng, Z., Laurent, L. G., & ChunYing, C. (2010). "Smart" nanomaterials for cancer therapy. *Science China, 53*(11). doi:10.1007/s11426-010-4122-9

Xiong, X., Campo, A. d., & Cui, J. (2019). Photoresponsive Polymers. Smart Polymers and Their Applications. Elsevier. doi:10.1016/B978-0-08-102416-4.00004-1

Yu, H., Shi, R., Zhao, Y., Waterhouse, G. I. N., Wu, L. Z., Tung, C. H., & Zhang, T. (2016, September 14). Smart Utilization of Carbon Dots in Semiconductor Photocatalysis. *Advanced Materials*, *28*(43), 9454–9477. doi:10.1002/adma.201602581 PMID:27623955

Yun, S., & Kim, J. (2009, August). Covalently bonded multi-walled carbon nanotubes-cellulose electro-active paper actuator. *Sensors and Actuators. A, Physical*, *154*(1), 73–78. doi:10.1016/j.sna.2009.06.011

Yun, S., & Kim, J. (2011, February). Mechanical, electrical, piezoelectric and electro-active behavior of aligned multi-walled carbon nanotube/cellulose composites. *Carbon*, *49*(2), 518–527. doi:10.1016/j.carbon.2010.09.051

Yun, S., Kim, J., & Ounaies, Z. (2006, April 25). Single-walled carbon nanotube/polyaniline coated cellulose based electro-active paper (EAPap) as hybrid actuator. *Smart Materials and Structures*, *15*(3), N61–N65. doi:10.1088/0964-1726/15/3/N02

Zhai, S., Jiang, W., Wei, L., Karahan, H. E., Yuan, Y., Ng, A. K., & Chen, Y. (2015). All-carbon solid-state yarn supercapacitors from activated carbon and carbon fibers for smart textiles. *Materials Horizons*, *2*(6), 598–605. doi:10.1039/C5MH00108K

Zhang, C., Lv, W., Tao, Y., & Yang, Q. H. (2015). Towards superior volumetric performance: Design and preparation of novel carbon materials for energy storage. *Energy & Environmental Science*, *8*(5), 1390–1403. doi:10.1039/C5EE00389J

Zhang, Z., Demir, K. G., & Gu, G. X. (2019, March 19). Developments in 4D-printing: A review on current smart materials, technologies, and applications. *International Journal of Smart and Nano Materials*, *10*(3), 205–224. doi:10.1080/19475411.2019.1591541

Zheng, Y., Huang, K., You, X., Huang, B., Wu, J., & Gu, Z. (2017, November). Hybrid hydrogels with high strength and biocompatibility for bone regeneration. *International Journal of Biological Macromolecules*, *104*, 1143–1149. doi:10.1016/j.ijbiomac.2017.07.017 PMID:28687387

Chapter 14
Analyzing the Properties of Zinc Oxide (ZnO) Thin Film Grown on Silicon (Si) Substrate, ZnO/Si Using RF Magnetron Sputtering Approach

Sumitra Choudhary
Department of Physics, JECRC University, India

Abhishek Sharma
Department of Physics, JECRC University, India

Vikas Sharma
Institute of Chemistry, The Hebrew University of Jerusalem, Israel

Ajay Kumar
Department of Mechanical Engineering, JECRC University, India

Parveen Kumar
Mechanical Engineering, Rawal Institute of Engineering and Technology, India

ABSTRACT

In this work, ZnO single layer thin film of 100 nm is deposited on a Si substrate using RF magnetron sputtering. Base pressure of 1.0×10−5 mbar, RF power of 100W at Ar flow 15 sccm and room temperature were process parameters. The average crystallite size of ZnO layer, deposited on Si substrate, using Scherrer's formula was 108.16 nm. XRD verifies the crystalline nature of ZnO with various peaks at (002), (101) and (103) planes. Using AFM technique, ZnO had an Average Surface Roughness

DOI: 10.4018/978-1-6684-9224-6.ch014

(Ra) of 2.75 nm and RMS roughness (Rq) of 3.70 nm. From Hall measurements at room temperature, the authors determined that ZnO is a n-type semiconductor having a resistivity of 1 to 100 Ωcm. The layer's sheet resistance was 7.05×103 Ω/sq, and its resistivity was 7.05×10−2 Ωcm. The Raman spectra analysis confirmed the presence of Raman active modes in the sample, confirming the existence of certain vibrational modes. In PL spectra, an emission peak was observed at 380.30 nm, which closely resembled pure ZnO. These results collectively shows that the ZnO/Si thin films grown on Si demonstrated excellent quality.

1. INTRODUCTION

In recent years, there has been a notable upsurge in the research and investigations concerning Zinc Oxide (ZnO). This heightened interest can be attributed to this material's unique properties and potential applications. Researchers from diverse fields have been actively exploring the various aspects of ZnO due to its intriguing characteristics and wide range of possible uses. As a result, the scientific community has been increasingly intrigued by the prospects of utilizing ZnO in innovative technologies and applications.

According to research conducted by Jandow et al., 2010, Zinc Oxide (ZnO) exhibits distinctive properties, with its wide band gap of 3.37 eV standing out as a key feature. This unique characteristic makes ZnO a promising material with significant potential for various applications in optoelectronics and piezoelectric devices. As a result of these findings, ZnO has garnered considerable attention as a viable candidate for advancing technologies in these specific fields.

According to research by (Umar et al., 2010), ZnO has a significant exciting binding energy of 60 eV at room temperature, which is helpful for a variety of applications. Additionally guarantees its suitability for UV photodetectors, room-temperature LEDs, solar cells, and gas monitors (R. Romero et al., 2004, R. Ghosh, D. & Basak, 2007, S. Sharma & C. Periasamy, 2014, S. Sharma et al., 2014). In several transparent conducting films (TCFs), ZnO is a commonly used transparent conducting substance. ZnO thin films have been documented to be produced using a variety of techniques, which includes sol-gel, thermal vapour deposition, pulsed laser deposition, e-beam deposition, and RF sputtering (L. Znaidi, 2010, pp.18-30, B.D. Yao et al., 2002, S.H. Bae et al., 2001, R. Al Asmar et al., 2006). Among these methods, RF sputtering is a very useful technique for the deposition of ZnO. Because in the RF sputtering process, various parameters like pressure, temperature, deposition time, gas flow rate, and RF power can regulate the characteristics of thin films that are being grown. The selection of the substrate material is a crucial factor in the deposition process of ZnO thin films. The various substrates used to deposit ZnO thin films include PPC plastic (Jandow et al., 2010), sapphire (Al_2O_3) (Pant P et al., 2009), soda lime glass (Jia G et al., 2012), Si (Rouhi J et al., 2014), GaAs (Mauricio MR et al., 2009), and polyimide (Li TC et al., 2012). Here, we have deposited ZnO on Si (silicon) substrate and further we are going to fabricate metal-dielectric multilayers for various optoelectronic applications (Vikas Sharma et al., 2017, Satyavir Singh et al., 2018, V. Sharma et al., 2016, Vikas Sharma et al., 2015, D. L. Raimondi & E. Kay, 1970, SH Ribut et al., 2019). This study aims to investigate the impact of ZnO thin film growth on a silicon substrate, focusing on its structural, electrical, and optical properties. X-ray diffraction (XRD) was employed to analyze the structural characteristics, while atomic force microscopy (AFM) was utilized to study the surface properties. Hall measurements were conducted to assess the electrical properties, and Raman spectroscopy was employed to analyze the optical properties of the ZnO thin films. The objective of this work is to study the effect of structural, electrical, and optical properties of ZnO thin films grown on the silicon substrate. The structural properties were characterized by X-ray diffraction (XRD), surface analysis was done by using atomic force microscopy (AFM), electrical properties were by Hall measurements, and optical properties were analysed by using Raman spectroscopy.

2. EXPERIMENTAL PROCEDURE

2.1 ZnO Thin Film Preparation

ZnO thin film of 100 nm was deposited on silicon substrates using RF magnetron sputtering technique (MNIT/MRC/65/RF-DC system). Prior to loading into the sputtering chamber, the substrates were cleaned with acetone. Additionally, the sputtering chamber was also pre-sputtered for 20 minutes to eliminate any impurities before the actual deposition process took place. Under conditions of room temperature and a base pressure of 1.0×10^{-5} mbar, a silicon substrate measuring 10×10 mm was coated with a 99.99% pure Zinc Oxide (ZnO) target. With a radio frequency (RF) power supply of 100 W and an Ar flow of 15 sccm, ZnO was deposited for 30 minutes at a working pressure of 1.86×10^{-2} mbar. During the sputtering process, the substrate was kept in continuous rotation at a speed of 12 rpm. Positioned at a distance of 7 cm from the target, the substrate experienced a deposition rate of approximately 0.5 Angstroms per second.

2.2 Thin Film Characterization

An atomic force microscope (MNIT/MRC/04/AFM) and X-ray diffractometer (PANalytical X'Pert Pro) were used to characterize the crystal structure and morphology of the ZnO deposited on the silicon substrate. Electrical properties were characterized by the Hall measurement system. Optical properties were characterized using Raman spectroscopy (STR-MPR-80 Techno Instruments).

3. RESULTS AND DISCUSSION

3.1 Structural/ XRD Measurements

The XRD patterns of the ZnO thin film on Si substrate which were prepared by the RF magnetron sputtering method are shown in Figure 1. The diffraction pattern has peaks at 2θ=25.62°, 2θ=34.30° and 2θ=62.58° that correspond to the (0 0 2), (1 0 1), and (1 0 3) planes respectively. The peaks observed in the diffraction pattern confirmed ZnO's wurtzite hexagonal crystal structure. This result is in agreement with the research done by Hu et al. [12]. A higher intensity is obtained on the peak corresponding to (0 0 2) plane, which indicates the extent of the diffracted plane's crystallinity. Using Scherrer's method, crystal size was determined.

$$D = \frac{\check{s}\ \lambda}{\beta Cos\theta}$$

Where K is the Scherrer constant (expected to be 0.89), λ is X-ray wavelength, β is full-width-half-maximum (FWHM) of the diffraction peak in radians and θ is the Bragg diffraction angle. The parameter "D" represents the average crystallite dimension perpendicular to the diffracting planes. We have tabulated the result of the structural information of the diffraction pattern for ZnO/Si in Table 1.

From the above relation, the average crystallite size was calculated as 108.16 nm for ZnO.

3.2 Surface Characterization/AFM Analysis

The surface morphology of the ZnO layer deposited on the silicon substrate was examined by using AFM (Multimode Scanning Probe Microscope, Bruker). The surface topography of the ZnO thin layer was observed in the results. The scan area for the measurement was taken 2 × 2 μm^2 and the results are presented in the following Figure 2 (a-h).

Figure 1. XRD pattern of deposited ZnO thin layer

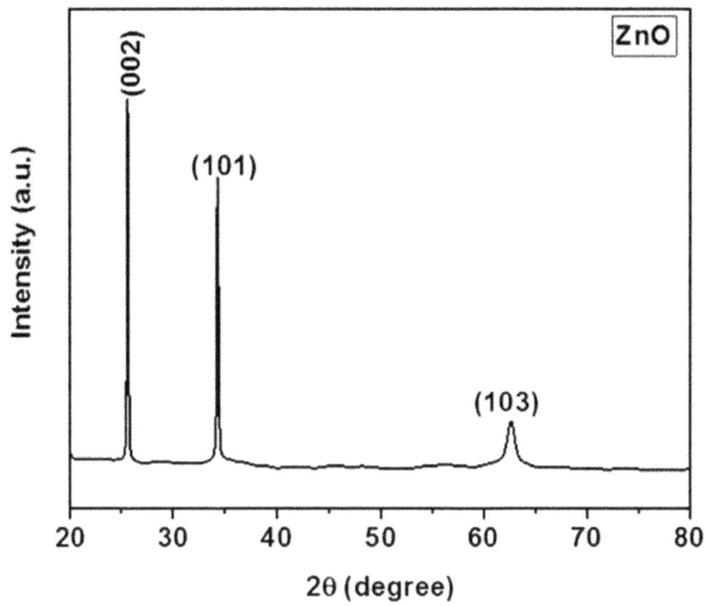

Table 1. Structural details obtained from the diffraction pattern analysis of ZnO thin films grown on a Si substrate

Pos. [2Theta]	FWHM Left Values	Diffraction planes
25.62	0 .1181	(0 0 2)
34.30	0 .1574	(1 0 1)
62.58	0.7872	(1 0 3)

The Average Surface Roughness (Ra) and Root Mean Square (RMS) Roughness (Rq) values of the ZnO/Si films are tabulated in Table 2. They were calculated with the help of Gwyddion software. The average roughness of the deposited ZnO thin film was 2.75nm as shown in Table 2. The surface of the ZnO/Si sample is slightly rough. The microstructural strain has a significant impact on the surface roughness of the ZnO thin film, which is partially determined by the growth environment, the temperature used to grow thin films, and residual thermal strain, which also rises with growth temperature (R. Romero et al., 2004).

Figure 2. Two and three- dimensional AFM images of ZnO/Si depicted at different micrometers, (a) 2D, 1μm (b) 2D, 2μm (c) 2D, 5μm (d) 2D, 500 nm (e) 3D, 1μm (f) 3D, 2μm (g) 3D, 5μm (h) 3D, 500nm

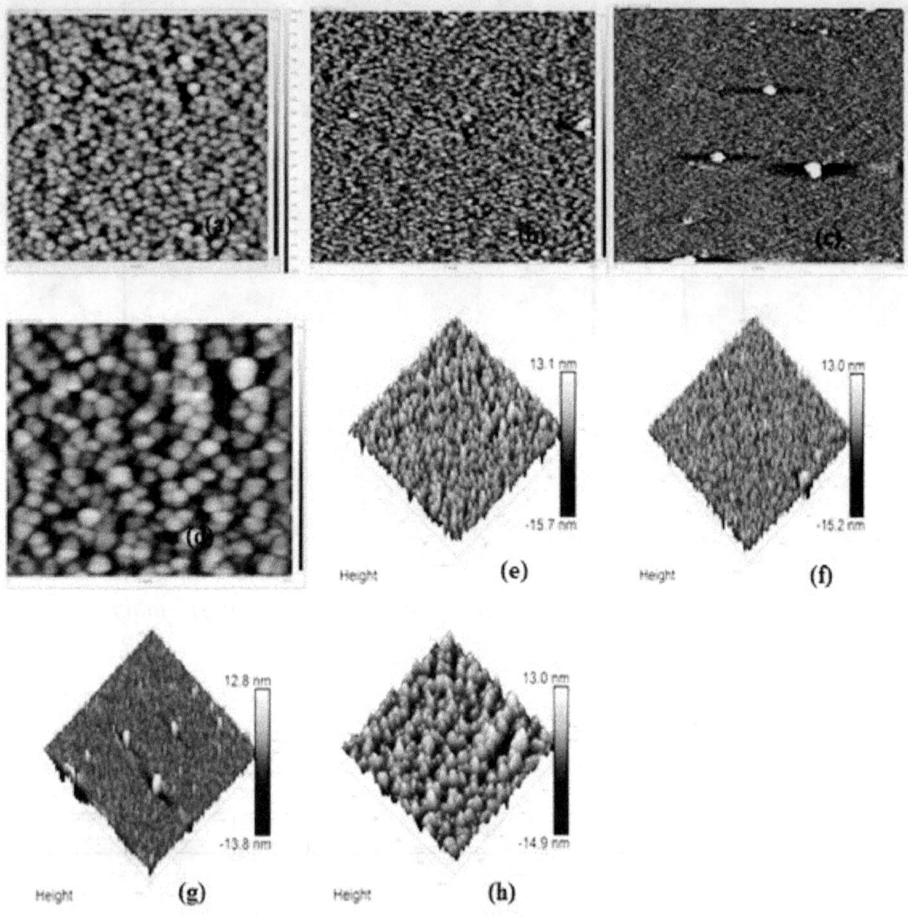

Table 2. Average surface roughness (Ra) and root mean square (RMS) roughness (Rq) values of the ZnO/Si films

Substrate	Root Mean Square Roughness, Rq	Average Roughness, Ra
Silicon (Si)	3.70 nm	2.75 nm

3.3 Electrical Properties/Hall Measurement

In the analysis of electric properties, the resistivity and sheet resistance of the ZnO layer was characterized by the Hall Measurement technique using HMS-5000 series, Hall Effect measurement system (at MNIT MRC, Jaipur) at room temperature. Under typical conditions, Zinc Oxide (ZnO) exhibits a slight oxygen deficiency in its composition. Due to this chemical deficiency, ZnO behaves as an n-type semiconductor with a typical resistivity ranging from 1 to 100 Ωcm (SH Ribut et al., 2019). The sheet resistance of the layer was measured to be 7.05×10^{3} Ω/sq and resistivity was 7.05×10^{-2} Ωcm, which is closer to the earlier reported results as shown below in Table 3.

Table 3. Shows the results of Hall measurements

Temp (SV)	Temp (PV)	Current (A)	Bulk Concentration	Sheet Resistance (Ω/□)	Resistivity (Ω cm)	Conductivity ($\Omega^{-1}m^{-1}$)	Mobility (cm^2/v.s)
300	300	5×10^{-3}	1.26×10^{19}	7.05×10^{3}	7.05×10^{-2}	1.51×10^{1}	1.60×10^{1}

Figure 3. Raman spectra of ZnO/Si thin film

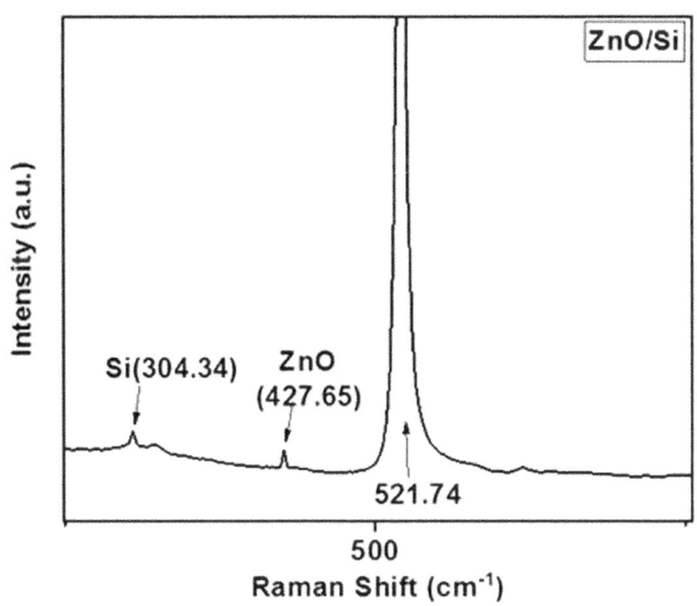

3.4 Optical Properties/Raman Spectroscopy

The analysis of optical properties of ZnO/Si thin film was examined using Raman spectroscopy. The Raman bands of the film are shown in Figure 3. The existence of Raman active modes was discovered using Raman spectroscopy.

Table 4. Shows the phonon frequency for the ZnO deposited on the silicon substrate

Sample	Phonons	Raman shift (cm^{-1})
ZnO/Si	E_2 (High)	427.65
	A_1 (Low)	521.74

In this case, only the A_1 low and E_2 high modes are observed and permitted, while the identification of other modes remains unidentified in the vertically aligned sample. The phonon frequency for the ZnO deposited on the silicon substrate is shown in the findings above shown in Table 4. These values of the ZnO/Si sample which are detected at 427.65 cm^{-1} for E_2 (high) and 521.74 cm^{-1} present A_1 (low) are more compatible with the findings given by (SH Ribut et al., 2019). The peak observed at approximately 427.65 cm^{-1} in the E_2 high mode confirms the characteristic wurtzite structure of ZnO.

The PL spectrum of the thin ZnO film deposited on the silicon substrate is shown in Figure 4. The optical characteristics of the ZnO thin film were also determined using photoluminescence spectroscopy. At room temperature, the measurement was carried out. Xenon lamp light with an excitation wavelength of 300 nm was used as the source of illumination to stimulate the sample.

The findings reported by (Haga et al., 2003 and Umar et al., 2010) are more consistent with the emission peak of the ZnO deposited on the silicon substrate detected by the PL signal, which corresponds to 380.30 nm (Eg = 3.26 eV). The Photoluminescence (PL) spectra from the ZnO deposited on the silicon substrate exhibited band emission that aligned with the findings of previous researchers. A distinct and sharp line at 380.30 nm in the ultraviolet spectrum is visible in the PL spectra.

Figure 4. PL spectra of ZnO thin film deposited on Si substrate

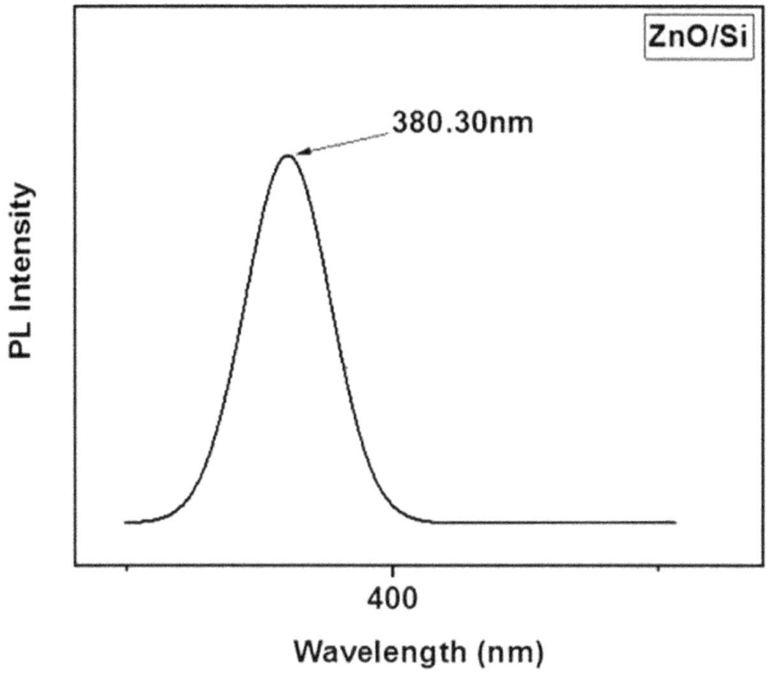

4. CONCLUSION

A simple method RF Magnetron sputtering is used to fabricate Zinc Oxide (ZnO) layer on Silicon substrates is presented. Various techniques, including X-ray diffraction (XRD), Atomic Force Microscopy (AFM), Hall measurement, Raman Spectroscopy, and photoluminescence, were used to explore and analyze different properties of the ZnO thin film on the silicon substrate.

The XRD results confirmed the wurtzite structure of the ZnO layer grown on Si substrate. XRD showed peaks at different angles 25.62°, 34.30°, and 62.5° that correspond to the (0 0 2), (1 0 1) and (1 0 3) planes respectively. The distinct and well-defined diffracted peaks observed in the XRD pattern provide strong evidence for the crystalline nature of the deposited ZnO thin film. Moreover, the analysis yielded an average crystallite size of 108.16 nm. In correlation with higher values of Root Mean Square (RMS), the samples exhibit increased roughness on their surfaces.

The average roughness was 2.75 nm, calculated using AFM results. Sheet resistance and resistivity were calculated using Hall measurements. The sheet resistance of the layer was measured to be 7.05×10^{3} Ω/sq and resistivity was 7.05×10^{-2} Ωcm, which confirms ZnO is an n-type semiconductor with a typical resistivity of 1–100 Ωcm.

Optical analysis was done by using Raman spectroscopy. Raman spectra detected the presence of Raman active modes for the deposited ZnO thin film. In PL analysis, the PL signal detects the emission peak of the ZnO deposited on silicon substrate corresponds to 380.30 nm (E_g = 3.26 eV). Finally, the results of the different characterizations confirmed the ZnO layer on Si substrate and the quality of ZnO/Si thin film is good.

5. FUTURE DIRECTION OF WORK

In light of the promising results obtained in this study, future research could focus on the fabrication of ZnO/Ag/ZnO trilayer structure. After the successful deposition, we will try to modify this trilayer to enhance its properties so that it can be used in various optoelectronic applications.

ACKNOWLEDGMENT

The characterization work was supported by Material Research Centre (MRC) - Malaviya National Institute of Technology, Jaipur, and Techno Instruments Pvt. Ltd, Jaipur.

REFERENCES

Al Asmar, R., Zaouk, D., Bahouth, Ph., Podleki, J., & Foucaran, A. (2006). Characterization of electron beam evaporated ZnO thin films and stacking ZnO fabricated by e-beam evaporation and rf magnetron sputtering for the realization of resonators. *Microelectronic Engineering*, *83*(3), 393–398. doi:10.1016/j.mee.2005.10.010

Bae, S. H., Lee, S. Y., Jin, B. J., & Im, S. (2001). Growth and characterization of ZnO thin films grown by pulsed laser deposition. *Applied Surface Science*, *169*, 525–528. doi:10.1016/S0169-4332(00)00752-2

Ghosh, R., & Basak, D. (2007). Electrical and ultraviolet photoresponse properties of quasialigned ZnO nanowires/p-Si heterojunction. *Applied Physics Letters*, *90*(24), 243106–3. doi:10.1063/1.2748333

Haga, K., Suzuki, T., Kashiwaba, Y., Watanabe, H., Zhang, B. P., & Segawa, Y. (2003). High-quality ZnO films prepared on Si wafers by low-pressure MO-CVD. *Thin Solid Films*, *433*(1), 131–134. doi:10.1016/S0040-6090(03)00327-4

Jandow, N. N., Yam, F. K., Thahab, S. M., Ibrahim, K., & Hassan, H. A. (2010). The characteristics of ZnO deposited on PPC plastic substrate. *Materials Letters*, *64*(21), 2366–2368. doi:10.1016/j.matlet.2010.07.025

Jia, G., Wang, Y., & Yao, J. (2012).. . *Digest Journal of Nanomaterials and Biostructures*, *7*(1), 261–267.

Li, T. C., Han, C. F., Wu, B. H., Hsieh, P. T., & Lin, J. F. (2012). Effects of Prestrain Applied to a Polyethylene Terephthalate Substrate Before the Coating of Al-Doped ZnO Film on Film Quality, Electrical Properties, and Pop-In Behavior During Nanoindentation. *Journal of Microelectromechanical Systems*, *21*(5), 1059–1070. doi:10.1109/JMEMS.2012.2205135

Lupan, O., Emelchenko, G. A., Ursaki, V. V., Chai, G., Redkin, A. N., Gruzintsev, A. N., Tiginyanu, I. M., Chow, L., Ono, L. K., Roldan Cuenya, B., Heinrich, H., & Yakimov, E. E. (2010). Synthesis and characterization of ZnO nanowires for nanosensor applications. *Materials Research Bulletin*, *45*(8), 1026–1032. doi:10.1016/j.materresbull.2010.03.027

Mauricio, M. R., Carvalho, G. M., Radovanovic, E., Muniz, E. C., & Rubira, A. F. (2009). Analysis of poly(N-isopropylacrylamide) grafted onto the surface of PET films by SI-ATRP technique. *Materials Science and Engineering C*, *29*(2), 594–598. doi:10.1016/j.msec.2008.10.017

Pant, P., Budai, J. D., Aggarwal, R., Narayan, R. J., & Narayan, J. (2009). Structural characterization of two-step growth of epitaxial ZnO films on sapphire substrates at low temperatures. *Journal of Physics. D, Applied Physics*, *42*(10), 105409. doi:10.1088/0022-3727/42/10/105409

Raimondi, D. L., & Kay, E. (1970). High Resistivity Transparent ZnO Thin Films. *Journal of Vacuum Science and Technology*, *7*(1), 96–99. doi:10.1116/1.1315841

Ribut, S. H., Abdullah, C., & Yusoff, M. (2019). Investigations of structural and optical properties of zinc oxide thin films growth on various substrates. *Results in Physics*, *13*, 102146. doi:10.1016/j.rinp.2019.02.082

Romero, R., Lopez, M. C., Leinen, D., Martin, F., & Ramos-Barrado, J. R. (2004). Electrical properties of the n-ZnO/c-Si heterojunction prepared by chemical spray pyrolysis. *Materials Science and Engineering B*, *110*(1), 87–93. doi:10.1016/j.mseb.2004.03.010

Rouhi, J., Alimanesh, M., Dalvand, R., Ooi, C. R., Mahmud, S., & Mahmood, M. R. (2014). Optical properties of well-aligned ZnO nanostructure arrays synthesized by an electric field-assisted aqueous solution method. *Ceramics International*, *40*(7), 11193–11198. doi:10.1016/j.ceramint.2014.03.157

Sharma, S., & Periasamy, C. (1633-1636). J. Electron Dev. 2014. *Superlattices and Microstructures, 19.*

Sharma, S., & Periasamy, C. (2014). A study on the electrical characteristic of n-ZnO/p-Si heterojunction diode prepared by vacuum coating technique. *Superlattices and Microstructures*, *73*, 12–21. doi:10.1016/j.spmi.2014.05.011

Sharma, V., Kumar, P., Kumar, A., Surbhi, Asokan, K., & Sachdev, K. (2017). High-performance radiation stable ZnO/Ag/ZnO multilayer transparent conductive electrode. *Solar Energy Materials and Solar Cells*, *169*, 122–131. doi:10.1016/j.solmat.2017.05.009

Sharma, V., Singh, S., Asokan, K., & Sachdev, K. (2015). A study on 100 MeV O7+ irradiated SnO2/Ag/SnO2 multilayer as transparent electrode for flat panel display application. *Nuclear Instruments & Methods in Physics Research. Section B, Beam Interactions with Materials and Atoms*, *379*, 141–145. doi:10.1016/j.nimb.2016.04.059

Sharma, V., Vyas, R., Bazylewski, P., Chang, G. S., Asokan, K., & Sachdeva, K. (2016). Probing the highly transparent and conducting SnO_x/Au/SnO_x structure for futuristic TCO applications. *RSC Advances*, *6*(35), 29135–29141. doi:10.1039/C5RA24422F

Singh, S., Sharma, V., Saini, D., Shekhawat, S., Asokan, K., & Sachdev, K. (2018). Influence of 100 keV Ar+ implantation on electrical and optical properties of TiO2/Ag/TiO2 multilayer films. *Materials Science in Semiconductor Processing*, *75*, 18–25. doi:10.1016/j.mssp.2017.11.016

Yao, B. D., Chan, Y. F., & Wang, N. (2002). Formation of ZnO nanostructures by a simple way of thermal evaporation. *Applied Physics Letters*, *81*(4), 757–759. doi:10.1063/1.1495878

Znaidi, L. (2010). Sol–gel-deposited ZnO thin films: A review. *Materials Science and Engineering B*, *174*(1-3), 18–30. doi:10.1016/j.mseb.2010.07.001

Conclusion

Recent developments in smart materials from various engineering and medical technology perspective are discussed and how these materials interact with the environment. It highlights the recent and advancements involved in proper utilization of smart materials, including chemical and biological approaches. Written in a didactic style, it offers a guide and insights into smart materials. Each chapter provides in-depth technical information on the smart materials theory and its applications. The book shows how smart materials may help to solve human and environmental issues in the future and suggests where current research may lead.

This book presents a collection of studies on state-of-art techniques developed specifically for smart materials with an emphasis on using various analytical strategies, computational and simulation approach, artificial intelligence approach to develop innovative smart materials. The book offers an ideal reference guide for academic researchers and industrial engineers in the fields of different smart materials. It can also be used as a comprehensive reference source for university students in chemical engineering, material science and environmental engineering.

The following points define major features and strengths of this book, which make it unique:

- This compilation brings to its readers the latest developments/advancements in smart materials, which is currently not available at one place.
- This book covers topics like hybrid techniques to process different smart materials that are difficult to process by conventional techniques.
- This book will benefit almost all stakeholders – industries, professionals, academics, research scholars, senior graduate students and human healthcare.
- The latest research and development of technology related to processing and characterization techniques, industry oriented case studies, applications in different sector of smart materials are available to professionals and engineers. The researchers in the area will get a comprehensive update on the state-of-art in the area and ample ideas for further research.

Compilation of References

Abdullah, B. J., Jiang, Q., & Omar, M. S. (2016). Effects of size on mass density and its influence on mechanical and thermal properties of ZrO2 nanoparticles in different structures. *Bulletin of Materials Science*, *39*(5), 1295–1302. doi:10.100712034-016-1244-5

Abdullah, B. J., Omar, M. S., & Jiang, Q. (2017). Size Effects on Cohesive Energy, Debye Temperature and Lattice Heat Capacity from First-Principles Calculations of Sn Nanoparticles. *Proceedings of the National Academy of Sciences. India. Section A, Physical Sciences*, *88*(4), 629–632. doi:10.100740010-017-0417-y

Abrahamsson, I., Zitzmann, N. U., Berglundh, T., Wennerberg, A., & Lindhe, J. (2001). Bone and soft tissue integration to titanium implants with different surface topography: An experimental study in the dog. *The International Journal of Oral & Maxillofacial Implants*, *16*(3). PMID:11432652

Adetunji, O. R., Adegbola, A. O., & Afolalu, S. A. (2015). Comparative study of case-hardening and water-quenching of mild steel rod on its mechanical properties. *International Journal of Advanced Research*, *3*(6), 1–9.

Adu, K. W., Humberto, R. G., & Eklund, P. (2007). *Phonons in Silicon Nanowires. Nanosilicon: new photonic and electronic material at the transition between solid and molecule*. Elsevier Ltd. https://doi.org/. doi:10.1016/B978-0-08-044528-1.50008-7

Aflori, M. (2021). Smart Nanomaterials for Biomedical Applications—A Review. *Nanomaterials (Basel, Switzerland)*, *11*(2), 396. doi:10.3390/nano11020396 PMID:33557177

Afolalu, S. A., Soetan, S. B., Ongbali, S. O., Abioye, A. A., & Oni, A. S. (2019, November). Impact of activated–flux tungsten inert gas (a-tig) welding on weld joint of a metal–Review. []. IOP Publishing.]. *IOP Conference Series. Materials Science and Engineering*, *640*(1), 012064. doi:10.1088/1757-899X/640/1/012064

Aggelis, D. G. (2015). Structural health monitoring in aerospace engineering: A review. *Journal of Intelligent Material Systems and Structures*, *26*(3), 350–374.

Aguilar, M. R. J. S. (2019). Introduction to Smart Polymers and Their Applications. Smart Polymers and their Applications. Elsevier.

Aguilar, M. R., & Román, J. S. (2019). Introduction to Smart Polymers and Their Applications. Smart Polymers and their Applications. Elsevier. doi:10.1016/B978-0-08-102416-4.00001-6

Ahamed, R., Choi, S.-B., & Ferdaus, M. M. (2018). A state of art on magneto-rheological materials and their potential applications. *Journal of Intelligent Material Systems and Structures*, *29*(10), 2051–2095. doi:10.1177/1045389X18754350

Ahmadi, S., Hedayati, R., Jain, R. A. K., Li, Y., Leeflang, S., & Zadpoor, A. (2017). Effects of laser processing parameters on the mechanical properties, topology, and microstructure of additively manufactured porous metallic biomaterials: A vector-based approach. *Materials & Design*, *134*, 234–243. doi:10.1016/j.matdes.2017.08.046

Aizenberg, J., & Fratzl, P. (2009). Biological and Biomimetic Materials. *Advanced Materials*, *21*(4), 387–388. doi:10.1002/adma.200803699

Ajay, P. A. (2023). Handbook of Sustainable Materials: Modelling, Characterization, and Optimization. CRC PPress.

Ajay, P. Ahmad, S., Sharma, J., & Gambhir, V. (2023). Handbook of Sustainable Materials: Modelling, Characterization, and Optimization. CRC Press. doi:10.1201/9781003297772

Akahori, T. (2002). Biomaterials: Journal of Japanese Society for Biomaterials 28 (3), 155-156, 2010-07-30. *Metallurgical and Materials Transactions. A, Physical Metallurgy and Materials Science*, *33*, 477–486.

Aktas Eken, G., & Hayri Acar, M. (2020). *PVDF-Based Shape Memory Materials, Oppor*. Fluoropolymers. doi:10.1016/B978-0-12-821966-9.00009-2

Al Asmar, R., Zaouk, D., Bahouth, Ph., Podleki, J., & Foucaran, A. (2006). Characterization of electron beam evaporated ZnO thin films and stacking ZnO fabricated by e-beam evaporation and rf magnetron sputtering for the realization of resonators. *Microelectronic Engineering*, *83*(3), 393–398. doi:10.1016/j.mee.2005.10.010

Alexander-Webber, J. A., Groschner, C. K., Sagade, A. A., Tainter, G., Gonzalez-Zalba, M. F., Di Pietro, R., Wong-Leung, J., Tan, H. H., Jagadish, C., Hofmann, S., & Joyce, H. J. (2017). 'Supporting Information for Engineering the Photoresponse of InAs Nanowires'. *ACS Applied Materials & Interfaces*, *9*(50), 43993–44000. doi:10.1021/acsami.7b14415 PMID:29171260

Algamaiah, H., Silikas, N., & Watts, D. C. (2021). Polymerization shrinkage and shrinkage stress development in ultra-rapid photo-polymerized bulk fill resin composites. *Dental materials: official publication of the Academy of Dental Materials*, *37*(4), 559–567. doi:10.1016/j.dental.2021.02.012

Allizond, V., Comini, S., Cuffini, A. M., & Banche, G. (2022). Current Knowledge on Biomaterials for Orthopedic Applications Modified to Reduce Bacterial Adhesive Ability. *Antibiotics (Basel, Switzerland)*, *11*(4), 529. doi:10.3390/antibiotics11040529 PMID:35453280

Almeida, H., Amaral, M. H., & Lobão, P. (2012). Temperature and pH stimuli-responsive polymers and their applications in controlled and selfregulated drug delivery. *Journal of Applied Pharmaceutical Science*, 01-10.

Alsamhan, A., Ragab, A. E., Dabwan, A., Nasr, M. M., & Hidri, L. (2019). Prediction of formation force during single-point incremental sheet metal forming using artificial intelligence techniques. *PLoS One*, *14*(8), e0221341. doi:10.1371/journal.pone.0221341 PMID:31437217

Al-Tabbaa, A., Lark, B., Paine, K., Jefferson, T., Litina, C., Gardner, D., & Embley, T. (2018). Biomimetic cementitious construction materials for next-generation infrastructure. *Proceedings of the Institution of Civil Engineers, Smart Infrastructure and Construction, 171*(2), 67–76, 10.1680/jsmic.18.00005

Al-Thobity, A. M. (2020). The impact of polymerization technique and glass-fiber reinforcement on the flexural properties of denture base resin material. *European journal of dentistry, 14*(01), 092-099.

Amalraju, D., & Dawood, A. S. (2012). Mechanical strength evaluation analysis of stainless steel and titanium locking plate for femur bone fracture. *Engineering Science and Technology: An International Journal*, *2*(3), 381–388.

Anselme, K. (2000). Osteoblast adhesion on biomaterials. *Biomaterials*, *21*(7), 667–681. doi:10.1016/S0142-9612(99)00242-2 PMID:10711964

Arab Hassani, F., Shi, Q., Wen, F., He, T., Haroun, A., Yang, Y., Feng, Y., & Lee, C. (2020). Smart materials for smart healthcare– moving from sensors and actuators to self-sustained nanoenergy nanosystems. *Smart Materials in Medicine*, *1*, 92–124. doi:10.1016/j.smaim.2020.07.005

Araujo, A. B. (2015). A Brief Review on Smart Materials. *Materials Research*, *18*(3), 529–538.

Aravindan, S., & Krishnamurthy, R. (1999). Joining of ceramic composites by microwave heating. *Materials Letters*, *38*(4), 245–249. doi:10.1016/S0167-577X(98)00166-9

Arden, G. P., Taylor, A. R., & Ansell, B. M. (1970). Total hip replacement using the McKee-Farrar prosthesis. In rheumatoid arthritis, Still's disease, and ankylosing spondylitis. *Annals of the Rheumatic Diseases*, *29*(1), 1–5. doi:10.1136/ard.29.1.1 PMID:5416097

Arnold, R. M., Huddleston, N. E., & Locklin, J. (2012). Utilizing click chemistry to design functional interfaces through post-polymerization modification. *Journal of Materials Chemistry*, *22*(37), 19357–19365. doi:10.1039/c2jm31708g

Asghari, F., Samiei, M., Adibkia, K., Akbarzadeh, A., & Davaran, S. (2017). Biodegradable and biocompatible polymers for tissue engineering application: A review. *Artificial Cells, Nanomedicine, and Biotechnology*, *45*(2), 185–192. doi:10.3109/21691401.2016.1146731 PMID:26923861

Asha, A. B., Srinivas, S., Hao, X., & Narain, R. (2019). Enzyme-Responsive Polymers: Classifications, Properties, Synthesis Strategies, and Applications. Smart Polymers and Their Applications. Elsevier.

Asmatulu, R. (2016). *Smart Materials: Science and Technology*. Springer.

Attar, H., Löber, L., Funk, A., Calin, M., Zhang, L. C., Prashanth, K. G., Scudino, S., Zhang, Y. S., & Eckert, J. (2015). Mechanical behavior of porous commercially pure Ti and Ti–TiB composite materials manufactured by selective laser melting. *Materials Science and Engineering A*, *625*, 350–356. doi:10.1016/j.msea.2014.12.036

Avinash Gudimetla, D. L. (2021). Modelling and optimization of wear parameters of Al4032 reinforced with coal ash using Taguchi and RSM approach. *Composites theory and practice, 21*, 3-11.

Avinash Gudimetla, D. L. (2020). Investigation of mechanical and tribological behavior of al 4032-SiHGM MMC. *Composites Theory and Practice*, *20*(3-4), 142–156.

Avinash Gudimetla, S. S. (2020). Influence of RHA reinforcements on mechanical and wear behavior of Al 4032 composites. *Emerging Materials Research*, *9*(4), 1237–1249. doi:10.1680/jemmr.20.00104

Azimi, B., Maleki, H., Zavagna, L., De la Ossa, J. G., Linari, S., Lazzeri, A., & Danti, S. (2020). Bio-based electrospun fibers for wound healing. *Journal of Functional Biomaterials*, *11*(3), 67. doi:10.3390/jfb11030067 PMID:32971968

Badami, V., & Ahuja, B. (2014). Biosmart Materials: Breaking New Ground in Dentistry. *TheScientificWorldJournal*, *2014*, 1–7. doi:10.1155/2014/986912 PMID:24672407

Badiger, R. I., Narendranath, S., & Srinath, M. S. (2015). Joining of Inconel-625 alloy through microwave hybrid heating and its characterization. *Journal of Manufacturing Processes*, *18*, 117–123. doi:10.1016/j.jmapro.2015.02.002

Badiger, R. I., Narendranath, S., & Srinath, M. S. (2018). Microstructure and mechanical properties of Inconel-625 welded joint developed through microwave hybrid heating. *Proceedings of the Institution of Mechanical Engineers. Part B, Journal of Engineering Manufacture*, *232*(14), 2462–2477. doi:10.1177/0954405417697350

Badiger, R. I., Narendranath, S., Srinath, M. S., & Hebbale, A. M. (2019). Effect of power input on metallurgical and mechanical characteristics of Inconel-625 welded joints processed through microwave hybrid heating. *Transactions of the Indian Institute of Metals*, *72*(3), 811–824. doi:10.100712666-018-1537-z

Bae, S. H., Lee, S. Y., Jin, B. J., & Im, S. (2001). Growth and characterization of ZnO thin films grown by pulsed laser deposition. *Applied Surface Science*, *169*, 525–528. doi:10.1016/S0169-4332(00)00752-2

Bagha, L., Shegal, S., & Thakur, A. (2016). Comparative analysis of microwave based joining/welding of SS304-SS304 using different interfacing materials. In *MATEC Web of Conferences* (*Vol. 57*, p. 03001). EDP Sciences.

Bagha, L., Sehgal, S., Thakur, A., & Kumar, H. (2017). Effects of powder size of interface material on selective hybrid carbon microwave joining of SS304–SS304. *Journal of Manufacturing Processes*, *25*, 290–295. doi:10.1016/j.jmapro.2016.12.013

Bagheri, H., Ayazi, Z., & Aghakhani, A. (2011). A novel needle trap sorbent based on carbon nanotube-sol–gel for microextraction of polycyclic aromatic hydrocarbons from aquatic media. *Analytica Chimica Acta*, *683*(2), 212–220. doi:10.1016/j.aca.2010.10.026 PMID:21167973

Bagudanch, I., Centeno, G., Vallellano, C., & Garcia-Romeu, M. L. (2019). Towards the manufacturing of near net shape medical prostheses in polymeric sheet by incremental sheet forming. *Near Net Shape Manufacturing Processes*, 1-33.

Bai, G., Liu, Z., Lin, J., Yu, Z., Hu, Y., & Wen, C. (2016). Effects of the addition of lanthanum and ultrasonic stirring on the microstructure and mechanical properties of the in situ Mg2Si/Al composites. *Materials & Design*, *90*, 424–432. doi:10.1016/j.matdes.2015.10.159

Bajpai, P. K., Singh, I., & Madaan, J. (2012). Joining of natural fiber reinforced composites using microwave energy: Experimental and finite element study. *Materials & Design*, *35*, 596–602. doi:10.1016/j.matdes.2011.10.007

Banerjee, A., Datta, S., Das, A., Chowdhury, A. R., & Datta, P. (2022). A Micro-Scale Non-Linear Finite Element Model to Optimize the Mechanical Behavior of Bioprinted Constructs. *3D printing and additive manufacturing, 9*(6), 490-502.

Bansal, A., Sharma, A. K., Das, S., & Kumar, P. (2016). On microstructure and strength properties of microwave welded Inconel 718/stainless steel (SS-316L). *Proceedings of the Institution of Mechanical Engineers, Part L: Journal of Materials: Design and Applications, 230*(5), 939-948.

Bansal, A., Sharma, A. K., & Das, S. (2013). Metallurgical and mechanical characterization of mild steel-mild steel joint formed by microwave hybrid heating process. *Sadhana*, *38*(4), 679–686. doi:10.100712046-013-0142-4

Bansal, A., Sharma, A. K., Kumar, P., & Das, S. (2012). Joining of mild steel plates using microwave energy. *Advanced Materials Research*, *585*, 465–469. doi:10.4028/www.scientific.net/AMR.585.465

Bao, Q., Zhang, D., & Qi, P. (2011). Synthesis and characterization of silver nanoparticle and graphene oxide nanosheet composites as a bactericidal agent for water disinfection. *Journal of Colloid and Interface Science*, *360*(2), 463–470. doi:10.1016/j.jcis.2011.05.009 PMID:21628064

Barbarino, S., Ameduri, S., Lecce, L., & Concilio, A. (2008). Wing Shape Control through an SMA-Based Device. *Journal of Intelligent Material Systems and Structures*, *20*(3), 283–296. doi:10.1177/1045389X08093825

Barker, S. L., Ross, D., Tarlov, M. J., Gaitan, M., & Locascio, L. E. (2000). Control of flow direction in microfluidic devices with polyelectrolyte multilayers. *Analytical Chemistry*, *72*(24), 5925–5929. doi:10.1021/ac0008690 PMID:11140758

Baroni, S., de Gironcoli, S., Dal Corso, A., & Giannozzi, P. (2001). Phonons and related crystal properties from density-functional perturbation theory. *Reviews of Modern Physics*, *73*(2), 515–562. doi:10.1103/RevModPhys.73.515

Barua, R., & Datta, S. RoyChowdhury, A., & Datta, P. (2022). Study of the surgical needle and biological soft tissue interaction phenomenon during insertion process for medical application: A Survey. *Proceedings of the Institution of Mechanical Engineers, Part H: Journal of Engineering in Medicine, 236*(10), 1465-1477.

Barua, R., Datta, P., & Chowdhury, A. R. (2021). The Impact of 3D Printing Technology on the COVID-19 Pandemic. *Quality of Healthcare in the Aftermath of the COVID-19 Pandemic*, 135.

Barua, R., Datta, S., Datta, P., & Chowdhury, A. R. (2019). Scaffolds and tissue engineering applications by 3D bio-printing process: a new approach. In Design, Development, and Optimization of Bio-Mechatronic Engineering Products (pp. 78-99). IGI global.

Barua, R., Datta, S., Datta, P., & RoyChowdhury, A. (2020). Study the Bioprinting Process in Biomedical and Tissue Engineering Fields. *Proceedings of Industry Interactive Innovations in Science, Engineering & Technology (I3SET2K19).*

Barua, R., Datta, S., Roychowdhury, A., & Datta, P. (2021). Importance of 3D printing technology in medical fields. In Research Anthology on Emerging Technologies and Ethical Implications in Human Enhancement (pp. 704-717). IGI Global.

Barua, R., Sarkar, A., & Datta, S. (2023). Emerging Advancement of 3D Bioprinting Technology in Modern Medical Science and Vascular Tissue Engineering Education. In Handbook of Research on Instructional Technologies in Health Education and Allied Disciplines (pp. 153-175). IGI Global.

Barua, R., Das, S., RoyChowdhury, A., & Datta, P. (2023). Simulation and experimental investigation of the surgical needle deflection model during the rotational and steady insertion process. *The International Journal of Artificial Organs*, *46*(1), 40–51. doi:10.1177/03913988221136154 PMID:36397288

Barua, R., Datta, S., Datta, P., & Roychowdhury, A. (2022). Study and Application of Machine Learning Methods in Modern Additive Manufacturing Processes. In S. Salunkhe, H. Hussein, & J. Davim (Eds.), *Applications of Artificial Intelligence in Additive Manufacturing* (pp. 75–95). IGI Global. doi:10.4018/978-1-7998-8516-0.ch004

Barua, R., Giria, H., Datta, S., Roy Chowdhury, A., & Datta, P. (2020). Force modeling to develop a novel method for fabrication of hollow channels inside a gel structure. *Proceedings of the Institution of Mechanical Engineers. Part H, Journal of Engineering in Medicine*, *234*(2), 223–231. doi:10.1177/0954411919891654 PMID:31774361

Barua, R., Roychowdhury, A., & Datta, P. (2022). Study of Different Additive Manufacturing Processes and Emergent Applications in Modern Healthcare. In *Advanced Manufacturing Techniques for Engineering and Engineered Materials* (pp. 239–259). IGI Global. doi:10.4018/978-1-7998-9574-9.ch014

Basu, B., Gowtham, N. H., Xiao, Y., Kalidindi, S. R., & Leong, K. W. (2022). Biomaterialomics: Data science-driven pathways to develop fourth-generation biomaterials. *Acta Biomaterialia*, *143*, 1–25. doi:10.1016/j.actbio.2022.02.027 PMID:35202854

Bawa, P., Pillay, V., Choonara, Y. E., & Du Toit, L. C. (2009). Stimuli-responsive polymers and their applications in drug delivery. *Biomedical Materials (Bristol, England)*, *4*(2), 022001. doi:10.1088/1748-6041/4/2/022001 PMID:19261988

Bazban-Shotorbani, S., Hasani-Sadrabadi, M. M., Karkhaneh, A., Serpooshan, V., Jacob, K. I., Moshaverinia, A., & Mahmoudi, M. (2017). Revisiting structure-property relationship of pH-responsive polymers for drug delivery applications. *Journal of Controlled Release*, *253*, 46–63. doi:10.1016/j.jconrel.2017.02.021 PMID:28242418

Bearat, H.H. & Vernon, B. L. (2011). Environmentally responsive injectable materials, Inject. *Biomater*, 263–297. . doi:10.1533/9780857091376.3.263

Bekas, D. G., Tsirka, K., Baltzis, D., & Paipetis, A. S. (2016). Self-healing materials: A review of advances in materials, evaluation, characterization and monitoring techniques. *Composites. Part B, Engineering*, *87*, 92–119. doi:10.1016/j.compositesb.2015.09.057

Bera, S. S. (2022). Review of defect engineering in perovskites for photovoltaic application. *Materials Advances* (*Vol.* 3, Issue 13). Royal Society of Chemistry (RSC). doi:10.1039/D2MA00194B

Bergbreiter, D. E., Zhang, L., & Mariagnanam, V. M. (1993). Smart ligands that regulate homogeneously catalyzed reactions. *Journal of the American Chemical Society*, *115*(20), 9295–9296. doi:10.1021/ja00073a058

Berman, D., Erdemir, A., & Sumant, A. V. (2014). Graphene: A new emerging lubricant. *Materials Today*, *17*(1), 31–42. doi:10.1016/j.mattod.2013.12.003

Berman, D., Erdemir, A., & Sumant, A. V. (2018). Approaches for achieving superlubricity in two-dimensional materials. *ACS Nano*, *12*(3), 2122–2137. doi:10.1021/acsnano.7b09046 PMID:29522673

Binyamin, G., Shafi, B. M., & Mery, C. M. (2006, November). Biomaterials: A primer for surgeons. [). WB Saunders.]. *Seminars in Pediatric Surgery*, *15*(4), 276–283. doi:10.1053/j.sempedsurg.2006.07.007 PMID:17055958

Biswas, P., Karn, K., Balasubramanian, P., & Kale, P. (2017). 'Biosensors and Bioelectronics Biosensor for Detection of Dissolved Chromium in Potable Water : A Review'. *Biosensors and Bioelectronic*. . doi:10.1016/j.bios.2017.03.043

Biswas, P., Mandal, D., & Mondal, M. K. (2020). Failures analysis of in-situ Al–Mg_2Si composites using actual microstructure based model. *Materials Science and Engineering A*, *797*, 140155. doi:10.1016/j.msea.2020.140155

Bobbert, F. S. L., & Zadpoor, A. A. (2017). Effects of bone substitute architecture and surface properties on cell response, angiogenesis, and structure of new bone. *Journal of Materials Chemistry. B, Materials for Biology and Medicine*, *5*(31), 6175–6192. doi:10.1039/C7TB00741H PMID:32264433

Bosshardt, D. D., Chappuis, V., & Buser, D. (2017). Osseointegration of titanium, titanium alloy and zirconia dental implants: Current knowledge and open questions. *Periodontology 2000*, *73*(1), 22–40. doi:10.1111/prd.12179 PMID:28000277

Bououdina, M. (2015). Smart Materials and Their Applications: A Review. *Journal of Materials Science Research*, *4*(4), 70–89.

Bourell, D., Kruth, J. P., Leu, M., Levy, G., Rosen, D., Beese, A. M., & Clare, A. (2017). Materials for additive manufacturing. *CIRP Annals - Manufacturing Technology, 66*(2), 659-681. https:// doi.org/ doi:10.1016/j.cirp.2017.05.009

Bozkurt, Y., Uzun, H., & Salman, S. (2011). Microstructure and mechanical properties of friction stir welded particulate reinforced AA2124/SiC/25p–T4 composite. *Journal of Composite Materials*, *45*(21), 2237–2245. doi:10.1177/0021998311401067

Bülbül, F., Efeoğlu, İ., & Arslan, E. (2007). The effect of bias voltage and working pressure on S/Mo ratio at MoS2–Ti composite films. *Applied Surface Science*, *253*(9), 4415–4419. doi:10.1016/j.apsusc.2006.09.059

Bussmann-Holder, A. (2020). Ferroelectrics: Past, present and future. *Ferroelectrics*, *569*(1), 1–8. doi:10.1080/00150193.2020.1791652

Byambaa, B., Annabi, N., Yue, K., Trujillo-de Santiago, G., Alvarez, M. M., Jia, W., Kazemzadeh-Narbat, M., Shin, S. R., Tamayol, A., & Khademhosseini, A. (2017). Bioprinted Osteogenic and Vasculogenic Patterns for Engineering 3D Bone Tissue. *Advanced healthcare materials, 6*(16). . doi:10.1002/adhm.201700015

Cano, E. E., Aguilar, M. R., Vázquez, B., & Román, J. S. (2019). Inflammation-Responsive Polymers. Smart Polymers and Their Applications, Elsevier, 219-254.

Cao, L., Kiely, J., Piano, M., & Luxton, R. (2019). *A Copper Oxide / Zinc Oxide Composite Nano-Surface for Use in a Biosensor*. MDPI. . doi:10.3390/ma12071126

Cao, W., & Hench, L. L. (1996, January). Bioactive materials. *Ceramics International*, *22*(6), 493–507. doi:10.1016/0272-8842(95)00126-3

Care, A., Bergquist, P. L., & Sunna, A. (2015, May). Solid-binding peptides: Smart tools for nanobiotechnology. *Trends in Biotechnology*, *33*(5), 259–268. doi:10.1016/j.tibtech.2015.02.005 PMID:25796487

Cargnello, M., & Thomas, R. (2014, October 08). Gordon, and Christopher B Murray. 2014. 'Solution-Phase Synthesis of Titanium Dioxide Nanoparticles and Nanocrystals'. *Chemical Reviews*, *114*(19), 9319–9345. doi:10.1021/cr500170p PMID:25004056

Casaletto, M. P., Ingo, G. M., Kaciulis, S., Mattogno, G., Pandolfi, L., & Scavia, G. (2001). Surface studies of in vitro biocompatibility of titanium oxide coatings. *Applied Surface Science*, *172*(1-2), 167–177. doi:10.1016/S0169-4332(00)00844-8

Centeno, G., Bagudanch, I., Martínez-Donaire, A. J., Garcia-Romeu, M. L., & Vallellano, C. (2014). Critical analysis of necking and fracture limit strains and forming forces in single-point incremental forming. *Materials & Design*, *63*, 20–29. doi:10.1016/j.matdes.2014.05.066

Chang, M. C., Hung, C. C., Chen, W. C., Tseng, S. C., Chen, Y. C., & Wang, J. C. (2019). Effects of pontic span and fiber reinforcement on fracture strength of multi-unit provisional fixed partial dentures. *Journal of Dental Sciences*, *14*(3), 309–317. doi:10.1016/j.jds.2018.11.008 PMID:31528260

Chassenieux, C., & Tsitsilianis, C. (2016). Recent trends in pH/thermo-responsive self-assembling hydrogels: From polyions to peptide-based polymeric gelators. *Soft Matter*, *12*(5), 1344–1359. doi:10.1039/C5SM02710A PMID:26781351

Chen, M., Guan, R., & Yang, S. (2018). Hybrids of Fullerenes and 2D Nanomaterials. *Advanced Science, 6*(1), 1800941. Portico. . doi:10.1002/advs.201800941

Chen, J., Qiu, Q., Han, Y., & Lau, D. (2019). Piezoelectric materials for sustainable building structures: Fundamentals and applications. *Renewable & Sustainable Energy Reviews*, *101*, 14–25. doi:10.1016/j.rser.2018.09.038

Chen, S., Guan, S., Li, W., Wang, H., Chen, J., Wang, Y., & Wang, H. (2012). In vivo degradation and bone response of a composite coating on Mg–Zn–Ca alloy prepared by microarc oxidation and electrochemical deposition. *Journal of Biomedical Materials Research. Part B, Applied Biomaterials*, *100*(2), 533–543. doi:10.1002/jbm.b.31982 PMID:22120974

Chen, S., Hu, J., Zhuo, H., & Zhu, Y. (2008). Two-way shape memory effect in polymer laminates. *Materials Letters*, *62*(25), 4088–4090. doi:10.1016/j.matlet.2008.05.073

Chen, Y. M., & Hong, W. J. (2014). *Synthesis and Characterization of ZnO Nanoparticles Templated by SBS-OH. Advanced Materials Research* (Vol. 864–867). Elsevier Ltd. doi:10.4028/www.scientific.net/AMR.864-867.694

Choi, S.-B., Li, W., Yu, M., Du, H., Fu, J., & Do, P. X. (2016). State of the art of control schemes for smart systems featuring magneto-rheological materials. *Smart Materials and Structures*, *25*(4), 043001. doi:10.1088/0964-1726/25/4/043001

Cho, J. P. (2020). Scalable roll-to-roll processing of smart materials and structures for IoT devices. *Smart Materials and Structures*, *29*(8).

Chorny, V., Yatsun, Y., & Golovakha, M. (2014). Historical aspects of using of biodegradable magnesium-based alloys for osteosynthesis (literature review). *Orthopaedics. Ortopediia Travmatologiia i Protezirovanie*, *0*(1), 105–109. doi:10.15674/0030-598720141105-109

Chowdhury, S. M., Chen, D. L., Bhole, S. D., Powidajko, E., Weckman, D. C., & Zhou, Y. (2011). Microstructure and mechanical properties of fiber-laser-welded and diode-laser-welded AZ31 magnesium alloy. *Metallurgical and Materials Transactions. A, Physical Metallurgy and Materials Science*, *42*(7), 1974–1989. doi:10.100711661-010-0574-y

Chrétien, S., & Salahub, D. R. (2001). Density Functional Theory, Methods, Techniques, and Applications. In C. Guet, P. Hobza, F. Speigelman, & F. David (Eds.), *Atomic clusters and nanoparticles. Agregats atomiques et nanoparticules: Les Houches Session LXXIII 2–28 July 2000* (pp. 105–160). Springer. doi:10.1007/3-540-45621-X_4

Ciocca, L., Fantini, M., De Crescenzio, F., Corinaldesi, G., & Scotti, R. (2011). Direct metal laser sintering (DMLS) of a customized titanium mesh for prosthetically guided bone regeneration of atrophic maxillary arches. *Medical & Biological Engineering & Computing*, *49*(11), 1347–1352. doi:10.100711517-011-0813-4 PMID:21779902

Cizaire, L., Vacher, B., Le Mogne, T., Martin, J. M., Rapoport, L., Margolin, A., & Tenne, R. (2002). Mechanisms of ultra-low friction by hollow inorganic fullerene-like MoS2 nanoparticles. *Surface and Coatings Technology*, *160*(2-3), 282–287. doi:10.1016/S0257-8972(02)00420-6

Clark, A. E. (2004). Magnetostrictive materials and their use in high-precision positioning systems. *Journal of Intelligent Material Systems and Structures*, *15*(10), 797–807.

Colnik, M., Knez-Hrncic, M., Skerget, M., & Knez, Z. (2020). Biodegradable polymers, current trends of research and their applications, a review. *Chemical Industry & Chemical Engineering Quarterly*, *26*(4), 401–418. doi:10.2298/CICEQ191210018C

Cornaton, Y., Ringholm, M., Louant, O., & Ruud, K. (2016). Analytic calculations of anharmonic infrared and Raman vibrational spectra. *Physical Chemistry Chemical Physics*, *18*(5), 4201–4215. doi:10.1039/C5CP06657C PMID:26784673

Crăciunescu, C., & Ercuta, A. (2015). Modulated interaction in double-layer shape memory-based micro-designed actuators. *Science and Technology of Advanced Materials*, *16*(6), 065003. doi:10.1088/1468-6996/16/6/065003 PMID:27877845

Cross, L. E. (1999). *Ceramics for electro-optics, piezoelectrics, and pyroelectrics*. Springer Science & Business Media.

CTTC. (n.d.). *Additive manufacturing of ceramics*. CTTC. https://www.cttc.fr/en/videos-downloads/additive-manufacturing-of-ceramics/

Cui, C., Hu, B., Zhao, L., & Liu, S. (2011). Titanium alloy production technology, market prospects and industry development. *Materials & Design*, *32*(3), 1684–1691. doi:10.1016/j.matdes.2010.09.011

Cui, J. F. (2020). Development of high-performance smart materials: Progress and challenges. *Journal of Materials Chemistry. A, Materials for Energy and Sustainability*, *8*(30), 14975–15000.

Cunliffe, D., de las Heras Alarcón, C., Peters, V., Smith, J. R., & Alexander, C. (2003). Thermoresponsive surface-grafted poly (N− isopropylacrylamide) copolymers: Effect of phase transitions on protein and bacterial attachment. *Langmuir*, *19*(7), 2888–2899. doi:10.1021/la026358l

Curry, J. F., Wilson, M. A., Luftman, H. S., Strandwitz, N. C., Argibay, N., Chandross, M., Sidebottom, M. A., & Krick, B. A. (2017). Impact of microstructure on MoS2 oxidation and friction. *ACS Applied Materials & Interfaces*, *9*(33), 28019–28026. doi:10.1021/acsami.7b06917 PMID:28758391

Dagani, R. (1995). Polymeric smart materials respond to changes in their environment. *Chemical & Engineering News*, (*Vol. 73*, Issue 38). American Chemical Society (ACS). . doi:10.1021/cen-v073n038.p030

Dagdelen, F., Kok, M., & Qader, I. N. (2019). Effects of Ta Content on Thermodynamic Properties and Transformation Temperatures of Shape Memory NiTi Alloy. *Metals and Materials International*, *25*(6), 1420–1427. doi:10.100712540-019-00298-z

Dai, H., Fan, Q., & Wang, C. (2022). Recent applications of immunomodulatory biomaterials for disease immunotherapy. *Exploration*, *2*(6), 20210157. doi:10.1002/EXP.20210157 PMID:37324799

Dallavalle, M., Sändig, N., & Zerbetto, F. (2012). Stability, dynamics, and lubrication of MoS2 platelets and nanotubes. *Langmuir*, *28*(19), 7393–7400. doi:10.1021/la300871q PMID:22530739

Damjanovic, D., & Newnham, R. E. (1992, April). Electrostrictive and Piezoelectric Materials for Actuator Applications. *Journal of Intelligent Material Systems and Structures*, *3*(2), 190–208. doi:10.1177/1045389X9200300201

Dämmer, G., Bauer, H., Neumann, R., & Major, Z. (2022). Design, additive manufacturing and component testing of pneumatic rotary vane actuators for lightweight robots. *Rapid Prototyping Journal*, *28*(11), 20–32. doi:10.1108/RPJ-03-2021-0052

Darabi, M. A., Khosrozadeh, A., Mbeleck, R., Liu, Y., Chang, Q., Jiang, J., Cai, J., Wang, Q., Luo, G., & Xing, M. (2017). Skin-inspired multifunctional autonomic-intrinsic conductive self-healing hydrogels with pressure sensitivity, stretchability, and 3D printability. *Advanced Materials*, *29*(31), 1700533. doi:10.1002/adma.201700533 PMID:28640439

Datta, S. (2023). Advantage of Alginate Bioinks in Biofabrication for Various Tissue Engineering Applications. *International Journal of Polymer Science*.

Datta, S., Barua, R., & Das, J. (2020). Importance of alginate bioink for 3D bioprinting in tissue engineering and regenerative medicine. *Alginates—Recent Uses of This Natural Polymer*, 1-11.

Datta, S., Barua, R., & Das, J. (2020). *Importance of alginate bioink for 3D bioprinting in tissue engineering and regenerative medicine*. Alginates—Recent Uses of This Natural Polymer. doi:10.5772/intechopen.90426

Datta, S., Barua, R., Sarkar, R., Barui, A., Chowdhury, A. R., & Datta, P. (2018, September). Design and development of alginate: Poly-l-lysine scaffolds by 3D bio printing and studying their mechanical, structural and cell viability properties. []. IOP Publishing.]. *IOP Conference Series. Materials Science and Engineering*, *402*, 012113. doi:10.1088/1757-899X/402/1/012113

Datta, S., Das, A., Chowdhury, A. R., & Datta, P. (2019). Bioink formulations to ameliorate bioprinting-induced loss of cellular viability. *Biointerphases*, *14*(5), 051006. doi:10.1116/1.5111392 PMID:31610661

Datta, S., Jana, S., Das, A., Chakraborty, A., Chowdhury, A. R., & Datta, P. (2020). Bioprinting of radiopaque constructs for tissue engineering and understanding degradation behavior by use of Micro-CT. *Bioactive Materials*, *5*(3), 569–576. doi:10.1016/j.bioactmat.2020.04.015 PMID:32373763

Dauda, K. T., Majebi, C., Ayoola, A. A., Agboola, O., Popoola, A. P., Fayomi, O. S. I., Banjo, S. O., Sojobi, J., & Ogunsanya, A. O. (2021). Biosmart Materials and Its Innovative Characteristics in Protective Composite Coating Application. *IOP Conference Series. Materials Science and Engineering*, *1107*(1), 012216. doi:10.1088/1757-899X/1107/1/012216

Davidson, J. A., Mishra, A. K., Kovacs, P., & Poggie, R. A. (1994). New surface-hardened, low-modulus, corrosion-resistant Ti-13Nb-13Zr alloy for total hip arthroplasty. *Bio-Medical Materials and Engineering*, *4*(3), 231–243. doi:10.3233/BME-1994-4310 PMID:7950871

Dawson, E., Mapili, G., Erickson, K., Taqvi, S., & Roy, K. (2008). Biomaterials for stem cell differentiation. *Advanced Drug Delivery Reviews*, *60*(2), 215–228. doi:10.1016/j.addr.2007.08.037 PMID:17997187

Dayyoub, T., Maksimkin, A. V., Filippova, O. V., Tcherdyntsev, V. V., & Telyshev, D. V. (2022). Shape Memory Polymers as Smart Materials: A Review. *Polymers*, *14*(17), 3511. doi:10.3390/polym14173511 PMID:36080587

De Caluwé, T. V. (2017). Addition of bioactive glass to glass ionomer cements: Effect on the physico-chemical properties and biocompatibility. *Dental Materials* (*Vol. 33*, Issue 4). Elsevier BV.

de Formanoir, C., Brulard, A., Vivès, S., Martin, G., Prima, F., Michotte, S., Rivière, E., Dolimont, A., & Godet, S. (2017). A strategy to improve the work-hardening behavior of Ti–6Al–4V parts produced by additive manufacturing. *Materials Research Letters*, *5*(3), 201–208. doi:10.1080/21663831.2016.1245681

Deb, S. K. (2014). Organic electrochromic materials and devices: Recent advances and future prospects. *Journal of Materials Chemistry. C, Materials for Optical and Electronic Devices*, *2*(15), 2656–2675.

DeRossi, D., Kajiwara, K., Osada, Y., & Yamauchi, A. (1991). *Polymer gels*. Fundamentals and Biomedical Applications. doi:10.1007/978-1-4684-5892-3

Deuis, R. L., Subramanian, C., & Yellup, J. M. (1997). Dry sliding wear of aluminium composites—A review. *Composites Science and Technology*, *57*(4), 415–435. doi:10.1016/S0266-3538(96)00167-4

Dhanasekaran, R., Sreenatha Reddy, S., Girish Kumar, B., & Anirudh, A. S. (2018). Shape Memory Materials for Bio-medical and Aerospace Applications. *Materials Today: Proceedings*, *5*(10), 21427–21435. doi:10.1016/j.matpr.2018.06.551

Dhanopia, A., & Bhargava, M. (2017). Finite element analysis of human fractured femur bone implantation with PMMA thermoplastic prosthetic plate. *Procedia Engineering*, *173*, 1658–1665. doi:10.1016/j.proeng.2016.12.190

Di Martino, S. F., Faulkner, R. G., Hogg, S. C., Vujic, S., & Tassa, O. (2014). Characterisation of microstructure and creep properties of alloy 617 for high-temperature applications. *Materials Science and Engineering A*, *619*, 77–86. doi:10.1016/j.msea.2014.09.046

Dierdorf, D. S. (2002). Smart materials: Some recent developments. *Annual Review of Materials Science*, *32*(1), 285–315.

Dietz, T. G., & Jaeger, H. (2013). *Magnetostrictive actuator of a medical ultrasound transducer assembly, and a medical ultrasound handpiece and a medical ultrasound system having such actuator*. Google Patents.

Ditler, E., & Luber, S. (2022). Vibrational spectroscopy by means of first-principles molecular dynamics simulations. *Wiley Interdisciplinary Reviews. Computational Molecular Science*, *12*(5), e1605. doi:10.1002/wcms.1605

Dobrescu, M., & Vasilescu, M. (2015). Processing and superelasticity of Ni-Ti shape memory alloys. *Key Engineering Materials*, *660*, 103–107.. doi:10.4028/www.scientific.net/KEM.660.103

Dodd, C. A., Hungerford, D. S., & Krackow, K. A. (1990). Total knee arthroplasty fixation. Comparison of the early results of paired cemented versus uncemented porous coated anatomic knee prostheses. *Clinical Orthopaedics and Related Research*, (260), 66–70. PMID:2225646

Dolez, P. I., & Mlynarek, J. (2016). *Smart materials for personal protective equipment:tendencies and recent developments, Smart Text.* Their Appl. doi:10.1016/B978-0-08-100574-3.00022-9

Donachie, M. J. (2000). *Titanium: a technical guide*. ASM international. doi:10.31399/asm.tb.ttg2.9781627082693

Donnet, C., Martin, J. M., Le Mogne, T., & Belin, M. (1996). Super-low friction of MoS2 coatings in various environments. *Tribology International*, *29*(2), 123–128. doi:10.1016/0301-679X(95)00094-K

Drossel, W. G., Kunze, H., Bucht, A., Weisheit, L., & Pagel, K. (2015). Smart3 - smart materials for smart applications. *Procedia CIRP*, *36*, 211–216. doi:10.1016/j.procir.2015.01.055

Duan, C., Yuan, D., Yang, Z., Li, S., Tao, L., Wang, Q., & Wang, T. (2018). High wear-resistant performance of thermosetting polyimide reinforced by graphitic carbon nitride (g-C3N4) under high temperature. *Composites. Part A, Applied Science and Manufacturing*, *113*, 200–208. doi:10.1016/j.compositesa.2018.07.008

Duarte Campos, D. F., Blaeser, A., Korsten, A., Neuss, S., Jäkel, J., Vogt, M., & Fischer, H. (2015). The stiffness and structure of three-dimensional printed hydrogels direct the differentiation of mesenchymal stromal cells toward adipogenic and osteogenic lineages. *Tissue Engineering. Part A*, *21*(3-4), 740–756. doi:10.1089/ten.tea.2014.0231 PMID:25236338

Duarte, A. R. C., Mano, J. F., & Reis, R. L. (2011, February). Thermosensitive polymeric matrices for three-dimensional cell culture strategies. *Acta Biomaterialia*, *7*(2), 526–529. doi:10.1016/j.actbio.2010.09.032 PMID:20883833

Dušek, K., & Patterson, D. (1968). Transition in swollen polymer networks induced by intramolecular condensation. *Journal of Polymer Science. Part A-2, Polymer Physics*, *6*(7), 1209–1216. doi:10.1002/pol.1968.160060701

Dutta, S. M., & Ghorbel, F. H. (2005). Differential hysteresis modeling of a shape memory alloy wire actuator. *IEEE/ASME Transactions on Mechatronics*, *10*(2), 189–197. doi:10.1109/TMECH.2005.844709

Elianov, O., Garusi, S., Rosentsveig, R., Cohen, S. R., Feldman, Y., Pinkas, I., Bendikov, T., Kaplan-Ashiri, I., Moshkovich, A., Perfilyev, V., Rapoport, L., Moshonov, J., Tenne, R., & Shay, B. (2018). Deposition of metal coatings containing fullerene-like MoS2 nanoparticles with reduced friction and wear. *Surface and Coatings Technology*, *353*, 116–125. doi:10.1016/j.surfcoat.2018.08.084

Elias, C. N., Lima, J. H. C., Valiev, R., & Meyers, M. (2008). Biomedical Applications of Titanium and its Alloys. *J. Miner. Met. Mater. Soc*, 46-9.

Emami, B., Lyman, J., Brown, A., Cola, L., Goitein, M., Munzenrider, J. E., & Wesson, M. (1991). Tolerance of normal tissue to therapeutic irradiation. *International Journal of Radiation Oncology* Biology* Physics, 21*(1), 109-122.

Emmelmann, C., Scheinemann, P., Munsch, M., & Seyda, V. (2011). Laser additive manufacturing of modified implant surfaces with osseointegrative characteristics. *Physics Procedia*, *12*, 364–368. doi:10.1016/j.phpro.2011.03.046

Erbel, R., Di Mario, C., Bartunek, J., Bonnier, J., de Bruyne, B., Eberli, F. R., Erne, P., Haude, M., Heublein, B., Horrigan, M., Ilsley, C., Böse, D., Koolen, J., Lüscher, T. F., Weissman, N., & Waksman, R. (2007). Temporary scaffolding of coronary arteries with bioabsorbable magnesium stents: A prospective, non-randomised multicentre trial. *Lancet*, *369*(9576), 1869–1875. doi:10.1016/S0140-6736(07)60853-8 PMID:17544767

Falahati, M., Ahmadvand, P., Safaee, S., Chang, Y.-C., Lyu, Z., Chen, R., Li, L., & Lin, Y. (2020). Smart polymers and nanocomposites for 3D and 4D printing. *Materials Today*, *40*, 215–245. doi:10.1016/j.mattod.2020.06.001

Fallahiarezoudar, E., Ahmadipourroudposht, M., Idris, A., & Yusof, N. M. (2015). A review of: Application of synthetic scaffold in tissue engineering heart valves. *Materials Science and Engineering C*, *48*, 556–565. doi:10.1016/j.msec.2014.12.016 PMID:25579957

Farahany, S., Ghandvar, H., Nordin, N. A., & Ourdjini, A. (2017). Microstructure Characterization, Mechanical, and Tribological Properties of Slow-Cooled Sb-Treated Al-20Mg 2 Si-Cu In Situ Composites. *Journal of Materials Engineering and Performance*, *26*(4), 1685–1700. doi:10.100711665-017-2624-8

Farahany, S., Ghandvar, H., Nordin, N. A., Ourdjini, A., & Idris, M. H. (2016). Effect of primary and eutectic Mg2Si crystal modifications on the mechanical properties and sliding wear behaviour of an Al–20Mg2Si–2Cu–xBi composite. *Journal of Materials Science and Technology*, *32*(11), 1083–1097. doi:10.1016/j.jmst.2016.01.014

Farid, M. I., Wu, W., Liu, X., & Wang, P. (2021). Additive manufacturing landscape and materials perspective in 4D printing. *International Journal of Advanced Manufacturing Technology*, *115*(9-10), 2973–2988. doi:10.100700170-021-07233-w PMID:34092882

Fazel, M., Salimijazi, H., Shamanian, M., Apachitei, I., & Zadpoor, A. (2019). Influence of hydrothermal treatment on the surface characteristics and electrochemical behavior of Ti-6Al-4V bio-functionalized through plasma electrolytic oxidation. *Surface and Coatings Technology*, *374*, 222–231. doi:10.1016/j.surfcoat.2019.05.088

Felix, S., Kollu, P., & Grace, A. N. (2019). Electrochemical Performance of Ag–CuO Nanocomposites towards Glucose Sensing. *Materials Research Innovations*, *23*(1), 27–32. doi:10.1080/14328917.2017.1358507

Fernández, L. G., Boza, A. M., & Ortega, F. R. (2019). pH-Responsive Polymers: Properties, Synthesis, and Applications. *Smart Polymers and Their Applications*. Elsevier. . doi:10.1016/B978-0-08-102416-4.00003-X

Fiorentino, A, Ceretti, E., Attanasio, A., Mazzoni, L. U. C. A., & Giardini, C. (2009). Analysis of forces, accuracy and formability in positive die sheet incremental forming. *International Journal of Material Forming*, *2*(S1), 805–808. doi:10.100712289-009-0467-z

Fischer, M., Joguet, D., Robin, G., Peltier, L., & Laheurte, P. (2016). In situ elaboration of a binary Ti-26Nb alloy by selective laser melting of elemental titanium and niobium mixed powders. *Materials Science and Engineering C*, *62*, 852–859. doi:10.1016/j.msec.2016.02.033 PMID:26952492

Flores-Ruiz, F. J., Enriquez-Flores, C. I., Chiñas-Castillo, F., & Espinoza-Beltrán, F. J. (2014). Nanotribological performance of fullerene-like carbon nitride films. *Applied Surface Science*, *314*, 193–198. doi:10.1016/j.apsusc.2014.06.168

Fukada, E. (1955, February 15). Piezoelectricity of Wood. *Journal of the Physical Society of Japan*, *10*(2), 149–154. doi:10.1143/JPSJ.10.149

Fu, Y. Q., Luo, J. K., Flewitt, A. J., Huang, W. M., Zhang, S., Du, H. J., & Milne, W. I. (2009). Thin film shape memory alloys and microactuators. *International Journal of Computational Materials Science and Surface Engineering*, 2(3/4), 208. doi:10.1504/IJCMSSE.2009.027483

Galaev, I. (1999, August 1). "Smart" polymers and what they could do in biotechnology and medicine. *Trends in Biotechnology*, *17*(8), 335–340. doi:10.1016/S0167-7799(99)01345-1 PMID:10407406

Ganjian, M., Modaresifar, K., Zhang, H., Hagedoorn, P.-L., Fratila-Apachitei, L. E., & Zadpoor, A. A. (2019). Reactive ion etching for fabrication of biofunctional titanium nanostructures. *Scientific Reports*, *9*(1), 18815. doi:10.103841598-019-55093-y PMID:31827149

Gao, W. D. (2015). Self-Powered and Self-Healing Soft Robots. *Advanced Materials*, *27*(43), 7289–7295.

García-Moreno, F. (2018). Self-healing smart materials. *Smart Materials and Structures*, *27*(4).

García-Vallejo, D. R.-L.-L. (2021). Challenges and opportunities for smart materials in the construction industry. *Applied Sciences (Basel, Switzerland)*, *11*(1), 251.

Garner, L. J., Wilson, L. N., Lagoudas, D. C., & Rediniotis, O. K. (2000, September). Development of a shape memory alloy actuated biomimetic vehicle. *Smart Materials and Structures*, *9*(5), 673–683. doi:10.1088/0964-1726/9/5/312

Garnett, E., Brongersma, M., Cui, Y., & Mcgehee, M. (2011). 'Nanowire Solar Cells'.. doi:10.1146/annurev-matsci-062910-100434

Gautam, R., Singh, R. D., Sharma, V. P., Siddhartha, R., Chand, P., & Kumar, R. (2012). Biocompatibility of polymethylmethacrylate resins used in dentistry. *Journal of Biomedical Materials Research. Part B, Applied Biomaterials*, *100*(5), 1444–1450. doi:10.1002/jbm.b.32673 PMID:22454327

Gaviria, L., Salcido, J. P., Guda, T., & Ong, J. L. (2014). Current trends in dental implants. *Journal of the Korean Association of Oral and Maxillofacial Surgeons*, *40*(2), 50. doi:10.5125/jkaoms.2014.40.2.50 PMID:24868501

Gawande, M. B., Shelke, S. N., Zboril, R., & Varma, R. S. (2014). Microwave-assisted chemistry: Synthetic applications for rapid assembly of nanomaterials and organics. *Accounts of Chemical Research*, *47*(4), 1338–1348. doi:10.1021/ar400309b PMID:24666323

Ge, J., Sun, L., Zhang, F. R., Zhang, Y., Shi, L. A., Zhao, H. Y., Zhu, H.-W., Jiang, H.-L., & Yu, S. H. (2016). A stretchable electronic fabric artificial skin with pressure-, lateral strain-, and flexion-sensitive properties. *Advanced Materials*, *28*(4), 722–728. doi:10.1002/adma.201504239 PMID:26618615

Genabai, N. K., Ahmad, S., Zhang, Z., Jiang, X., Gabaldon, C. A., & Gangwani, L. (2015). Genetic inhibition of JNK3 ameliorates spinal muscular atrophy. *Human Molecular Genetics*, *24*(24), 6986–7004. doi:10.1093/hmg/ddv401 PMID:26423457

Genchi, G. G., Marino, A., Tapeinos, C., & Ciofani, G. (2017). Smart Materials Meet Multifunctional Biomedical Devices: Current and Prospective Implications for Nanomedicine. *Frontiers in Bioengineering and Biotechnology*, *5*, 80. doi:10.3389/fbioe.2017.00080 PMID:29326928

Ge, Q., Liu, X., Qiao, A., & Mu, Y. (2022). Compressive Properties and Degradable Behavior of Biodegradable Porous Zinc Fabricated with the Protein Foaming Method. *Journal of Functional Biomaterials*, *13*(3), 151. doi:10.3390/jfb13030151 PMID:36135585

Ghosh, R., & Basak, D. (2007). Electrical and ultraviolet photoresponse properties of quasialigned ZnO nanowires/p-Si heterojunction. *Applied Physics Letters*, *90*(24), 243106–3. doi:10.1063/1.2748333

Ghosh, S., Sanghavi, S., & Sancheti, P. (2018). Metallic biomaterial for bone support and replacement. In *Fundamental biomaterials: metals* (pp. 139–165). Woodhead Publishing. doi:10.1016/B978-0-08-102205-4.00006-4

Giannatsis, J., & Dedoussis, V. (2009). Additive fabrication technologies applied to medicine and health care: A review. *International Journal of Advanced Manufacturing Technology*, *40*(1), 116–127. doi:10.100700170-007-1308-1

Gil, E. S., & Hudson, S. M. (2004). Stimuli-reponsive polymers and their bioconjugates. *Progress in Polymer Science*, *29*(12), 1173–1222. doi:10.1016/j.progpolymsci.2004.08.003

Giurgiutiu, V., & Zagrai, A. N. (2000). Use of smart materials technologies in radiation environments and nuclear industry. *Smart Structures and Materials 2000: Smart Structures and Integrated Systems*. SPIE. . doi:10.1117/12.388812

Goh, E. G., Xu, X., & McCormick, P. G. (2014). 'Effect of Particle Size on the UV Absorbance of Zinc Oxide Nanoparticles'. *Scripta Materialia*, *78–79*, 49–52. doi:10.1016/j.scriptamat.2014.01.033

Gokuldoss, P. K., Kolla, S., & Eckert, J. (2017). Additive Manufacturing Processes: Selective Laser Melting, Electron Beam Melting and Binder Jetting-Selection Guidelines. *Materials (Basel)*, *10*(6), 672. doi:10.3390/ma10060672 PMID:28773031

Gorgin Karaji, Z., Speirs, M., Dadbakhsh, S., Kruth, J. P., Weinans, H., Zadpoor, A. A., & Amin Yavari, S. (2017). Additively Manufactured and Surface Biofunctionalized Porous Nitinol. *ACS Applied Materials & Interfaces*, *9*(2), 1293–1304. doi:10.1021/acsami.6b14026 PMID:28001358

Gottardo, S., Mech, A., Drbohlavová, J., Małyska, A., Bøwadt, S., Riego Sintes, J., & Rauscher, H. (2021, January). Towards safe and sustainable innovation in nanotechnology: State-of-play for smart nanomaterials. *NanoImpact*, *21*, 100297. doi:10.1016/j.impact.2021.100297 PMID:33738354

Granqvist, C. G. (1995). Electrochromics for smart windows: Thin films of tungsten oxide and nickel oxide, and devices based on these. *Journal of Materials Chemistry*, *5*(4), 547–565.

Gray, H. N., & Bergbreiter, D. E. (1997). Applications of Polymeric Smart Materials to Environmental Problems. *Environmental Health Perspectives*, *105*, 55. doi:10.2307/3433397 PMID:9114277

Gulati, V., & Kumar, A. (2018). Investigation of some process parameters on forming force in single point incremental forming. *Int Res J Eng Technol*, *4*(4), 784–791.

Gupta, P., Adhikary, M., Kumar, M., Bhardwaj, N., & Mandal, B. B. (2016). Biomimetic, osteoconductive non-mulberry silk fiber reinforced tricomposite scaffolds for bone tissue engineering. *ACS Applied Materials & Interfaces*, *8*(45), 30797–30810. doi:10.1021/acsami.6b11366 PMID:27783501

Gupta, P., & Kumar, S. (2014). Investigation of stainless steel joint fabricated through microwave energy. *Materials and Manufacturing Processes*, *29*(8), 910–915. doi:10.1080/10426914.2014.892975

Gupta, R. K., Kumar, V. A., Mathew, C., & Rao, G. S. (2016). Strain hardening of titanium alloy Ti6Al4V sheets with prior heat treatment and cold working. *Materials Science and Engineering A*, *662*, 537–550. doi:10.1016/j.msea.2016.03.094

Gupta, T. K., Srivastava, A. K., & Srivastava, V. S. (2021). Microstructural and tensile behaviour of hybrid MMC Al7075/SiC/B4C produced by mechanical stir casting. *Materials Today: Proceedings*, *47*, 4107–4113. doi:10.1016/j.matpr.2021.06.430

Guvendiren, M., Molde, J., Soares, R. M., & Kohn, J. (2016). Designing biomaterials for 3D printing. *ACS Biomaterials Science & Engineering*, *2*(10), 1679–1693. doi:10.1021/acsbiomaterials.6b00121 PMID:28025653

Gu, X., Zheng, Y., Cheng, Y., Zhong, S., & Xi, T. (2009). In vitro corrosion and biocompatibility of binary magnesium alloys. *Biomaterials*, *30*(4), 484–498. doi:10.1016/j.biomaterials.2008.10.021 PMID:19000636

Hadjiloizi, E. (2020). Modelling and simulation of fatigue behaviour in smart materials and structures. *Fatigue & Fracture of Engineering Materials & Structures*, *43*(7), 1636–1653.

Haga, K., Suzuki, T., Kashiwaba, Y., Watanabe, H., Zhang, B. P., & Segawa, Y. (2003). High-quality ZnO films prepared on Si wafers by low-pressure MO-CVD. *Thin Solid Films*, *433*(1), 131–134. doi:10.1016/S0040-6090(03)00327-4

Haghniaz, R. (2018). Smart materials: A review of applications and technologies. *Renewable & Sustainable Energy Reviews*, *91*, 1295–1307.

Halonen, N., Pálvölgyi, P. S., Bassani, A., Feorentini, C., Nair, R., Spigno, G., & Kordas, K. (2020). Bio-Based Smart Materials for Food Packaging and Sensors – A Review. *Frontiers in Materials*, *7*, 82. doi:10.3389/fmats.2020.00082

Hamdi, O., Mighri, F., & Rodrigue, D. (2018). Piezoelectric cellular polymer films: Fabrication, properties and applications. *AIMS Materials Science*, *5*(5), 845–869. doi:10.3934/matersci.2018.5.845

Hamilton, R. F., Sehitoglu, H., Chumlyakov, Y., & Maier, H. J. (2004). Stress dependence of the hysteresis in single crystal NiTi alloys. *Acta Materialia*, *52*(11), 3383–3402. doi:10.1016/j.actamat.2004.03.038

Han, B., Wang, Y., Ding, S., Yu, X., Zhang, L., Li, Z., & Ou, J. (2016, July 28). Self-sensing cementitious composites incorporated with botryoid hybrid nano-carbon materials for smart infrastructures. *Journal of Intelligent Material Systems and Structures*, *28*(6), 699–727. doi:10.1177/1045389X16657416

Han, C. D. (2007). *Experimental Methods for Measurement of the Rheological Properties of Polymeric Fluids. Rheology and Processing of Polymeric Materials* (Vol. 1). Polymer Rheology. doi:10.1093/oso/9780195187823.001.0001

Han, M. D. (2017). Smart materials for energy storage systems: A review of recent advances. *Materials Today. Energy*, *6*, 169–185.

Hao, Y. L., Li, S. J., & Yang, R. (2016). Biomedical titanium alloys and their additive manufacturing. *Rare Metals*, *35*(9), 661–671. doi:10.100712598-016-0793-5

Hartl, D. J., Mooney, J. T., Lagoudas, D. C., Calkins, F. T., & Mabe, J. H. (2009). Use of a Ni60Ti shape memory alloy for active jet engine chevron application: II. Experimentally validated numerical analysis. *Smart Materials and Structures*, *19*(1), 015021. doi:10.1088/0964-1726/19/1/015021

Harun, W. S. W., Asri, R. I. M., Alias, J., Zulkifli, F. H., Kadirgama, K., Ghani, S. A. C., & Shariffuddin, J. H. M. (2018). A comprehensive review of hydroxyapatite-based coatings adhesion on metallic biomaterials. *Ceramics International*, *44*(2), 1250–1268. doi:10.1016/j.ceramint.2017.10.162

Harvey, J. A. (2000, December). Smart materials, Kirk-Othmer Encycl. *Chem. Technol.*, *1*. doi: 10.1002/0471238961.1913011808011822.a01

Hassan, A. M., Almomani, M., Qasim, T., & Ghaithan, A. (2012). Effect of processing parameters on friction stir welded aluminum matrix composites wear behavior. *Materials and Manufacturing Processes*, *27*(12), 1419–1423. doi:10.1080/10426914.2012.700156

Hawkins, M. J., Soon-Shiong, P., & Desai, N. (2008). Protein nanoparticles as drug carriers in clinical medicine. *Advanced Drug Delivery Reviews*, *60*(8), 876–885. doi:10.1016/j.addr.2007.08.044 PMID:18423779

Hench, L. L. (1971). Bonding mechanisms at the interface of ceramic prosthetic materials. *Journal of Biomedical Materials Research* (*Vol. 5*, Issue 6). Wiley. doi:10.1002/jbm.820050611

Hench, L. L. (2006). The story of Bioglass®. In Journal of Materials Science: Materials in Medicine (Vol. 17, Issue 11). Springer Science and Business Media LLC. doi:10.100710856-006-0432-z

Herod-Schmidt, U., Floeth, E., Last, B., Schafer, W., Zaglauer Daimler-Benz, H., & Dornier, A. (1997), Qualification of smart composites for the use in aerospace applications. In J. M. Sater (Ed.), *SPIE Proceedings*. SPIE. 10.1117/12.274660

Higgins, S. G & Stevens, M. (2021). 'Europe PMC Funders Group High-Aspect-Ratio Nanostructured Surfaces as Biological Metamaterials' 32 (9): 1–100. https://doi.org/. doi:10.1002/adma.201903862.High-Aspect-Ratio

Hoa, N. D., Van Quy, N., Jung, H., Kim, D., Kim, H., & Hong, S. K. (2010). 'Synthesis of Porous CuO Nanowires and Its Application to Hydrogen Detection'. *Sensors and Actuators. B, Chemical*, *146*(1), 266–272. doi:10.1016/j.snb.2010.02.058

Hockaday, L. A., Kang, K. H., Colangelo, N. W., Cheung, P. Y. C., Duan, B., Malone, E., Wu, J., Girardi, L. N., Bonassar, L. J., Lipson, H., Chu, C. C., & Butcher, J. T. (2012). Rapid 3D printing of anatomically accurate and mechanically heterogeneous aortic valve hydrogel scaffolds. *Biofabrication*, *4*(3), 035005. doi:10.1088/1758-5082/4/3/035005 PMID:22914604

Hoffman, A. S. (2012). Hydrogels for biomedical applications. *Advanced Drug Delivery Reviews*, *64*, 18–23. doi:10.1016/j.addr.2012.09.010 PMID:11755703

Hoffman, A. S., Stayton, P. S., Bulmus, V., Chen, G., Chen, J., Cheung, C., Chilkoti, A., Ding, Z., Dong, L., Fong, R., Lackey, C. A., Long, C. J., Miura, M., Morris, J. E., Murthy, N., Nabeshima, Y., Park, T. G., Press, O. W., Shimoboji, T., & Miyata, T. (2000). Really smart bioconjugates of smart polymers and receptor proteins. *Journal of Biomedical Materials Research*, *52*(4), 577–586. doi:10.1002/1097-4636(20001215)52:4<577::AID-JBM1>3.0.CO;2-5 PMID:11033539

Honarpisheh, M., Keimasi, M., & Alinaghian, I. (2018). Numerical and experimental study on incremental forming process of Al/Cu bimetals: Influence of process parameters on the forming force, dimensional accuracy and thickness variations. *Journal of Mechanics of Materials and Structures*, *13*(1), 35–51. doi:10.2140/jomms.2018.13.35

Hoogenboom, R. (2019). Temperature-Responsive Polymers: Properties, Synthesis, and Applications. *Smart Polymers and Their Applications*. Elsevier. . doi:10.1016/B978-0-08-102416-4.00002-8

Hosseini, E. S., Dervin, S., Ganguly, P., & Dahiya, R. (2021). Biodegradable Materials for Sustainable Health Monitoring Devices. *ACS Applied Bio Materials*, *4*(1), 163–194. doi:10.1021/acsabm.0c01139 PMID:33842859

Hsieh, C. T., Liao, C. Y., Dai, N. T., Tseng, C. S., Yen, B. L., & Hsu, S. H. (2018). 3D printing of tubular scaffolds with elasticity and complex structure from multiple waterborne polyurethanes for tracheal tissue engineering. *Applied Materials Today*, *12*, 330–341. doi:10.1016/j.apmt.2018.06.004

Huai-Ling Gao, W. Z.-Y., Cui, C., Bao, J.-Z., Zhu, Y.-B., Xia, J., Wen, S.-M., & Yu, S.-H. (2021). A Highly Compressible and Stretchable Carbon Spring for Smart Vibration and Magnetism Sensors. *Advanced Materials*, *33*(39), 1–9. doi:10.1002/adma.202102724 PMID:34387379

Huang, H. Y. (2010). Effects of activating flux on the welded joint characteristics in gas metal arc welding. *Materials & Design (1980-2015)*, *31*(5), 2488-2495.

Hu, J. L. (2016). *Introduction to Active Coatings for Smart Textiles, Act.* Coatings Smart Text. doi:10.1016/B978-0-08-100263-6.00001-0

Ibrahim Nazem, QADER, M. K. (2019). A Review of Smart Materials: Researches and Applications. *El-Cezerî Journal of Science and Engineering*, *6*(3), 755–788.

ISO. (2015). *IAAF. Additive Manufacturing—General Principles—Terminology*. ISO.

Ista, L. K., & Lopez, G. P. (1998). Lower critical solubility temperature materials as biofouling release agents. *Journal of Industrial Microbiology & Biotechnology*, *20*(2), 121–125. doi:10.1038j.jim.2900490

J., H. & Y., H. (2017). Recent Progress in Smart Materials Research. *Journal of Materials Science*, *52*(22), 12947–12966.

Jagger, D. C., Harrison, A., & Al-Marzoug, K. (2000). Effect of the addition of poly (methyl methacrylate) beads on some properties of acrylic resin. *The International Journal of Prosthodontics*, *13*(5). PMID:11203657

Jaishree Damodharan, A. S. (2018). A Review on Smart Materials, Types and Applications. *International Journal of Engineering Technology Science and Research, 5*(3).

Jandow, N. N., Yam, F. K., Thahab, S. M., Ibrahim, K., & Hassan, H. A. (2010). The characteristics of ZnO deposited on PPC plastic substrate. *Materials Letters*, *64*(21), 2366–2368. doi:10.1016/j.matlet.2010.07.025

Javidani, M., & Larouche, D. (2014). Application of cast Al-Si alloys in internal combustion engine components. *International Materials Reviews*, *59*(3), 132–158. doi:10.1179/1743280413Y.0000000027

Jeong, H., Noh, Y., Young, G., Lee, H., & Lee, D. (2022). 'Roll-to-Roll Processed Silver Nanowire / Silicon Dioxide Microsphere Composite for High-Accuracy Flexible Touch Sensing Application'. *Surfaces and Interfaces*, *30*(April), 101976. doi:10.1016/j.surfin.2022.101976

Jeswiet, J., Micari, F., Hirt, G., Bramley, A., Duflou, J., & Allwood, J. (2005). Asymmetric single point incremental forming of sheet metal. *CIRP Annals*, *54*(2), 88–114. doi:10.1016/S0007-8506(07)60021-3

Jia, G., Wang, Y., & Yao, J. (2012).. . *Digest Journal of Nanomaterials and Biostructures*, *7*(1), 261–267.

Jin, H., Jin, Q., & Jian, J. (2018). Smart Materials for Wearable Healthcare Devices. In *Wearable Technologies*. InTech. doi:10.5772/intechopen.76604

Ji, Y. B., Park, J. Y., Kang, Y., Lee, S., Ju, H. J., Choi, S., Lee, B. Y., & Kim, M. S. (2021). Scaffold printing using biodegradable poly(1,4-butylene carbonate) ink: Printability, *in vivo* physicochemical properties, and biocompatibility. *Materials Today. Bio*, *12*, 100129. doi:10.1016/j.mtbio.2021.100129 PMID:34604731

John, A. (2015). Preparation and characterization of cellulose–ZnOhybrid film by blending method. *Macromolecular Research*, *23*(9), 814–818. doi:10.100713233-015-3111-4

John, A., Ko, H.-U., Kim, D.-G., & Kim, J. (2011). Preparation of cellulose–ZnOhybrid films by a wet chemical method and their characterization. *Cellulose (London, England)*, *18*(3), 675–680. doi:10.100710570-011-9523-1

Johnson, W. W. (1959). The history of prosthetic dentistry. *The Journal of Prosthetic Dentistry*, *9*(5), 841–846. doi:10.1016/0022-3913(59)90049-6

Joshi, G. P., Tyagi, A., Meena, S. L., Chourasia, S., Saxena, N. S., & Kumar, A. (2021). Optical Band Gap Determination of Ni-Al Doped Polyaniline at Room Temperature and Different Annealing Temperatures. *Journal of Physics: Conference Series*, *1950*(1), 012091. doi:10.1088/1742-6596/1950/1/012091

Jung, C., Son, A., Her, N., Zoh, K.-D., Cho, J., & Yoon, Y. (2015). Removal of endocrine disrupting compounds, pharmaceuticals, and personal care products in water using carbon nanotubes: A review. *Journal of Industrial and Engineering Chemistry*, *27*, 1–11. doi:10.1016/j.jiec.2014.12.035

Jurczyk, M. (Ed.). (2012). *Bionanomaterials for dental applications*. CRC Press. doi:10.1201/b13255

Kajtez, J., Nilsson, F., Fiorenzano, A., Parmar, M., & Emnéus, J. (2021). 3D biomaterial models of human brain disease. *Neurochemistry International*, *147*, 105043. doi:10.1016/j.neuint.2021.105043 PMID:33887378

Kambale, B. (2021). *Which Additive Manufacturing Process Is Right for You? | manufacturing*. REPORT.

Kamila, S. (2013). Introduction, classification and applications of smart materials: Anoverview. *American Journal of Applied Sciences*, *10*(8), 876–880. doi:10.3844/ajassp.2013.876.880

Kandagal, Z. B., Akkimardi, V. G., & Sonnad, A. N. (2021). Investigation on microstructural and mechanical properties of microwave welded Al6061-SiC-graphite hybrid metal matrix composites. *Metallography, Microstructure, and Analysis*, *10*(3), 334–346. doi:10.100713632-021-00745-7

Kandola, B. K. (2019). *Smart Textiles and Their Applications*. Woodhead Publishing.

Kaner, R. B. (2017). Nanotechnology for the energy challenge. *Nanotechnology*, *28*(11).

Kanjana, S., & Nalankilli, G. (2018). Smart, Waterproof, Breathable Sportswear- AReview. *Journal of Textile and Apparel, Technology and Management*, *10*(3).

Kaplan, K. M., Spivak, J. M., & Bendo, J. A. (2005). Embryology of the spine and associated congenital abnormalities. *The Spine Journal*, *5*(5), 564–576. doi:10.1016/j.spinee.2004.10.044 PMID:16153587

Kawamura, H., Matsusaka, S., Nomoto, K., Kodaka, H., Hidai, H., Chiba, A., & Morita, N. (2020). 'Improvement in Etching Efficiency of Borosilicate Glass by Dissolving Internal Silver Precipitates'. *Precision Engineering*, *64*(April), 108–112. doi:10.1016/j.precisioneng.2020.03.016

Kedjarune, U., Charoenworaluk, N., & Koontongkaew, S. (1999). Release of methyl methacrylate from heat-curved and autopolymerized resins: Cytotoxicity testing related to residual monomer. *Australian Dental Journal*, *44*(1), 25–30. doi:10.1111/j.1834-7819.1999.tb00532.x PMID:10217017

Kereiakes, D. J., Cox, D. A., Hermiller, J. B., Midei, M. G., Bachinsky, W. B., Nukta, E. D., Leon, M. B., Fink, S., Marin, L., & Lansky, A. J.Guidant Multi-Link Vision Stent Registry Investigators. (2003). Usefulness of a cobalt chromium coronary stent alloy. *The American Journal of Cardiology*, *92*(4), 463–466. doi:10.1016/S0002-9149(03)00669-6 PMID:12914881

Ke, X., Zhang, L., Yang, X., Wang, J., Zhuang, C., Jin, Z., Liu, A., Zhao, T., Xu, S., Gao, C., Gou, Z., & Yang, G. (2018, January 8). Low-melt bioactive glass-reinforced 3D printing akermanite porous cages with highly improved mechanical properties for lumbar spinal fusion. *Journal of Tissue Engineering and Regenerative Medicine*, *12*(5), 1149–1162. doi:10.1002/term.2624 PMID:29222837

Khademhosseini, A., & Langer, R. (2016). A decade of progress in tissue engineering. *Nature Protocols*, *11*(10), 1775–1781. doi:10.1038/nprot.2016.123 PMID:27583639

Khan, F., & Tanaka, M. (2017). Designing Smart Biomaterials for Tissue Engineering. *International Journal of Molecular Sciences*, *19*(1), 17. doi:10.3390/ijms19010017 PMID:29267207

Khan, F., Tanaka, M., & Ahmad, S. R. (2015). Fabrication of polymeric biomaterials: A strategy for tissue engineering and medical devices. *Journal of Materials Chemistry. B, Materials for Biology and Medicine*, *3*(42), 8224–8249. doi:10.1039/C5TB01370D PMID:32262880

Khan, F., Tare, R. S., Oreffo, R. O., & Bradley, M. (2009). Versatile biocompatible polymer hydrogels: Scaffolds for cell growth. *Angewandte Chemie*, *121*(5), 996–1000. doi:10.1002/ange.200804096 PMID:19115339

Khayat, S., Fanaei, H., & Ghanbarzehi, A. (2017). Minerals in pregnancy and lactation: A review article. *Journal of Clinical and Diagnostic Research : JCDR*, *11*(9), QE01. doi:10.7860/JCDR/2017/28485.10626 PMID:29207789

Khindria, S. K., Mittal, S., & Sukhija, U. (2009). Evolution of denture base materials. *Journal of Indian Prosthodontic Society*, *9*(2), 64. doi:10.4103/0972-4052.55246

Khorasani, M., Ghasemi, A., Leary, M., Sharabian, E., Cordova, L., Gibson, I., Downing, D., Bateman, S., Brandt, M., & Rolfe, B. (2022). The effect of absorption ratio on meltpool features in laser-based powder bed fusion of IN718. *Optics & Laser Technology*, *153*, 108263. doi:10.1016/j.optlastec.2022.108263

Khorshidi, R., Honarbakhsh-Raouf, A., & Mahmudi, R. (2018). Effect of minor Gd addition on the microstructure and creep behavior of a cast Al–15Mg2Si in situ composite. *Materials Science and Engineering A*, *718*, 9–18. doi:10.1016/j.msea.2018.01.081

Kikuchi, A., & Okano, T. (2002). Pulsatile drug release control using hydrogels. *Advanced Drug Delivery Reviews*, *54*(1), 53–77. doi:10.1016/S0169-409X(01)00243-5 PMID:11755706

Kikuchi, A., & Okano, T. (2002, July). Intelligent thermoresponsive polymeric stationary phases for aqueous chromatography of biological compounds. *Progress in Polymer Science*, *27*(6), 1165–1193. doi:10.1016/S0079-6700(02)00013-8

Kim, J. (2017). Multifunctional smart biopolymer composites as actuators. *Biopolym. Compos. Electron*, 311–331. . doi:10.1016/B978-0-12-809261-3.00011-5

Kim, B. S., Baez, C. E., & Atala, A. (2000). Biomaterials for tissue engineering. *World Journal of Urology*, *18*(1), 2–9. doi:10.1007003450050002 PMID:10766037

Kim, D. A., Lee, J. H., Jun, S. K., Kim, H. W., Eltohamy, M., & Lee, H. H. (2017, July). Sol–gel-derived bioactive glass nanoparticle-incorporated glass ionomer cement with or without chitosan for enhanced mechanical and biomineralization properties. *Dental Materials*, *33*(7), 805–817. doi:10.1016/j.dental.2017.04.017 PMID:28535954

Kim, J. H., Seol, Y. J., Ko, I. K., Kang, H. W., Lee, Y. K., Yoo, J. J., Atala, A., & Lee, S. J. (2018). 3D Bioprinted Human Skeletal Muscle Constructs for Muscle Function Restoration. *Scientific Reports*, *8*(1), 12307. doi:10.103841598-018-29968-5 PMID:30120282

Kim, J., Bae, S. H., & Lim, H. G. (2006, May 17). Micro transfer printing on cellulose electro-active paper. *Smart Materials and Structures*, *15*(3), 889–892. doi:10.1088/0964-1726/15/3/026

Kim, J., Yun, S., & Lee, S. K. (2007, October 17). Cellulose Smart Material: Possibility and Challenges. *Journal of Intelligent Material Systems and Structures*, *19*(3), 417–422. doi:10.1177/1045389X07083140

Kim, J., Yun, S., & Ounaies, Z. (2006, May 19). Discovery of Cellulose as a Smart Material. *Macromolecules*, *39*(12), 4202–4206. doi:10.1021/ma060261e

Kim, P. (2018). Smart hydrogels for biomedical applications: A review of recent developments. *Smart Materials and Structures*, *27*(4).

Kim, P., & Lieber, C. M. (1999). Nanotube Nanotweezers. *Science*, *286*(5447), 2148–2150. doi:10.1126cience.286.5447.2148 PMID:10591644

Kim, S. X. (2017). Piezoelectric energy harvesters for powering wireless electronics: State-of-the-art and future challenges. *Micro and Nano Systems Letters*, *5*(1), 1–11.

Klemm, D., Heublein, B., Fink, H. P., & Bohn, A. (2005, May 30). Cellulose: Fascinating Biopolymer and Sustainable Raw Material. *Angewandte Chemie International Edition*, *44*(22), 3358–3393. doi:10.1002/anie.200460587 PMID:15861454

Ko, H. U., Kim, H. C., Kim, J. W., Zhai, L., & Kim, J. (2017, June 2). Fabrication Method Study of ZnO Nanocoated Cellulose Film and Its Piezoelectric Property. *Materials (Basel)*, *10*(6), 611. doi:10.3390/ma10060611 PMID:28772971

KÖK, M., Qader, İ. N., Dağdelen, F., & Aydoğdu, Y. (2019). Akıllı Malzemeler üzerine derleme: araştırmalar ve uygulamaları. *El-Cezeri Fen ve Mühendislik Dergisi*. doi:10.31202/ecjse.562177

Kong, L., Valdivia, D. O., Simon, C. M., Hassinan, C. W., Delestrée, N., Ramos, D. M., Park, J. H., Pilato, C. M., Xu, X., Crowder, M., Grzyb, C. C., King, Z. A., Petrillo, M., Swoboda, K. J., Davis, C., Lutz, C. M., Stephan, A. H., Zhao, X., Weetall, M., & Sumner, C. J. (2021). Impaired prenatal motor axon development necessitates early therapeutic intervention in severe SMA. *Science Translational Medicine*, *13*(578), eabb6871. doi:10.1126citranslmed.abb6871 PMID:33504650

Kong, X. C. (2017). A review of smart materials in tire applications. *Smart Materials and Structures*, *26*(2).

Kornilova, A. V., Gorbachevskii, M. V., Kuralbayeva, G. A., Jana, S., Novikov, A. A., Eliseev, A. A., Vasiliev, A. N., & Timoshenko, V. Y. (2019). 'Plasmonic Properties of Halloysite Nanotubes with Immobilized Silver Nanoparticles for Applications in Surface-Enhanced Raman Scattering'. *Physica Status Solidi (A). Physica Status Solidi. A, Applications and Materials Science*, *1800886*(14), 1–5. doi:10.1002/pssa.201800886

Koshy, J., Samuel, M. S., Chandran, A., George, K. C., Predeep, P., Thakur, M., & Varma, M. K. R. (2011). 'Optical Properties of CuO Nanoparticles'. *AIP Conference Proceedings*, *1391*(1), 576–578. doi:10.1063/1.3643615

Kost, J., & Langer, R. (2012). Responsive polymeric delivery systems. *Advanced Drug Delivery Reviews*, *64*, 327–341. doi:10.1016/j.addr.2012.09.014 PMID:11259837

Kothari, M., & Hung, W. N. (2018). Suppressing aluminum carbide in welding aluminum silicon carbide composite. *International journal of engineering materials and manufacture, 3*(1), 41-54.

Kulkarni, H., Zohaib, K., Khusru, A., & Shravan Aiyappa, K. (2018). Application of piezoelectric technology in automotive systems. *Materials Today: Proceedings*, *5*(10), 21299–21304. doi:10.1016/j.matpr.2018.06.532

Kumar, A., Gulati, V., & Kumar, P. (2019c). Experimental investigation of forming forces in single point incremental forming. In *Advances in Industrial and Production Engineering: Select Proceedings of FLAME 2018* (pp. 423-430). Springer Singapore. 10.1007/978-981-13-6412-9_41

Kumar, A., Thakre, G. D., Arya, P. K., & Jain, A. K. (2017, December). 2D Structured Nano-Sheets of Octadecylamine Grafted Graphitic-Carbon Nitride (g-C3N4) as Lubricant Additives. In Macromolecular Symposia, 376(1), p. 1700009. doi:10.1002/masy.201700009

Kumar, A. K. (2023). Materials processed by additive manufacturing techniques. In *Advances in Additive Manufacturing Artificial Intelligence, Nature-Inspired, and Biomanufacturing* (pp. 217–233). Elsevier. doi:10.1016/B978-0-323-91834-3.00014-4

Kumar, A. M. (2022). *Advances in Additive Manufacturing: Artificial Intelligence, Nature-Inspired, and Biomanufacturing*. Elsevier.

Kumar, A., & Gulati, V. (2018). Experimental investigations and optimization of forming force in incremental sheet forming. *Sadhana*, *43*(10), 1–15. doi:10.100712046-018-0926-7

Kumar, A., & Gulati, V. (2019). Experimental investigation and optimization of surface roughness in negative incremental forming. *Measurement*, *131*, 419–430. doi:10.1016/j.measurement.2018.08.078

Kumar, A., & Gulati, V. (2021). Optimization and investigation of process parameters in single point incremental forming. [IJEMS]. *Indian Journal of Engineering and Materials Sciences*, *27*(2), 246–255.

Kumar, A., Gulati, V., & Kumar, P. (2018a). Investigation of surface roughness in incremental sheet forming. *Procedia Computer Science*, *133*, 1014–1020. doi:10.1016/j.procs.2018.07.074

Kumar, A., Gulati, V., & Kumar, P. (2018b). Effects of process parameters on surface roughness in incremental sheet forming. *Materials Today: Proceedings*, *5*(14), 28026–28032. doi:10.1016/j.matpr.2018.10.043

Kumar, A., Gulati, V., & Kumar, P. (2018c). Investigation of process variables on forming forces in incremental sheet forming. *Int J Eng Technol*, *10*(3), 680–684. doi:10.21817/ijet/2018/v10i3/181003021

Kumar, A., Gulati, V., Kumar, P., & Singh, H. (2019b). Forming force in incremental sheet forming: A comparative analysis of the state of the art. *Journal of the Brazilian Society of Mechanical Sciences and Engineering*, *41*(6), 1–45. doi:10.100740430-019-1755-2

Kumar, A., Gulati, V., Kumar, P., Singh, H., Singh, V., Kumar, S., & Haleem, A. (2020). Parametric investigation of forming forces in single point incremental forming. *Materials Today: Proceedings*, *24*, 611–617. doi:10.1016/j.matpr.2020.04.315

Kumar, A., Gulati, V., Kumar, P., Singh, V., Kumar, B., & Singh, H. (2019a). Parametric effects on formability of AA2024-O aluminum alloy sheets in single point incremental forming. *Journal of Materials Research and Technology*, *8*(1), 1461–1469. doi:10.1016/j.jmrt.2018.11.001

Kumar, A., Kumar, M., Goyat, M. S., & Avasthi, D. K. (2022). A Review of the Latest Developments in the Production and Applications of Ag-Nanowires as Transparent Electrodes. *Materials Today. Communications*, *33*(September), 104433. doi:10.1016/j.mtcomm.2022.104433

Kumar, A., Kumar, P., Mittal, R. K., & Gambhir, V. (2023). Materials processed by additive manufacturing techniques. In A. Kumar, R. K. Mittal, & A. Haleem (Eds.), *Advances in Additive Manufacturing* (pp. 217–233). Elsevier. doi:10.1007/978-3-031-34563-0

Kumar, A., Srivastava, A., Galaev, I. Y., & Mattiasson, B. (2007). Smart polymers: Physical forms and bioengineering applications. *Progress in Polymer Science*, *32*(10), 1205–1237. doi:10.1016/j.progpolymsci.2007.05.003

Kumar, N., Arora, N., & Goel, S. K. (2020). Effect of base metal solution annealing on mechanical and metallurgical properties of GMA welded nitronic steel. *Materials Science and Engineering A*, *771*, 138542. doi:10.1016/j.msea.2019.138542

Kumar, S. N., Devarajaiah, R. M., & Prabhu, T. R. (2021). Review on aluminium based functionally graded composites. *Materials Today: Proceedings*, *39*, 1743–1749. doi:10.1016/j.matpr.2020.06.307

Kumar, S., Bhalla, A., & Cross, L. (1994, September). Smart Ceramics for Broadband Vibration Control. *Journal of Intelligent Material Systems and Structures*, *5*(5), 673–677. doi:10.1177/1045389X9400500511

Kumar, V. P. (2017). *Future Perspectives of Bioactive Glasses for the Clinical Applications. Bioactive Glasses*. Springer International Publishing., doi:10.1007/978-3-319-45716-1_11

Kunchur, M. N., Dean, C. L., Moghadam, N. S., Knight, J. M., He, Q. L., Liu, H., Wang, J., Lortz, R., Sou, I. K., & Gurevich, A. (2015). Current-induced depairing in the Bi 2 Te 3/FeTe interfacial superconductor. *Physical Review B: Condensed Matter and Materials Physics*, *92*(9), 094502. doi:10.1103/PhysRevB.92.094502

Kunchur, M. N., Dean, C., Liang, M., Moghaddam, N. S., Guarino, A., Nigro, A., Grimaldi, G., & Leo, A. (2013). Depairing current density of Nd2− xCexCuO4− δ superconducting films. *Physica. C, Superconductivity*, *495*, 66–68. doi:10.1016/j.physc.2013.08.005

Kuroda, D., Niinomi, M., Morinaga, M., Kato, Y., & Yashiro, T. (1998). Design and mechanical properties of new _ type titanium alloys for implant materials. *Materials Science and Engineering A*, *243*(1-2), 244–249. doi:10.1016/S0921-5093(97)00808-3

Kusaka, K. (2016). Application of smart materials for improving efficiency and reliability of mechanical systems. *Journal of Mechanical Science and Technology*, *30*(9), 3939–3946.

Kuzyk, M., & Dawson, N. (2020). Photomechanical materials and applications: A tutorial. *Advances in Optics and Photonics*, *12*(4), 847. doi:10.1364/AOP.387366

Kwon, I. C., Bae, Y. H., & Kim, S. W. (1991). Electrically erodible polymer gel for controlled release of drugs. *Nature*, *354*(6351), 291–293. doi:10.1038/354291a0 PMID:1956379

Lamprecht, C. P. (2020). Smart materials: Challenges and approaches in production and applications. *Journal of Materials Processing Technology*, 275.

Lee, A. Y., An, J., Chua, C. K., & Zhang, Y. (2019). Preliminary investigation of the reversible 4D printing of a dual-layer component. *Engineering (Beijing)*, *5*(6), 1159–1170. doi:10.1016/j.eng.2019.09.007

Lee, D. B., & Kim, S. K. (2005). High temperature oxidation of a (Cr 2 N− Mo 2 S 3)/Cr double-layered coating deposited on steel by DC magnetron sputtering. *Metals and Materials International*, *11*(3), 227–232. doi:10.1007/BF03027447

Lee, H.-S., & Cho, C. (2008). Magnetoelastic Characteristics of the Multilayered Magnetostrictive Thin Film With Polyimide Substrate for Micro Actuator. *IEEE Transactions on Magnetics*, *44*(11), 4049–4052. doi:10.1109/TMAG.2008.2003070

Lee, J. Y., An, J., & Chua, C. K. (2017). Fundamentals and applications of 3D printing for novel materials. *Applied Materials Today*, *7*, 120–133. doi:10.1016/j.apmt.2017.02.004

Lee, J., Mahendra, S., & Alvarez, P. J. J. (2010, July). Nanomaterials in the Construction Industry: A Review of Their Applications and Environmental Health and Safety Considerations. *ACS Nano*, *4*(7), 3580–3590. doi:10.1021/nn100866w PMID:20695513

Lee, S. J. (2020). Application of shear thickening fluid (STF) in sports equipment. *Polymers*, *12*(9).

Lexcellent, C., Leclercq, S., Gabry, B., & Bourbon, G. (2000). Two way shape memory effect of shape memory alloys: an experimental study and a phenomenological model. *Int.J. Plast. 16*(10), 1155–1168. https://doi.org/ 00005-X. doi:10.1016/S0749-6419(00)

Liang, S. X., Wang, X., Zhang, W., Liu, Y. J., Wang, W., & Zhang, L. C. (2020). Selective laser melting manufactured porous Fe-based metallic glass matrix composite with remarkable catalytic activity and reusability. *Applied Materials Today*, *19*, 100543. doi:10.1016/j.apmt.2019.100543

Li, C. (2003). Elastic moduli of multi-walled carbon nanotubes and the effect of van der Waals forces. *Composites Science and Technology*, *63*(11), 1517–1524. doi:10.1016/S0266-3538(03)00072-1

Li, C., Thostenson, E. T., & Chou, T.-W. (2008). Sensors and actuators based on carbon nanotubes and their composites: A review. *Composites Science and Technology*, *68*(6), 1227–1249. doi:10.1016/j.compscitech.2008.01.006

Li, D., Wu, Y., Kim, P., Shi, L., Yang, P., & Majumdar, A. (2003). Thermal conductivity of individual silicon nanowires. *Applied Physics Letters*, *83*(14), 2934–2936. doi:10.1063/1.1616981

Li, H., Wang, J., Gao, S., Chen, Q., Peng, L., Liu, K., & Wei, X. (2017). Superlubricity between MoS2 monolayers. *Advanced Materials*, *29*(27), 1701474. doi:10.1002/adma.201701474 PMID:28497859

Li, J., Guan, X., Fan, Z., Ching, L.-M., Li, Y., Wang, X., Cao, W.-M., & Liu, D.-X. (2020). Non-Invasive Biomarkers for Early Detection of Breast Cancer. *Cancers (Basel)*, *12*(10), 2767. doi:10.3390/cancers12102767 PMID:32992445

Lim, K. S., Levato, R., Costa, P. F., Castilho, M. D., Alcala-Orozco, C. R., Van Dorenmalen, K. M., Melchels, F. P. W., Gawlitta, D., Hooper, G. J., Malda, J., & Woodfield, T. B. (2018). Bio-resin for high resolution lithography-based biofabrication of complex cell-laden constructs. *Biofabrication*, *10*(3), 034101. doi:10.1088/1758-5090/aac00c PMID:29693552

Li, N., & Zheng, Y. (2013). Novel magnesium alloys developed for biomedical application: A review. *Journal of Materials Science and Technology*, *29*(6), 489–502. doi:10.1016/j.jmst.2013.02.005

Li, S. K., & D'Emanuele, A. (2001). On–off transport through a thermoresponsive hydrogel composite membrane. *Journal of Controlled Release*, *75*(1-2), 55–67. doi:10.1016/S0168-3659(01)00365-0 PMID:11451497

Li, T. (2019). Metallic wood: A perspective on its synthesis, properties, and applications. *Materials Today*, *27*, 56–69.

Li, T. C., Han, C. F., Wu, B. H., Hsieh, P. T., & Lin, J. F. (2012). Effects of Prestrain Applied to a Polyethylene Terephthalate Substrate Before the Coating of Al-Doped ZnO Film on Film Quality, Electrical Properties, and Pop-In Behavior During Nanoindentation. *Journal of Microelectromechanical Systems*, *21*(5), 1059–1070. doi:10.1109/JMEMS.2012.2205135

Liu, Y. (2000). *Shape memory implants, Shape Mem.* Implant. . doi:10.1007/978-3-642-59768-8

Liu, H., Duan, C., Yang, C., Chen, X., Shen, W., & Zhu, Z. (2015). 'A Novel Nitrite Biosensor Based on the Direct Electron Transfer Hemoglobin Immobilized in the WO 3 Nanowires with High Length – Diameter Ratio'. *Materials Science and Engineering C*, *53*, 43–49. doi:10.1016/j.msec.2015.04.016 PMID:26042689

Liu, J. (2021). Thermochromic VO2 thin films for smart windows: A review. *Journal of Materials Chemistry. C, Materials for Optical and Electronic Devices*, *9*(4), 1102–1118.

Liu, J. H. (2021). Advances in piezoelectric materials for biomedical applications. *Journal of Materials Chemistry. B, Materials for Biology and Medicine*, *9*(25), 4957–4975. PMID:34008692

Liu, X. (2015). Magnetostrictive sensors: A review of materials and devices. *Sensors (Basel)*, *15*(3), 4655–4686.

Liu, Y. (2018). A review of smart materials in aerospace. nternational. *Journal of Aerospace Engineering*.

Liu, Y. S. (2018). Smart materials and devices for harsh environment sensing and control. *Smart Materials and Structures*, *27*(1).

Liu, Y. W. (2021). Fatigue properties of Ni–Mn–Ga/CNTs composite wire for smart actuator applications. *Journal of Alloys and Compounds*, 873.

Liu, Y., & Hsu, S. H. (2018). Synthesis and biomedical applications of self-healing hydrogels. *Frontiers in Chemistry*, *6*, 449. doi:10.3389/fchem.2018.00449 PMID:30333970

Li, W. H., Yao, G. Z., Chen, G., Yeo, S. H., & Yap, F. F. (2000). Testing and steady state modeling of a linear MR damper under sinusoidal loading. *Smart Materials and Structures*, *9*(1), 95–102. doi:10.1088/0964-1726/9/1/310

Li, W. J., Shi, E. W., Ko, J. M., Chen, Z. Z., Ogino, H., & Fukuda, T. (2003). Hydrothermal synthesis of MoS2 nanowires. *Journal of Crystal Growth*, *250*(3-4), 418–422. doi:10.1016/S0022-0248(02)02412-0

Li, X. L., & Li, Y. D. (2003). Formation of MoS2 inorganic fullerenes (IFs) by the reaction of MoO3 nanobelts and S. *Chemistry (Weinheim an der Bergstrasse, Germany)*, *9*(12), 2726–2731. doi:10.1002/chem.200204635 PMID:12866535

Li, Y. Z. (2020). 3D printing of smart materials: A review on recent progresses in additive manufacturing. *Materials Horizons*, *7*(4), 1041–1065.

Li, Y.-C., Zhang, Y. S., Akpek, A., Shin, S. R., & Khademhosseini, A. (2016). 4D bioprinting: The next-generation technology for biofabrication enabled by stimuli-responsive materials. *Biofabrication*, *9*(1), 012001. doi:10.1088/1758-5090/9/1/012001 PMID:27910820

Li, Y., Jahr, H., Lietaert, K., Pavanram, P., Yilmaz, A., Fockaert, L. I., Leeflang, M. A., Pouran, B., Gonzalez-Garcia, Y., Weinans, H., Mol, J. M. C., Zhou, J., & Zadpoor, A. A. (2018). Additively manufactured biodegradable porous iron. *Acta Biomaterialia*, *77*, 380–393. doi:10.1016/j.actbio.2018.07.011 PMID:29981948

Li, Y., Yang, C., Zhao, H., Qu, S., Li, X., & Li, Y. (2014). New developments of Ti-based alloys for biomedical applications. *Materials (Basel)*, *7*(3), 1709–1800. doi:10.3390/ma7031709 PMID:28788539

Li, Y., Zhou, J., Pavanram, P., Leeflang, M. A., Fockaert, L. I., Pouran, B., Tümer, N., Schröder, K. U., Mol, J. M. C., Weinans, H., Jahr, H., & Zadpoor, A. A. (2018). Additively manufactured biodegradable porous magnesium. *Acta Biomaterialia*, *67*, 378–392. doi:10.1016/j.actbio.2017.12.008 PMID:29242158

Llordés, A. G. (2013). Electrochromic materials for smart windows: Progress and prospects. *Chemical Society Reviews*, *42*(20), 8299–8311.

Loh, G. H., Pei, E., Harrison, D., & Monzón, M. D. (2018). An overview of functionally graded additive manufacturing. *Additive Manufacturing*, *23*, 34–44. doi:10.1016/j.addma.2018.06.023

Long, M., & Rack, H. J. (1998). Titanium alloys in total joint replacement—A materials science perspective. *Biomaterials*, *19*(18), 1621–1639. doi:10.1016/S0142-9612(97)00146-4 PMID:9839998

Luo, Y., Yang, L., & Tian, M. (2013). Application of biomedical-grade titanium alloys in trabecular bone and artificial joints. In *Biomaterials and Medical Tribology* (pp. 181–216). Woodhead Publishing. doi:10.1533/9780857092205.181

Lupan, O., Emelchenko, G. A., Ursaki, V. V., Chai, G., Redkin, A. N., Gruzintsev, A. N., Tiginyanu, I. M., Chow, L., Ono, L. K., Roldan Cuenya, B., Heinrich, H., & Yakimov, E. E. (2010). Synthesis and characterization of ZnO nanowires for nanosensor applications. *Materials Research Bulletin*, *45*(8), 1026–1032. doi:10.1016/j.materresbull.2010.03.027

Lu, T.-L., & Tsai, Y.-C. (2011). Sensitive electrochemical determination of acetaminophen in pharmaceutical formulations at multiwalled carbon nanotube-alumina-coated silica nanocomposite modified electrode. *Sensors and Actuators. B, Chemical*, *153*(2), 439–444. doi:10.1016/j.snb.2010.11.013

Lu, Y., Deshmukh, S., Jones, I., & Chiu, Y. L. (2021). Biodegradable magnesium alloys for orthopaedic applications. *Biomaterials Translational*, *2*(3), 214–235. doi:10.12336/biomatertransl.2021.03.005 PMID:35836650

Lyon, L. A., Meng, Z., Singh, N., Sorrell, C. D., & St. John, A. (2009). Thermoresponsive microgel-based materials. *Chemical Society Reviews*, *38*(4), 865. doi:10.1039/b715522k PMID:19421566

Ma, B., Wang, J., Lee, T. H., Dorris, S. E., Wen, J., & Balachandran, U. (2018). Microstructural characterization of Al 4 C 3 in aluminum–graphite composite prepared by electron-beam melting. *Journal of Materials Science*, *53*(14), 10173–10180. doi:10.100710853-018-2336-y

Machado, L. G., & Savi, M. A. (2003). Medical applications of shape memory alloys. *Brazilian Journal of Medical and Biological Research*, *36*(6), 683–691. doi:10.1590/S0100-879X2003000600001 PMID:12792695

Madangopal, K., Banerjee, S., & Lele, S. (1994). Thermal arrest memory effect. *Acta Metallurgica et Materialia*, *42*(6), 1875–1885. doi:10.1016/0956-7151(94)90012-4

Magin, C. M., Alge, D. L., & Anseth, K. S. (2016). Bio-inspired 3D microenvironments: A new dimension in tissue engineering. *Biomedical Materials (Bristol, England)*, *11*(2), 022001. doi:10.1088/1748-6041/11/2/022001 PMID:26942469

Magnusson, M., Paul, N., & De Nys R. (2012). *For Oil-Based Biofuel and Bioproducts Total Lipid and Fatty Acid Composition of Seaweeds for the Selection of Species for Oil-Based Biofuel and Bioproducts*. Wiley. . doi:10.1111/j.1757-1707.2012.01175.x

Magudeeswaran, G., Balasubramanian, V., Reddy, G. M., & Balasubramanian, T. S. (2008). Effect of welding processes and consumables on tensile and impact properties of high strength quenched and tempered steel joints. *Journal of Iron and Steel Research International*, *15*(6), 87–94. doi:10.1016/S1006-706X(08)60273-3

Maheswari, A. U. (2022). Smart Materials- Types & Applications. *International Journal for Research in Applied Science and Engineering Technology*, *10*(1), 1752–1755. doi:10.22214/ijraset.2022.40147

Mahinroosta, M., Jomeh Farsangi, Z., Allahverdi, A., & Shakoori, Z. (2018). Hydrogels as intelligent materials: A brief review of synthesis, properties and applications. *Materials Today. Chemistry*, *8*, 42–55. doi:10.1016/j.mtchem.2018.02.004

Malat, T. A., Glombitza, M., Dahmen, J., Hax, P. M., & Steinhausen, E. (2018, April). The Use of Bioactive Glass S53P4 as Bone Graft Substitute in the Treatment of Chronic Osteomyelitis and Infected Non-Unions – a Retrospective Study of 50 Patients. *Zeitschrift für Orthopädie und Unfallchirurgie*, *156*(02), 152–159. doi:10.1055-0043-124377 PMID:29665602

Mamand, S., Omar, M., & Muhammed, A. (2012). Calculation Of Lattice Thermal Conductivity Of Suspended GaAs nanobeams: Effect Of Size Dependent Parameters. *Advanced Materials Letters*, *3*(6), 449–458. doi:10.5185/amlett.2012.icnano.102

Ma, N. (2018). Recent advances in shape memory polymers and composites: A review. *Journal of Materials Science*, *53*(3), 1743–1775.

Mani, G., Feldman, M. D., Patel, D., & Agrawal, C. M. (2007). Coronary stents: A materials perspective. *Biomaterials*, *28*(9), 1689–1710. doi:10.1016/j.biomaterials.2006.11.042 PMID:17188349

Manju, V. G., & Kumar, A. (2020). Process parameters and thickness reduction in single point incremental forming. *International Research Journal of Engineering and Technology*, *7*(1), 473–476.

Mantovani, D. (2000). Shape memory alloys: Properties and biomedical applications, JOM. *Journal of the Minerals Metals & Materials Society*, *52*(10), 36–44. doi:10.100711837-000-0082-4

Marinopoulou, E., & Katakalos, K. (2022). Thermomechanical Fatigue Testing on Fe-Mn-Si Shape Memory Alloys in Prestress Conditions. *Materials (Basel)*, *16*(1), 237. doi:10.3390/ma16010237 PMID:36614575

Marrese, M., Guarino, V., & Ambrosio, L. (2017). 'Atomic Force Microscopy: A Powerful Tool to Address Scaffold Design in Tissue Engineering'. *Journal of Functional Biomaterials*, *8*(1), 7. doi:10.3390/jfb8010007 PMID:28208801

Mather, P. T. (2010). Shape memory polymer research. *Annual Review of Materials Research*, *40*, 179–210.

Matsumoto, K. (2016). Magnetostrictive energy harvesting using laminated Terfenol-D/PVDF composite. *Journal of Intelligent Material Systems and Structures*, *27*(10), 1249–1255.

Mauricio, M. R., Carvalho, G. M., Radovanovic, E., Muniz, E. C., & Rubira, A. F. (2009). Analysis of poly(N-isopropylacrylamide) grafted onto the surface of PET films by SI-ATRP technique. *Materials Science and Engineering C*, *29*(2), 594–598. doi:10.1016/j.msec.2008.10.017

Mauro, M. (2018). Multifunctional Smart Materials for Future Aerospace Applications. In Handbook of Smart Materials in Analytical Chemistry, 13-34.

Mazzocchi, A., Devarasetty, M., Huntwork, R., Soker, S., & Skardal, A. (2018). Optimization of collagen type I-hyaluronan hybrid bioink for 3D bioprinted liver microenvironments. *Biofabrication*, *11*(1), 015003. doi:10.1088/1758-5090/aae543 PMID:30270846

Melgarejo, Z. H., Suárez, O. M., & Sridharan, K. (2008). Microstructure and properties of functionally graded Al–Mg–B composites fabricated by centrifugal casting. *Composites. Part A, Applied Science and Manufacturing*, *39*(7), 1150–1158. doi:10.1016/j.compositesa.2008.04.002

Melly, S. K., Liu, L., Liu, Y., & Leng, J. (2020, July 10). On 4D printing as a revolutionary fabrication technique for smart structures. *Smart Materials and Structures*, *29*(8), 083001. doi:10.1088/1361-665X/ab9989

Meng, Y., Chen, G., & Huang, M. (2022). Piezoelectric Materials: Properties, Advancements, and Design Strategies for High-Temperature Applications. *Nanomaterials (Basel, Switzerland)*, *12*(7), 1171. doi:10.3390/nano12071171 PMID:35407289

Mercuri, E., Finkel, R. S., Muntoni, F., Wirth, B., Montes, J., Main, M., Mazzone, E. S., Vitale, M., Snyder, B., Quijano-Roy, S., Bertini, E., Davis, R. H., Meyer, O. H., Simonds, A. K., Schroth, M. K., Graham, R. J., Kirschner, J., Iannaccone, S. T., Crawford, T. O., & Szlagatys-Sidorkiewicz, A.SMA Care Group. (2018). Diagnosis and management of spinal muscular atrophy: Part 1: Recommendations for diagnosis, rehabilitation, orthopedic and nutritional care. *Neuromuscular Disorders*, *28*(2), 103–115. doi:10.1016/j.nmd.2017.11.005 PMID:29290580

Mishra, A. K. (Ed.). (2016). *Smart Materials for Waste Water Applications.*, doi:10.1002/9781119041214

Mishra, R. R., & Sharma, A. K. (2016). On mechanism of in-situ microwave casting of aluminium alloy 7039 and cast microstructure. *Materials & Design*, *112*, 97–106. doi:10.1016/j.matdes.2016.09.041

Mishra, R. R., & Sharma, A. K. (2018, April). Experimental investigation on in-situ microwave casting of copper. []. IOP Publishing.]. *IOP Conference Series. Materials Science and Engineering*, *346*(1), 012052. doi:10.1088/1757-899X/346/1/012052

Mittal, R. K. (2020). *Incremental sheet forming technologies: principles, merits, limitations, and applications*. CRC Press.

Miyazaki, S., & Sachdeva, R. L. (2009). Shape memory effect and superelasticity in Ti—Ni alloys. *Shape Memory Alloys for Biomedical Applications*, 3–19. Elsevier. . doi:10.1533/9781845695248.1.3

Mohd Jani, J. L., Leary, M., Subic, A., & Gibson, M. A. (2014). A review of shape memory alloy research, applications and opportunities. *Materials & Design*, *56*, 1078–1113. doi:10.1016/j.matdes.2013.11.084

Mondal, S., & Hu, J. L. (2006). A novel approach to excellent UV protecting cotton fabric with functionalized MWNT containing water vapor permeable PU coating. *Journal of Applied Polymer Science*, *103*(5), 3370–3376. doi:10.1002/app.25437

Montgomery, M., Ahadian, S., Davenport Huyer, L., Lo Rito, M., Civitarese, R. A., Vanderlaan, R. D., Wu, J., Reis, L. A., Momen, A., Akbari, S., Pahnke, A., Li, R.-K., Caldarone, C. A., & Radisic, M. (2017). Flexible shape-memory scaffold for minimally invasive delivery of functional tissues. *Nature Materials*, *16*(10), 1038–1046. doi:10.1038/nmat4956 PMID:28805824

Moon, R. J., Martini, A., Nairn, J., Simonsen, J., & Youngblood, J. (2011). Cellulose nanomaterials review: Structure, properties and nanocomposites. *Chemical Society Reviews*, *40*(7), 3941. doi:10.1039/c0cs00108b PMID:21566801

Morsiya, C. (2022). A review on parameters affecting properties of biomaterial SS 316L. *Australian Journal of Mechanical Engineering*, *20*(3), 803–813. doi:10.1080/14484846.2020.1752975

Mortimer, R. J. (1997). Electrochromic materials. *Chemical Society Reviews*, *26*(3), 147. doi:10.1039/cs9972600147

Mortimer, R. J. (2011). Electrochromic Materials. *Annual Review of Materials Research, 41*(1), 241–268. doi:10.1146/annurev-matsci-062910-100344

Moyez, S. A., Maitra, S., Mukherjee, K., Sengupta, A., & Roy, S. (2020, October 13). Structural Features and Optical Properties of CH3NH3Pb(1−x)SnxCl3 Thin-Film Perovskites for Photovoltaic Applications. *Journal of Electronic Materials, 49*(12), 7133–7143. doi:10.100711664-020-08529-5

Moyez, S. A., & Roy, S. (2018, October). Dual-step thermal engineering technique: A new approach for fabrication of efficient CH3NH3PbI3-based perovskite solar cell in open air condition. *Solar Energy Materials and Solar Cells, 185*, 145–152. doi:10.1016/j.solmat.2018.05.027

Mukherjee, B., Varghese, B., Zheng, M., Karthik, K. R. G., Mathews, N., Mhaisalkar, S. G., Tok, E. S., & Sow, C. H. (2012). 'Synthesis, Characterization and Electrical Properties of Hybrid Zn 2GeO 4-ZnO Beaded Nanowire Arrays'. *Journal of Crystal Growth, 346*(1), 32–39. doi:10.1016/j.jcrysgro.2012.02.008

Mulenga, C. (2019). A cross-section study to determine human height using femur length in Zambian population. *Acta Scientific Microbiology, 2*(11), 57–61. doi:10.31080/ASMI.2019.02.0399

Mun, S., Maniruzzaman, M., Ko, H. U., Kafy, A., & Kim, J. (2015). Preparation and characterisation of cellulose ZnO hybrid film by blending method and its glucose biosensor application. *Materials Technology, 30*(sup7), B150–B154. doi:10.1179/1753555714Y.0000000197

Murr, L. E., Gaytan, S. M., Medina, F., Lopez, H., Martinez, E., Machado, B. I., Hernandez, D. H., Martinez, L., Lopez, M. I., Wicker, R. B., & Bracke, J. (2010). Next-generation biomedical implants using additive manufacturing of complex, cellular and functional mesh arrays. *Philosophical transactions. Series A, Mathematical, physical, and engineering sciences, 368*(1917), 1999–2032. https://doi.org/ doi:10.1098/rsta.2010.0010

Nabiyouni, M., Brückner, T., Zhou, H., Gbureck, U., & Bhaduri, S. B. (2018). Magnesium-based bioceramics in orthopedic applications. *Acta Biomaterialia, 66*, 23–43. doi:10.1016/j.actbio.2017.11.033 PMID:29197578

Nag, A. (2015). Smart windows with an environment-sensitive material: A review. *Journal of Materials Chemistry. A, Materials for Energy and Sustainability, 3*(45), 22416–22429.

Nai, S. M. L., Gupta, M., & Lim, C. Y. H. (2003). Synthesis and wear characterization of Al based, free standing functionally graded materials: Effects of different matrix compositions. *Composites Science and Technology, 63*(13), 1895–1909. doi:10.1016/S0266-3538(03)00158-1

Najeeb, S., Khurshid, Z., Zafar, M., Khan, A., Zohaib, S., Martí, J., Sauro, S., Matinlinna, J., & Rehman, I. (2016, July 14). Modifications in Glass Ionomer Cements: Nano-Sized Fillers and Bioactive Nanoceramics. *International Journal of Molecular Sciences, 17*(7), 1134. doi:10.3390/ijms17071134 PMID:27428956

Nakayama, M., Okano, T., Miyazaki, T., Kohori, F., Sakai, K., & Yokoyama, M. (2006). Molecular design of biodegradable polymeric micelles for temperature-responsive drug release. *Journal of Controlled Release, 115*(1), 46–56. doi:10.1016/j.jconrel.2006.07.007 PMID:16920217

Nakhaei, P., Margiana, R., Bokov, D. O., Abdelbasset, W. K., Jadidi Kouhbanani, M. A., Varma, R. S., Marofi, F., Jarahian, M., & Beheshtkhoo, N. (2021). Liposomes: Structure, Biomedical Applications, and Stability Parameters With Emphasis on Cholesterol. *Frontiers in Bioengineering and Biotechnology*, *9*, 705886. doi:10.3389/fbioe.2021.705886 PMID:34568298

Nalwa, H. S. (2007). Handbook Of Nanostructured Biomaterials And Their Applications In Nanobiotechnology: : Biomaterials. American Scientific Publishers.

Nandwani, S., Vardhan, S., & Bagha, A. K. (2020). A literature review on the exposure time of microwave based welding of different materials. *Materials Today: Proceedings*, *27*, 2526–2528. doi:10.1016/j.matpr.2019.10.056

Naresh, C., Bose, P. S. C., & Rao, C. S. P. (2016). Shape memory alloys: a state of art review. *IOP Conf. Ser. Mater. Sci. Eng. 149* (1). 10.1088/1757-899X/149/1/012054

Narva, K. K., Lassila, L. V., & Vallittu, P. K. (2005). The static strength and modulus of fiber reinforced denture base polymer. *Dental Materials*, *21*(5), 421–428. doi:10.1016/j.dental.2004.07.007 PMID:15826698

Newnham, R. (1999, May). Ceramics into next millennium. *British Ceramic Transactions*, *98*(5), 251–255. doi:10.1179/096797899680534

Newnham, R. E., Xu, Q. C., Kumar, S., & Cross, L. E. (1990). Smart ceramics. *Ferroelectrics*, *102*(1), 259–266. doi:10.1080/00150199008221486

Nguyen, M. K., & Lee, D. S. (2010, March 1). Injectable Biodegradable Hydrogels. *Macromolecular Bioscience*, *10*(6), 563–579. doi:10.1002/mabi.200900402 PMID:20196065

Nguyen, T. H. (2018). A review of fatigue behavior in smart materials and structures. *Smart Materials and Structures*, *27*(11).

Nia, E. M., Zawawi, N. A. W. A., & Singh, B. S. M. (2017). A review of walking energy harvesting using piezoelectric materials. *IOP Conference Series. Materials Science and Engineering*, *291*, 012026. doi:10.1088/1757-899X/291/1/012026

Nierengarten, J.-F. (2017). Fullerene hexa-adduct scaffolding for the construction of giant molecules. *Chemical Communications (Cambridge)*, *53*(87), 11855–11868. doi:10.1039/C7CC07479D PMID:29051931

Niinomi, M., Nakai, M., & Hieda, J. (2012). Development of new metallic alloys for biomedical applications. *Acta Biomaterialia*, *8*(11), 3888–3903. doi:10.1016/j.actbio.2012.06.037 PMID:22765961

Nirmalraj, P. N. (2017). Smart textiles: A review on thermochromic materials and their applications. *Journal of Industrial Textiles*, *47*(8), 1745–1776.

Niu, J., Xiong, M., Guan, X., Zhang, J., Huang, H., Pei, J., & Yuan, G. (2016). The in vivo degradation and bone-implant interface of Mg-Nd-Zn-Zr alloy screws: 18 months post-operation results. *Corrosion Science*, *113*, 183–187. doi:10.1016/j.corsci.2016.10.009

Núez, G., & Carlos, A. F. (2020a). 'Single GaAs Nanowire Based Photodetector Fabricated by Dielectrophoresis'. *Nanotechnology*, *31*(22), 225604. doi:10.1088/1361-6528/ab76ee PMID:32187022

Odegard, G. M. (2018). 6.2 Computational Multiscale Modeling – Nanoscale to Macroscale. In P. W. R. Beaumont & C. H. Zweben (Eds.), *Comprehensive Composite Materials II* (pp. 28–51). Elsevier. doi:10.1016/B978-0-12-803581-8.10021-9

Ok, Y. S., Chang, S. X., Gao, B., & Chung, H. J. (2015, October). SMART biochar technology—A shifting paradigm towards advanced materials and healthcare research. *Environmental Technology & Innovation*, *4*, 206–209. doi:10.1016/j.eti.2015.08.003

Olabi, A. G., & Grunwald, A. (2008). Design and application of magnetostrictive materials. Materials &amp. *Materials & Design*, *29*(2), 469–483. doi:10.1016/j.matdes.2006.12.016

Ölander, A. (1932). An Electrochemical Investigation Of Solid Cadmium-Gold Alloys. *Journal of the American Chemical Society*, *54*(10), 3819–3833. doi:10.1021/ja01349a004

Oleksik, V., Pascu, A., Gavrus, A., & Oleksik, M. (2010). Experimental studies regarding the single point incremental forming process. *Academic Journal of Manufacturing Engineering*, *8*(2), 51–56.

Omar, M. S. (2016). Structural and Thermal Properties of Elementary and Binary Tetrahedral Semiconductor Nanoparticles. *International Journal of Thermophysics*, *37*(1), 11. doi:10.100710765-015-2026-9

Ortín, J., & Delaey, L. (2002). Hysteresis in shape-memory alloys. *International Journal of Non-linear Mechanics*, *37*(8), 1275–1281. doi:10.1016/S0020-7462(02)00027-6

Osada, Y., & Gong, J. P. (1998, August). Soft and Wet Materials: Polymer Gels. *Advanced Materials*, *10*(11), 827–837. doi:10.1002/(SICI)1521-4095(199808)10:11<827::AID-ADMA827>3.0.CO;2-L

Osada, Y., & Ross-Murphy, S. B. (1993). Intelligent gels. *Scientific American*, *268*(5), 82–87. doi:10.1038cientificamerican0593-82

Otsuka, K. &. (1998). *Shape memory materials.*

Otsuka, H., Tanahashi, H., Maruyama, T., Murakami, M., & Yamada, H. (1998). Development of Fe-Mn-Si shape memory alloys for pipe joints in steel construction. *Materia (Sendai)*, *37*(4), 283–285. doi:10.2320/materia.37.283

Ougizawa, T., & Inoue, T. (1986, July). UCST and LCST Behavior in Polymer Blends and Its Thermodynamic Interpretation. *Polymer Journal*, *18*(7), 521–527. doi:10.1295/polymj.18.521

Ounaies, Z. P. (2016). *Smart composite material.* CRC Press.

Ozbolat, I. T., & Hospodiuk, M. (2016). Current advances and future perspectives in extrusion-based bioprinting. *Biomaterials*, *76*, 321–343. doi:10.1016/j.biomaterials.2015.10.076 PMID:26561931

Padhy, M. K., & Saini, R. P. (2008). A review on silt erosion in hydro turbines. *Renewable & Sustainable Energy Reviews*, *12*(7), 1974–1987. doi:10.1016/j.rser.2007.01.025

Panthongsy, P., Isarakorn, D., Hamamoto, K., & Janphuang, P. (2019). Performance and Behavior Analysis of Piezoelectric Energy Harvesting Floor Tiles. *2019 5th International Conference on Engineering, Applied Sciences and Technology (ICEAST)*. IEEE. 10.1109/ICEAST.2019.8802555

Pant, P., Budai, J. D., Aggarwal, R., Narayan, R. J., & Narayan, J. (2009). Structural characterization of two-step growth of epitaxial ZnO films on sapphire substrates at low temperatures. *Journal of Physics. D, Applied Physics*, *42*(10), 105409. doi:10.1088/0022-3727/42/10/105409

Parham, S., Kharazi, A. Z., Bakhsheshi-Rad, H. R., Ghayour, H., Ismail, A. F., Nur, H., & Berto, F. (2020). Electrospun Nano-Fibers for Biomedical and Tissue Engineering Applications: A Comprehensive Review. *Materials (Basel)*, *13*(9), 2153. doi:10.3390/ma13092153 PMID:32384813

Park, J., Kim, J.-K., Park, S. A., & Lee, D.-W. (2018). Biodegradable polymer material based smart stent: Wireless Presor Sensor and 3D Printed stent. *Microelectronic Engineering*, *206*, 1–5. doi:10.1016/j.mee.2018.12.007

Parry, L., Ashcroft, I. A., & Wildman, R. D. (2016). Understanding the effect of laser scan strategy on residual stress in selective laser melting through thermo-mechanical simulation. *Additive Manufacturing*, *12*, 1–15. doi:10.1016/j.addma.2016.05.014

Parvizi, S., Hashemi, S., & Moein, S. (2022). NiTi shape memory alloys: Properties Nickel-Titanium Smart Hybrid Mater. From *Micro- to Nano-structured Alloy, Emerg. Appl.* 399–426. Elsevier. . doi:10.1016/B978-0-323-91173-3.00021-3

Patel, D. K., & Lim, K. T. (2021). Stimuli-responsive Polymers and Their Biomedical Applications. In Functionalized Polymers: Synthesis, Characterization and Applications (pp. 250-264). CRC Press. doi:10.1201/9780367821913-13

Patel, N. R., & Gohil, P. P. (2012). A review on biomaterials: Scope, applications & human anatomy significance. *International Journal of Emerging Technology and Advanced Engineering*, *2*(4), 91–101.

Pattanashetti, N. A., Heggannavar, G. B., & Kariduraganavar, M. Y. (2017). Smart biopolymers and their biomedical applications. *Procedia Manufacturing*, *12*, 263–279. doi:10.1016/j.promfg.2017.08.030

Peppas, N. A. (1997). Hydrogels and drug delivery. *Current Opinion in Colloid & Interface Science*, *2*(5), 531–537. doi:10.1016/S1359-0294(97)80103-3

Perez-Masia, R., Lopez-Rubio, A., Fabra, M. J., & Lagaron, J. M. (2013, June 14). Biodegradable polyester-based heat management materials of interest in refrigeration and smart packaging coatings. *Journal of Applied Polymer Science*, *130*(5), 3251–3262. doi:10.1002/app.39555

Peter, I., & Rosso, M. (2015). Study of Ti-enriched CoCrMo alloy for dental application. *IEEE Access : Practical Innovations, Open Solutions*, *3*, 73–80. doi:10.1109/ACCESS.2015.2398312

Peterson, I. S., Mazzella, A. J., Belter, L. T., Curry, M. A., Cruz, R. E., & Jarecki, J. (2022). The Cure SMA Clinical Trial Experience Survey: A Study of Trial Participant Perspectives on Clinical Trial Management and Patient-Centric Management Practices. *Neurology and Therapy*, *11*(3), 1167–1181. doi:10.100740120-022-00360-w PMID:35635603

Pethe, A. M., & Yadav, K. S. (2019). Polymers, responsiveness and cancer therapy. *Artificial Cells, Nanomedicine, and Biotechnology*, *47*(1), 395–405. doi:10.1080/21691401.2018.1559176 PMID:30688110

Petrini, L., & Migliavacca, F. (2011). Biomedical Applications of Shape Memory Alloys. *Journal of Metallurgy*, *2011*, 1–15. doi:10.1155/2011/501483

Phillies, G. D. J. (2013). *The VW transformation—A Simple Alternative to the Wilson GF Method* (arXiv:1306.5499). arXiv. https://doi.org//arXiv.1306.5499 doi:10.48550

Phillips, A., & Thelin, L. W. (1927). An x-ray study of the beta transformation in copper-zinc alloys. *Journal of the Franklin Institute*, *204*(3), 359–368. doi:10.1016/S0016-0032(27)92102-4

Poinern, G. E. J., Brundavanam, S., & Fawcett, D. (2012). Biomedical magnesium alloys: A review of material properties, surface modifications and potential as a biodegradable orthopaedic implant. *American Journal of Biomedical Engineering*, *2*(6), 218–240. doi:10.5923/j.ajbe.20120206.02

Popa, A. C., Stan, G., Husanu, M. A., Mercioniu, I., Santos, L., Fernandes, H., & Ferreira, J. (2017, January). Bioglass implant-coating interactions in synthetic physiological fluids with varying degrees of biomimicry. *International Journal of Nanomedicine*, *12*, 683–707. doi:10.2147/IJN.S123236 PMID:28176941

Pour, A. G., Lincoln, C. N., Perlík, V., Šanda, F., & Hauer, J. (2017). Anharmonic vibrational effects in linear and two-dimensional electronic spectra. *Physical Chemistry Chemical Physics*, *19*(36), 24752–24760. doi:10.1039/C7CP05189A PMID:28868559

Prabhu, T. R. (2017). Processing and properties evaluation of functionally continuous graded 7075 Al alloy/SiC composites. *Archives of Civil and Mechanical Engineering*, *17*(1), 20–31. doi:10.1016/j.acme.2016.08.004

Prasad, A. R., Ramji, K., & Datta, G. L. (2014). An experimental study of wire EDM on Ti-6Al- 4V alloy. *Procedia Materials Science*, *5*, 2567–2576. doi:10.1016/j.mspro.2014.07.517

Purushothaman, K. K., & Muralidharan, G. (2009). The effect of annealing temperature on the electrochromic properties of nanostructured NiO films. *Solar Energy Materials and Solar Cells*, *93*(8), 1195–1201. doi:10.1016/j.solmat.2008.12.029

Putra, N. E., Leeflang, M. A., Minneboo, M., Taheri, P., Fratila-Apachitei, L. E., Mol, J. M. C., Zhou, J., & Zadpoor, A. A. (2021). Extrusion-based 3D printed biodegradable porous iron. *Acta Biomaterialia*, *121*, 741–756. doi:10.1016/j.actbio.2020.11.022 PMID:33221501

Qader, I. N., & Omar, M. S. (2017). Carrier concentration effect and other structure-related parameters on lattice thermal conductivity of Si nanowires. *Bulletin of Materials Science*, *40*(3), 599–607. doi:10.100712034-017-1393-1

Qiu, Y., & Park, K. (2001). Environment-sensitive hydrogels for drug delivery. *Advanced Drug Delivery Reviews*, *53*(3), 321–339. doi:10.1016/S0169-409X(01)00203-4 PMID:11744175

Qi, Z. W. (2019). Recent progress in smart materials-based sports textiles. *Journal of Textile Research*, *40*(6), 107–116.

Radhikaa, C. (2020). A Study on various modes of vibration in spectroscopy. *Journal of Matematik*, (2), 1103-1105. https://doi.org/ doi:10.26637/MJM0S20/0285

Rahmani, M., Ahmadi, M. T., Shayesteh, N., Amin, N. A., Rahmani, K., & Ismail, R. (2011, September). Current-voltage modeling of bilayer graphene nanoribbon schottky diode. In *2011 IEEE Regional Symposium on Micro and Nano Electronics* (pp. 256-258). IEEE. 10.1109/RSM.2011.6088337

Raimondi, D. L., & Kay, E. (1970). High Resistivity Transparent ZnO Thin Films. *Journal of Vacuum Science and Technology*, *7*(1), 96–99. doi:10.1116/1.1315841

Rajan, P. K., Ganesan, V., Dennison, M. S., & Kalishwaran, P. (2018). Fabrication and characterisation of hybrid aluminium metal matrix composite with welding slag and flux by using stir casting technique. *International Research Journal of Engineering and Technology*, *25*, 24–27.

Ram, S. C., Chattopadhyay, K., & Chakrabarty, I. (2016). Dry Sliding Wear Behavior of A356 Alloy/Mg 2 Sip Functionally Graded in-situ Composites: Effect of Processing Conditions. *Tribology in Industry, 38*(3).

Ramasamy, T., Ruttala, H. B., Gupta, B., Poudel, B. K., Choi, H. G., Yong, C. S., & Kim, J. O. (2017). Smart chemistry-based nanosized drug delivery systems for systemic applications: A comprehensive review. *Journal of Controlled Release*, *258*, 226–253. doi:10.1016/j.jconrel.2017.04.043 PMID:28472638

Ram, S. C., Chattopadhyay, K., & Bhushan, A. (2023). A literature review on Al-Si alloy matrix based in situ Al-Mg_2Si FG-composites: Synthesis, microstructure features, and mechanical characteristics. *Proceedings of the Institution of Mechanical Engineers. Part C, Journal of Mechanical Engineering Science*, *237*(4), 919–940. doi:10.1177/09544062221124064

Ram, S. C., Chattopadhyay, K., & Chakrabarty, I. (2017). High temperature tensile properties of centrifugally cast in-situ Al-Mg_2Si functionally graded composites for automotive cylinder block liners. *Journal of Alloys and Compounds*, *724*, 84–97. doi:10.1016/j.jallcom.2017.06.306

Ram, S. C., Chattopadhyay, K., & Chakrabarty, I. (2018, July). Effect on microstructures, hardness and SDAS of primary Mg2Si/Al-Si eutectic phases of centrifugally cast functionally graded Al-(Mg2Si) p in-situ composites. [). IOP Publishing.]. *IOP Conference Series. Materials Science and Engineering*, *390*(1), 012012. doi:10.1088/1757-899X/390/1/012012

Ram, S. C., Chattopadhyay, K., & Chakrabarty, I. (2019). Microstructures and high temperature mechanical properties of A356-Mg2Si functionally graded composites in as-cast and artificially aged (T6) conditions. *Journal of Alloys and Compounds*, *805*, 454–470. doi:10.1016/j.jallcom.2019.07.075

Rapoport, L., Fleischer, N., & Tenne, R. (2005). Applications of WS 2 (MoS 2) inorganic nanotubes and fullerene-like nanoparticles for solid lubrication and for structural nanocomposites. *Journal of Materials Chemistry*, *15*(18), 1782–1788. doi:10.1039/b417488g

Ratner, B. D., Hoffman, A. S., Schoen, F. J., & Lemons, J. E. (2004). *Biomaterials science: an introduction to materials in medicine.*

Ravindra, P., Baral, M., Biswas, T., Nand, M., Jha, S. N., Athresh, E., Ranjan, R., Jain, M., Ganguli, T., & Avasthi, S. (2019). 'Electrical and Optical Properties of Low-Bandgap Oxide Zn 2 Mo 3 O 8 for Optoelectronic Applications'. *Thin Solid Films*, *677*(March), 95–102. doi:10.1016/j.tsf.2019.03.023

Razzi, F., Lovrak, M., Gruzdyte, D., Den Hartog, Y., Duncker, D. J., van Esch, J. H., van Steijn, V., & van Beusekom, H. M. M. (2022). An Implantable Artificial Atherosclerotic Plaque as a Novel Approach for Drug Transport Studies on Drug-Eluting Stents. *Advanced Healthcare Materials*, *11*(6), e2101570. doi:10.1002/adhm.202101570 PMID:34865315

Rebouillat, S., Lyons, M. E., Brandon, M. P., & Doyle, R. L. (2011, November). Paving the Way to The Integration of Smart Nanostructures: Part II: Nanostructured Microdispersed Hydrated Metal Oxides for Electrochemical Energy Conversion and Storage Applications. *International Journal of Electrochemical Science*, *6*(11), 5830–5917. doi:10.1016/S1452-3981(23)18448-3

Rempe, S. B., & Jonsson, H. (1998). A Computational Exercise Illustrating Molecular Vibrations and Normal Modes. *Chemical Educator (Boise, Idaho)*, *3*(4), 1. doi:10.100700897980231a

Ribut, S. H., Abdullah, C., & Yusoff, M. (2019). Investigations of structural and optical properties of zinc oxide thin films growth on various substrates. *Results in Physics*, *13*, 102146. doi:10.1016/j.rinp.2019.02.082

Robert, B. (2014). Smart materials: A review of capabilities. *Assembly Automation*, *34*(1), 16–22. doi:10.1108/AA-10-2013-094

Rock, M. (2012). *Temperature responsive smart textile.* Google Patents.

Rogers, C. A. (1988). U.S. Army Research Office Workshop on Smart Materials, Structures and Mathematical Issues Held in Blacksburg. *Defense Technical Information Center* (p. 5). Blacksburg: Technomic Publishing Co.

Rogers, C. A. (1989). *U.S. Army Research Office Workshop on Smart Materials.* Structures and Mathematical Issues Held in Blacksburg. 10.21236/ADA218528

Romero, R., Lopez, M. C., Leinen, D., Martin, F., & Ramos-Barrado, J. R. (2004). Electrical properties of the n-ZnO/c-Si heterojunction prepared by chemical spray pyrolysis. *Materials Science and Engineering B*, *110*(1), 87–93. doi:10.1016/j.mseb.2004.03.010

Rosenqvist, K., Gursoy, M., Kononen, E., Gursoy, U. K., & Juppo, A. M. (2017). A Pilot Study - Comparison between a Novel Combination of Bioactive Glass with Clodronate and Bioactive Glass Alone as a Treatment for Chronic Periodontitis. *Journal of Biotechnology & Biomaterials*, *07*(03). doi:10.4172/2155-952X.1000265

Rouhi, J., Alimanesh, M., Dalvand, R., Ooi, C. R., Mahmud, S., & Mahmood, M. R. (2014). Optical properties of well-aligned ZnO nanostructure arrays synthesized by an electric field-assisted aqueous solution method. *Ceramics International*, *40*(7), 11193–11198. doi:10.1016/j.ceramint.2014.03.157

Roy, D., Brooks, W. L. A., & Sumerlin, B. S. (2013). New directions in thermoresponsive polymers. *Chemical Society Reviews*, *42*(17), 7214. doi:10.1039/c3cs35499g PMID:23450220

Roy, S., & Botte, G. G. (2018). Perovskite solar cell for photocatalytic water splitting with a TiO2/Co-doped hematite electron transport bilayer. *RSC Advances*, *8*(10), 5388–5394. doi:10.1039/C7RA11996H PMID:35542422

Roytburd, A., Slutsker, J., & Wuttig, M. (2000). Smart composites with shape memory alloys. *Compr. Compos. Mater.*. doi:10.1016/B0-08-042993-9/00212-6

Ruel-Gariepy, E., & Leroux, J. C. (2004). In situ-forming hydrogels—Review of temperature-sensitive systems. *European Journal of Pharmaceutics and Biopharmaceutics*, *58*(2), 409–426. doi:10.1016/j.ejpb.2004.03.019 PMID:15296964

Rus, D., & Tolley, M. T. (2015). Design, fabrication and control of soft robots. *Nature*, *521*(7553), 467–475. doi:10.1038/nature14543 PMID:26017446

Rush, J. A. (1997). *Memory wire robotic hand*. Google Patents.

Ruskowitz, E. R., & DeForest, C. A. (2018). Photoresponsive biomaterials for targeted drug delivery and 4D cell culture. *Nature Reviews. Materials*, *3*(2), 1–17. doi:10.1038/natrevmats.2017.87

Safari, A. (2019). *Piezoelectric energy harvesting: Fundamentals, materials, and devices*. Springer.

Sahani, S., Roy, T., & Sharma, Y. C. (2020, May). Smart waste management of waste cooking oil for large scale high quality biodiesel production using Sr-Ti mixed metal oxide as solid catalyst: Optimization and E-metrics studies. *Waste Management (New York, N.Y.)*, *108*, 189–201. doi:10.1016/j.wasman.2020.04.036 PMID:32360999

Saini, N., Jindal, R., & Tripathi, A. (2022). Effect of M site cation ordering on the lattice dynamical properties of the tetragonal Sr2MO4 (M=Cr, Fe, Co, Mo, and Sn) crystals. *Philosophical Magazine*, *102*(11), 1055–1068. doi:10.1080/14786435.2022.2035006

Saini, N., Jindal, R., & Tripathi, A. (2023a). Influence of B-site cation in lattice dynamics of bilayered Ruddlesden–Popper Sr3B2O7 (B = Zr, Mo, Sn, Hf) compounds. *Journal of Raman Spectroscopy : JRS*, jrs.6561. doi:10.1002/jrs.6561

Saini, N., Jindal, R., & Tripathi, A. (2023b). Normal coordinate analysis to study the optical phonons in B-site cation disordered Ruddlesden-Popper Sr3B2O7 (B = V, Cr, Fe, Co) crystals. *Solid State Sciences*, *107209*, 107209. doi:10.1016/j.solidstatesciences.2023.107209

Saini, N., Jindal, R., & Tripathi, A. (2023c). Study of lattice dynamics of bilayered tetragonal Ruddlesden-Popper compounds (Ca, Sr)3Mn2O7. *Materials Today. Communications*, *34*, 105190. doi:10.1016/j.mtcomm.2022.105190

Saito, K. (2017). *Magnetic rheological fluid shock absorber*. Google Patents.

Salhi, B., Hossain, M. K., Mukhaimer, A. W., & Al-Sulaiman, F. A. (2016). 'Nanowires : A New Pathway to Nanotechnology-Based Applications'. *Journal of Electroceramics*, *37*(April), 34–49. doi:10.100710832-016-0037-y PMID:32214875

Salih, O. S., Ou, H., Sun, W., & McCartney, D. G. (2015). A review of friction stir welding of aluminium matrix composites. *Materials & Design*, *86*, 61–71. doi:10.1016/j.matdes.2015.07.071

Sanchez, A. H. M., Luthringer, B. J., Feyerabend, F., & Willumeit, R. (2015). Mg and Mg alloys: How comparable are in vitro and in vivo corrosion rates? A review. *Acta Biomaterialia*, *13*, 16–31. doi:10.1016/j.actbio.2014.11.048 PMID:25484334

Santos, G. (2017). The importance of metallic materials as biomaterials. *Adv Tissue Eng Regen Med Open Access*, *3*(1), 300–302. doi:10.15406/atroa.2017.03.00054

Sarkar, S., Guibal, E., Quignard, F., & SenGupta, A. K. (2012, January 25). Polymer-supported metals and metal oxide nanoparticles: Synthesis, characterization, and applications. *Journal of Nanoparticle Research*, *14*(2), 715. doi:10.100711051-011-0715-2

Sasaki, H., Hamanaka, I., Takahashi, Y., & Kawaguchi, T. (2016). Effect of long-term water immersion or thermal shock on mechanical properties of high-impact acrylic denture base resins. *Dental Materials Journal*, *35*(2), 204–209. doi:10.4012/dmj.2015-291 PMID:27041009

Sato, T., Katayama, K., Suzuki, T., & Shiomi, T. (1998, February). UCST and LCST behaviour in polymer blends containing poly(methyl methacrylate-statstyrene). *Polymer*, *39*(4), 773–780. doi:10.1016/S0032-3861(97)00339-X

Saveleva, M. S., Eftekhari, K., Abalymov, A., Douglas, T. E. L., Volodkin, D., Parakhonskiy, B. V., & Skirtach, A. G. (2019, April 4). Hierarchy of Hybrid Materials—The Place of Inorganics-in-Organics in it, Their Composition and Applications. *Frontiers in Chemistry*, *7*, 179. Advance online publication. doi:10.3389/fchem.2019.00179 PMID:31019908

Saxena, A., Dwivedi, S. P., Maurya, N. K., & Srivastava, A. K. (2021). Investigation on mechanical properties and deformation behavior of copper-based three-phase metal matrix composite: Experimental and micro-macro-mechanical finite element analysis. *Proceedings of the Institution of Mechanical Engineers, Part L: Journal of Materials: Design and Applications, 235*(8), 1850-1867. 10.1177/14644207211009254

Saxena, M., & Kirtania, S. (2016). Stiffness analysis of symmetric cross-ply laminated composite plates. *ADBU Journal of Engineering Technology, 4*. https://journals.dbuniversity.ac.in/ojs/index.php/AJET/article/view/153/pdf_32

Saxena, A., Dwivedi, S. P., Dixit, A., Sharma, S., Srivastava, A. K., & Maurya, N. K. (2021). Computational and experimental investigation on mechanical behavior of zirconia toughened alumina and nickel powder reinforced EN31 based composite material. *Materialwissenschaft und Werkstofftechnik*, *52*(5), 548–560. doi:10.1002/mawe.202000152

Saxena, M., Sharma, A. K., Srivastava, A. K., Singh, N., & Dixit, A. R. (2023). An Investigation for Minimizing the Wear Loss of Microwave-Assisted Synthesized g-C3N4/MoS2 Nanocomposite Coated Substrate. *Coatings*, *13*(1), 118. doi:10.3390/coatings13010118

Saxena, M., Sharma, A. K., Srivastava, A. K., Singh, R. K., Dixit, A. R., Nag, A., & Hloch, S. (2022). Microwave-Assisted Synthesis, Characterization and Tribological Properties of a g-C3N4/MoS2 Nanocomposite for Low Friction Coatings. *Coatings*, *12*(12), 1840. doi:10.3390/coatings12121840

Schetky, L. (1993). *A review of: "Engineering Aspects Of Shape Memory Alloys" by T.W. Duerig, K.N. Melton, D. Stockel.* Taylor & Francis. doi:10.1080/10426919308934859

Schmaljohann, D. (2006). Thermo-and pH-responsive polymers in drug delivery. *Advanced Drug Delivery Reviews*, *58*(15), 1655–1670. doi:10.1016/j.addr.2006.09.020 PMID:17125884

Schmalz, G., & Galler, K. M. (2017). Biocompatibility of biomaterials – Lessons learned and considerations for the design of novel materials. *Dental Materials*, *33*(4), 382–393. doi:10.1016/j.dental.2017.01.011 PMID:28236437

Sehrawat, S. K., Kumar, A., Grover, S., Dogra, N., Nindra, J., Rathee, S., Dahiya, M., & Kumar, A. (2022). Study of 3D scanning technologies and scanners in orthodontics. *Materials Today: Proceedings*, *56*, 186–193. doi:10.1016/j.matpr.2022.01.064

Sehrawat, S. K., Kumar, A., Prabhakar, M., & Nindra, J. (2022). The expanding domains of 3D printing pertaining to the speciality of orthodontics. *Materials Today: Proceedings*, *50*, 1611–1618. doi:10.1016/j.matpr.2021.09.124

Shafieizad, A. H., Zarei-Hanzaki, A., Abedi, H. R., & Al-Fadhalah, K. J. (2015). The Mg2Si phase evolution during thermomechanical processing of in-situ aluminum matrix macro-composite. *Materials Science and Engineering A*, *644*, 310–317. doi:10.1016/j.msea.2015.07.060

Shang, X., Pan, E., & Qin, L. (2008). Mathematical modeling and numerical computation for the vibration of a magnetostrictive actuator. *Smart Materials and Structures*, *17*(4), 045026. doi:10.1088/0964-1726/17/4/045026

Sharma, A. K., & Gupta, D. (2012). On microstructure and flexural strength of metal–ceramic composite cladding developed through microwave heating. *Applied Surface Science*, *258*(15), 5583–5592. doi:10.1016/j.apsusc.2012.02.019

Sharma, D., & Hussain, C. M. (2020, January). Smart nanomaterials in pharmaceutical analysis. *Arabian Journal of Chemistry*, *13*(1), 3319–3343. doi:10.1016/j.arabjc.2018.11.007

Sharma, G. D. (2018). Hybrid electrochromic materials and devices: A review. *Journal of Materials Chemistry. C, Materials for Optical and Electronic Devices*, *6*(31), 8217–8232.

Sharma, S., & Periasamy, C. (1633-1636). J. Electron Dev. 2014. *Superlattices and Microstructures*, *19*.

Sharma, S., & Periasamy, C. (2014). A study on the electrical characteristic of n-ZnO/p-Si heterojunction diode prepared by vacuum coating technique. *Superlattices and Microstructures*, *73*, 12–21. doi:10.1016/j.spmi.2014.05.011

Sharma, S., Sudhakara, P., Singh, J., Ilyas, R. A., Asyraf, M. R. M., & Razman, M. R. (2021). Critical Review of Biodegradable and Bioactive Polymer Composites for Bone Tissue Engineering and Drug Delivery Applications. *Polymers*, *13*(16), 2623. doi:10.3390/polym13162623 PMID:34451161

Sharma, V., Kumar, P., Kumar, A., Surbhi, Asokan, K., & Sachdev, K. (2017). High-performance radiation stable ZnO/Ag/ZnO multilayer transparent conductive electrode. *Solar Energy Materials and Solar Cells*, *169*, 122–131. doi:10.1016/j.solmat.2017.05.009

Sharma, V., Singh, S., Asokan, K., & Sachdev, K. (2015). A study on 100 MeV O7+ irradiated SnO2/Ag/SnO2 multilayer as transparent electrode for flat panel display application. *Nuclear Instruments & Methods in Physics Research. Section B, Beam Interactions with Materials and Atoms*, *379*, 141–145. doi:10.1016/j.nimb.2016.04.059

Sharma, V., Vyas, R., Bazylewski, P., Chang, G. S., Asokan, K., & Sachdeva, K. (2016). Probing the highly transparent and conducting SnO_x/Au/SnO_x structure for futuristic TCO applications. *RSC Advances*, *6*(35), 29135–29141. doi:10.1039/C5RA24422F

Shasshi Bahl, H. N. (2020). Smart materials types, properties and applications: A review. *Materials Today: Proceedings*, *28*(3), 1302–1306. doi:10.1016/j.matpr.2020.04.505

Shayan, M. M. (2019). Shape memory alloys for biomedical applications: A review. *Journal of Materials Science. Materials in Medicine*, *30*(7), 79. PMID:31240399

Shellaiah, M., & Sun, K. W. (2020, July 14). Review on Sensing Applications of Perovskite Nanomaterials. *Chemosensors (Basel, Switzerland)*, *8*(3), 55. doi:10.3390/chemosensors8030055

Shiga, T. (1999). Deformation and viscoelastic behavior of polymer gels in electric fields. *Neutron spin echo spectroscopy viscoelasticity rheology*, 131-163.

Shi, J. L., Yan, H. G., Su, B., Chen, J. H., Zhu, S. Q., & Chen, G. (2011). Preparation of a functionally gradient aluminum alloy metal matrix composite using the technique of spray deposition. *Materials and Manufacturing Processes*, *26*(10), 1236–1241. doi:10.1080/10426914.2010.544814

Shimanouchi, T., Tsuboi, M., & miyazawa, T. (1961). Optically Active Lattice Vibrations as Treated by the GF-Matrix Method. *The Journal of Chemical Physics*, *35*(5), 1597–1612. doi:10.1063/1.1732116

Shivaji S. & Aher, A. D. (2015). A Review On Smart Materials: Future Potentials In Engineering. *International journal of science technology and management, 10.*

Silinsh, E. A. (1980). Introduction: Characteristic Features of Organic Molecular Crystals. In E. A. Silinsh (Ed.), *Organic Molecular Crystals: Their Electronic States* (pp. 1–46). Springer. doi:10.1007/978-3-642-81464-8_1

Singh, I., Bajpai, P. K., Malik, D., Sharma, A. K., & Kumar, P. (2011). Feasibility Study on Microwave Joining of 'green composites'. *Akademeia, 1*(1), ea0101.

Singh, S., Singh, P., Gupta, D., Jain, V., Kumar, R., & Kaushal, S. (2019). Development and characterization of microwave processed cast iron joint. *Engineering Science and Technology, an International Journal, 22*(2), 569-577.

Singh, A. K. (2019). 3D printing of smart materials: A review on recent progresses in the field. *Materials Today: Proceedings*, *18*, 5017–5022.

Singh, A., & Elisseeff, J. (2010). Biomaterials for stem cell differentiation. *Journal of Materials Chemistry*, *20*(40), 8832–8847. doi:10.1039/c0jm01613f

Singh, D., Wang, S. B., Xia, T., Tainsh, L., Ghiassi-Nejad, M., Xu, T., Peng, S., Adelman, R. A., & Rizzolo, L. J. (2018). A biodegradable scaffold enhances differentiation of embryonic stem cells into a thick sheet of retinal cells. *Biomaterials*, *154*, 158–168. doi:10.1016/j.biomaterials.2017.10.052 PMID:29128844

Singh, S., Belokar, R. M., & Suri, N. M. (2019). XRD characterization of aluminium joint processed by microwave hybrid heating. *International Journal of Scientific Research and Reviews*, *7*(2).

Singh, S., Sharma, V., Saini, D., Shekhawat, S., Asokan, K., & Sachdev, K. (2018). Influence of 100 keV Ar+ implantation on electrical and optical properties of TiO2/Ag/TiO2 multilayer films. *Materials Science in Semiconductor Processing*, *75*, 18–25. doi:10.1016/j.mssp.2017.11.016

Singh, S., Suri, N. M., & Belokar, R. M. (2015). Characterization of joint developed by fusion of aluminum metal powder through microwave hybrid heating. *Materials Today: Proceedings*, *2*(4-5), 1340–1346. doi:10.1016/j.matpr.2015.07.052

Singh, V. G. (2023). Experimental investigations of thermo physical properties and convective heat transfer of Al2O3 and CuO nanofluids in a copper tube: Proposing New Correlations. *Biointerface Research in Applied Chemistry*, 229.

Singh, V. G., Gupta, M., & Kumar, A. (2019). Investigating the morphology and thermophysical properties of Al2O3 and CuO nanofluids. *IOP Conference Series. Materials Science and Engineering*, *577*(1), 012052. doi:10.1088/1757-899X/577/1/012052

Singh, V. K., Kumar, A., Alam, M., Kumar, A., Kumar, P., & Goyat, V. (2022). A study of morphology, UV measurements and zeta potential of zinc ferrite and Al2O3 nanofluids. *Materials Today: Proceedings*, *59*, 1034–1039. doi:10.1016/j.matpr.2022.02.371

Singla, M., Dwivedi, D. D., Singh, L., & Chawla, V. (2009). Development of aluminium based silicon carbide particulate metal matrix composite. *Journal of Minerals & Materials Characterization & Engineering*, *8*(06), 455–467. doi:10.4236/jmmce.2009.86040

Sireesha, M., Albert, S. K., Shankar, V., & Sundaresan, S. (2000). A comparative evaluation of welding consumables for dissimilar welds between 316LN austenitic stainless steel and Alloy 800. *Journal of Nuclear Materials*, *279*(1), 65–76. doi:10.1016/S0022-3115(99)00275-5

Snyder, J. E. (2013). Magnetostrictive materials for energy harvesting. *Journal of Applied Physics*, *113*(17).

Sobczak, J. J., & Drenchev, L. B. (2009). *Metal Based Functionally Graded Materials*. Bentham Science Publishers. doi:10.2174/97816080503831090101

Sodano, A., Inman, D., & Park, G. (2004). A Review of Power Harvesting from Vibration Using Piezoelectric Materials. *Shock Vib Digest, 36*(3), 197–205. . doi:10.1177/0583102404043275

Soltani, N., Nodooshan, H. J., Bahrami, A., Pech-Canul, M. I., Liu, W., & Wu, G. (2014). Effect of hot extrusion on wear properties of Al–15 wt.% Mg2Si in situ metal matrix composites. *Materials & Design*, *53*, 774–781. doi:10.1016/j.matdes.2013.07.084

Song, M., Lin, X., Peng, Z., Xu, S., Jin, L., Zheng, X., & Luo, H. (2021). Materials and Methods of Biosensor Interfaces With Stability. *Frontiers in Materials*, *7*, 583739. doi:10.3389/fmats.2020.583739

Soni, P., Sehgal, S., Kumar, H., & Singh, A. P. (2018). Joining of SS316-SS316 through microwave hybrid heating by using Nickel nano-powder. *International Journal of Applied Engineering Research: IJAER*, *8*(8), 6446–6449.

Soutis, C. (2010). ACM Special Issue: ICCM-17: Deformation and Fracture of Composites. *Applied Composite Materials*, *18*(1), 1–1. doi:10.100710443-010-9183-6 PMID:35095206

Spanner, K., & Koc, B. (2016). Piezoelectric Motors, an Overview. *Actuators*, *5*(1), 6. doi:10.3390/act5010006

Springer. (1995). Molecular Vibrations. In I. Hargittai & M. Hargittai (Eds.), Symmetry through the Eyes of a Chemist (pp. 207–225). Springer US. doi:10.1007/978-0-585-31234-7_5

Srinath, M. S., Sharma, A. K., & Kumar, P. (2011a). A novel route for joining of austenitic stainless steel (SS-316) using microwave energy. *Proceedings of the Institution of Mechanical Engineers. Part B, Journal of Engineering Manufacture*, *225*(7), 1083–1091. doi:10.1177/2041297510393451

Srinath, M. S., Sharma, A. K., & Kumar, P. (2011b). Investigation on microstructural and mechanical properties of microwave processed dissimilar joints. *Journal of Manufacturing Processes*, *13*(2), 141–146. doi:10.1016/j.jmapro.2011.03.001

Srinath, M. S., Sharma, A. K., & Kumar, P. (2011c). A new approach to joining of bulk copper using microwave energy. *Materials & Design*, *32*(5), 2685–2694. doi:10.1016/j.matdes.2011.01.023

Staiger, M. P., Pietak, A. M., Huadmai, J., & Dias, G. (2006). Magnesium and its alloys as orthopedic biomaterials: A review. *Biomaterials*, *27*(9), 1728–1734. doi:10.1016/j.biomaterials.2005.10.003 PMID:16246414

Stevens, M. M. (2008). Biomaterials for bone tissue engineering. *Materials Today*, *11*(5), 18–25. doi:10.1016/S1369-7021(08)70086-5

Sudha, P. N., Sangeetha, K., Vijayalakshmi, K., & Barhoum, A. (2018). *Nanomaterials History, Classification, Unique Properties, Production and Market. Emerging Applications of Nanoparticles and Architecture Nanostructures*. Elsevier Inc. doi:10.1016/B978-0-323-51254-1/00012-9

Su, J., & Tajitsu, Y. (2016). Piezoelectric and Electrostrictive Polymers as EAPs: Materials. In F. Carpi (Ed.), *Electromechanically Active Polymers: A Concise Reference* (pp. 509–531). Springer International Publishing. doi:10.1007/978-3-319-31530-0_22

Sujan, D., Yeo, C. W., Rahman, M. E., Reddy, M. M., Maleque, M. A., & Mohammad, Y. A. (2012). Aluminum-silicon carbide composites for enhanced physio-mechanical properties. *Advanced Materials Research*, *576*, 370–373. doi:10.4028/www.scientific.net/AMR.576.370

Sumaru, K., Kameda, M., Kanamori, T., & Shinbo, T. (2004). Characteristic phase transition of aqueous solution of poly (N-isopropylacrylamide) functionalized with spirobenzopyran. *Macromolecules*, *37*(13), 4949–4955. doi:10.1021/ma049661x

Sunderland, A., Ciegis, R., Henty, D., Kågström, B., & Žilinskas, J. (2009). *Parallel Diagonalization Performance on High-Performance Computers*., doi:10.1007/978-0-387-09707-7_5

Suresha, K. M., Yang, S. Y., Lee, M. H., Kim, J. H., & Kim, J. (2008, January). Actuator, sensor and MEMS devices based on regenerated cellulose. *Composite Interfaces*, *15*(7–9), 679–685. doi:10.1163/156855408786778393

Suresh, K., Bagade, S. D., & Regalla, S. P. (2015). Deformation behavior of extra deep drawing steel in single-point incremental forming. *Materials and Manufacturing Processes*, *30*(10), 1202–1209. doi:10.1080/10426914.2014.994755

Swait, T. J., Rauf, A., Grainger, R., Bailey, P. B. S., Lafferty, A. D., Fleet, E. J., Hand, R. J., & Hayes, S. A. (2012). Smart composite materials for self-sensing and self-healing. *Plastics, Rubber and Composites*, *41*(4–5), 215–224. doi:10.1179/1743289811Y.0000000039

Takagi, T. (1990). A Concept of Intelligent Materials. Journal of Intelligent Material Systems and Structures, 1(2), 149–156. doi:10.1177/1045389X9000100201

Takahashi, Y., Hamanaka, I., & Shimizu, H. (2013). Flexural properties of denture base resins subjected to long-term water immersion. *Acta Odontologica Scandinavica*, *71*(3-4), 716–720. doi:10.3109/00016357.2012.715196 PMID:22900745

Takhsha Ghahfarokhi, M., Nasi, L., Casoli, F., Fabbrici, S., Trevisi, G., Cabassi, R., & Albertini, F. (2020). Following the Martensitic Configuration Footprints in the Transition Route of Ni-Mn-Ga Magnetic Shape Memory Films: Insight into the Role of Twin Boundaries and Interfaces. *Materials (Basel)*, *13*(9), 2103. doi:10.3390/ma13092103 PMID:32370074

Tamang, S., & Aravindan, S. (2017). An investigation on joining of Al6061-T6 to AZ31B by microwave hybrid heating using active braze alloy as an interlayer. *Journal of Manufacturing Processes*, *28*, 94–100. doi:10.1016/j.jmapro.2017.05.027

Tang, G., Wang, Y., Chen, W., Tang, H., & Li, C. (2013). Hydrothermal synthesis and characterization of novel flowerlike MoS2 hollow microspheres. *Materials Letters*, *100*, 15–18. doi:10.1016/j.matlet.2013.02.103

Tang, L., Zheng, Y., Melo, M. B., Mabardi, L., Castaño, A. P., Xie, Y. Q., Li, N., Kudchodkar, S. B., Wong, H. C., Jeng, E. K., Maus, M. V., & Irvine, D. J. (2018). Enhancing T cell therapy through TCR-signaling-responsive nanoparticle drug delivery. *Nature Biotechnology*, *36*(8), 707–716. doi:10.1038/nbt.4181 PMID:29985479

Tan, L., Dong, J., Chen, J., & Yang, K. (2018). Development of magnesium alloys for biomedical applications: Structure, process to property relationship. *Materials Technology*, *33*(3), 235–243. doi:10.1080/10667857.2017.1405890

Tay, F. E. (2015). Smart materials: Recent advances and future directions. *Progress in Materials Science*, *74*, 163–239.

Taylor, D. L., & in het Panhuis, M. (2016). Self-healing hydrogels. *Advanced Materials*, *28*(41), 9060–9093. doi:10.1002/adma.201601613 PMID:27488822

Thakur, S. (2017). *Shape Memory Polymers for Smart Textile Applications*. Textiles for Advanced Applications., doi:10.5772/intechopen.69742

Tiginyanu, I., Ursaki, V., & Popa, V. (2011). *Nanoimprint Lithography (NIL) and Related Techniques for Electronics Applications. Nanocoatings and Ultra-Thin Films*. Woodhead Publishing Limited. doi:10.1533/9780857094902.2.280

Tofighy, M. A., & Mohammadi, T. (2011). Adsorption of divalent heavy metal ions from water using carbon nanotube sheets. *Journal of Hazardous Materials*, *185*(1), 140–147. doi:10.1016/j.jhazmat.2010.09.008 PMID:20926186

Toh, W. Q., Tan, X., Bhowmik, A., Liu, E., & Tor, S. B. (2017). Tribochemical characterization and tribocorrosive behavior of CoCrMo alloys: A review. *Materials (Basel)*, *11*(1), 30. doi:10.3390/ma11010030 PMID:29278375

Toozandehjani, M. (2018). Conventional and Advanced Composites in Aerospace Industry: Technologies Revisited. *American Journal of Aerospace Engineering*, *5*(1), 9. doi:10.11648/j.ajae.20180501.12

Truong, V. X., Ablett, M. P., Richardson, S. M., Hoyland, J. A., & Dove, A. P. (2015). Simultaneous orthogonal dual-click approach to tough, in-situ-forming hydrogels for cell encapsulation. *Journal of the American Chemical Society*, *137*(4), 1618–1622. doi:10.1021/ja511681s PMID:25590670

Tsang, V. L., & Bhatia, S. N. (2007). Fabrication of three-dimensional tissues. *Tissue Engineering Ii: Basics of Tissue Engineering and Tissue Applications*, 189-205.

Tsaturyants, M., Sheremetyev, V., Dubinskiy, S., Komarov, V., Polyakova, K., Korotitskiy, A., Prokoshkin, S., Borisov, E., Starikov, K., Kaledina, D., Popovich, A., & Brailovski, V. (2022). Structure and properties of Ti-50.2Ni alloy processed by laser powder bed fusion and subjected to a combination of thermal cycling and heat treatments. *Shape Mem. Superelasticity*, *8*(1), 16–32. doi:10.100740830-022-00363-4

Uchino, K. (1986). Piezoelectric and Electrostrictive Actuators. *Sixth IEEE International Symposium on Applications of Ferroelectrics*. IEEE. 10.1109/ISAF.1986.201218

Uchino, K. (2006). Piezoelectric Actuators and Devices: Recent Developments and Prospects. *Japanese Journal of Applied Physics*, *45*(10A), 7571–7579.

Uchino, K. (2019). *Piezoelectric actuators and ultrasonic motors*. Springer.

Undre, B, S., R. Pandya, S., Kumar, V., & Singh, M. (2016). Dendrimers As Smart Materials For Developing The Various Applications In The Field Of Biomedical Sciences. *Advanced Materials Letters*, *7*(7), 502–516. doi:10.5185/amlett.2016.6153

Uzun, G., Hersek, N., & Tincer, T. (1999). Effect of five woven fiber reinforcements on the impact and transverse strength of a denture base resin. *The Journal of prosthetic dentistry*, *81*(5), 616-620.

Vacanti, J. P., & Langer, R. (1999). Tissue engineering: The design and fabrication of living replacement devices for surgical reconstruction and transplantation. *Lancet*, *354*, S32–S34. doi:10.1016/S0140-6736(99)90247-7 PMID:10437854

Van de Velde, K., & Kiekens, P. (2002). Biopolymers: Overview of several properties and consequences on their applications. *Polymer Testing*, *21*(4), 433–442. doi:10.1016/S0142-9418(01)00107-6

van Hengel, I. A. J., Riool, M., Fratila-Apachitei, L. E., Witte-Bouma, J., Farrell, E., Zadpoor, A. A., Zaat, S. A. J., & Apachitei, I. (2017). Data on the surface morphology of additively manufactured Ti-6Al-4V implants during processing by plasma electrolytic oxidation. *Data in Brief*, *13*, 385–389. doi:10.1016/j.dib.2017.06.015 PMID:28664174

Vanderroost, M., Ragaert, P., Devlieghere, F., & De Meulenaer, B. (2014, September). Intelligent food packaging: The next generation. *Trends in Food Science & Technology*, *39*(1), 47–62. doi:10.1016/j.tifs.2014.06.009

Varaprasad, D. V. (1998). *Electrochromic mirrors and devices*. Google Patents.

Vashist, A., & Ahmad, S.VASHIST. (2013). Hydrogels: Smart Materials for Drug Delivery. *Oriental Journal of Chemistry*, *29*(03), 861–870. doi:10.13005/ojc/290303

Vattikuti, S. P., & Byon, C. (2015). Synthesis and characterization of molybdenum disulfide nanoflowers and nanosheets: Nanotribology. *Journal of Nanomaterials*, *2015*, 9–9. doi:10.1155/2015/710462

Vayssieres, L., Hagfeldt, A., & Lindquist, S. E. (2000, January 1). Purpose-built metal oxide nanomaterials. The emergence of a new generation of smart materials. *Pure and Applied Chemistry*, *72*(1–2), 47–52. doi:10.1351/pac200072010047

Vicente, A., Santano, D., Zubiate, P., Urrutia, A., Del, I., & Ruiz, C. (2019, April). 2020. 'Sensors and Actuators B : Chemical Lossy Mode Resonance Sensors Based on Nanocoated Multimode-Coreless- Multimode Fi Bre'. *Sensors and Actuators. B, Chemical, 304*, 126955. doi:10.1016/j.snb.2019.126955

Vieira, A. C., Sequeira, P. D., Gomes, J. R., & Rocha, L. A. (2009). Dry sliding wear of Al alloy/SiCp functionally graded composites: Influence of processing conditions. *Wear, 267*(1-4), 585–592. doi:10.1016/j.wear.2009.01.041

Vukojicic, A., Delestrée, N., Fletcher, E. V., Pagiazitis, J. G., Sankaranarayanan, S., Yednock, T. A., Barres, B. A., & Mentis, G. Z. (2019). The Classical Complement Pathway Mediates Microglia-Dependent Remodeling of Spinal Motor Circuits during Development and in SMA. *Cell Reports, 29*(10), 3087–3100.e7. doi:10.1016/j.celrep.2019.11.013 PMID:31801075

Wachesk, C. C., Seabra, S. H., Dos Santos, T. A. T., Trava-Airoldi, V. J., Lobo, A. O., & Marciano, F. R. (2021). In vivo biocompatibility of diamond-like carbon films containing TiO_2 nanoparticles for biomedical applications. *Journal of Materials Science. Materials in Medicine, 32*(9), 117. doi:10.100710856-021-06596-6 PMID:34460018

Wada, K., & Liu, Y. (2008). On the mechanisms of two-way memory effect and stress-assisted two-way memory effect in NiTi shape memory alloy. *Journal of Alloys and Compounds, 449*(1–2), 125–128. doi:10.1016/j.jallcom.2006.01.128

Wang, X. (2017). Smart Materials for Advanced Biomedical Applications. *Smart Materials for Advanced Environmental Applications,* 313-346.

Wang, J. L., Xu, J. K., Hopkins, C., Chow, D. H. K., & Qin, L. (2020). Biodegradable magnesium-based implants in orthopedics—A general review and perspectives. *Advancement of Science, 7*(8), 1902443. doi:10.1002/advs.201902443 PMID:32328412

Wang, K., Meng, Q., Wang, Q., Zhang, W., Guo, J., Cao, S., Elezzabi, A. Y., Yu, W. W., Liu, L., & Li, H. (2021, September 3). Advances in Energy-Efficient Plasmonic Electrochromic Smart Windows Based on Metal Oxide Nanocrystals. *Advanced Energy and Sustainability Research, 2*(12), 2100117. doi:10.1002/aesr.202100117

Wang, P., Yue, W., Lu, Z., Zhang, G., & Zhu, L. (2018). Friction and wear properties of MoS2-based coatings sliding against Cu and Al under electric current. *Tribology International, 127*, 379–388. doi:10.1016/j.triboint.2018.06.028

Wang, S., Lee, J. M., & Yeong, W. Y. (2015). Smart hydrogels for 3D bioprinting. *International Journal of Bioprinting, 1*(1).

Wang, X., Li, Y., Gao, Y., Yang, Z., Lu, C., & Zhu, T. (2018). A quantum mechanical computational method for modeling electrostatic and solvation effects of protein. *Scientific Reports, 8*(1), 5475. doi:10.103841598-018-23783-8 PMID:29615707

Wang, Y., Lany, S., Ghanbaja, J., Fagot-Revurat, Y., Chen, Y. P., Soldera, F., Horwat, D., Mücklich, F., & Pierson, J. F. (2016). 'Electronic Structures of C U2 O, C U4 O3, and CuO: A Joint Experimental and Theoretical Study'. *Physical Review. B*, *94*(24), 245418. Advance online publication. doi:10.1103/PhysRevB.94.245418

Wang, Z. L. (2017). Smart materials for energy efficiency in buildings: A review. *Renewable & Sustainable Energy Reviews*, *68*, 1074–1089.

Wanhill, R., & Barter, S. (2011). *Fatigue of beta processed and beta heat-treated titanium alloys*. Springer Science & Business Media.

Watanabe, Y., & Sato, H. (2011). Review fabrication of functionally graded materials under a centrifugal force. *Nanocomposites with unique properties and applications in medicine and industry*, *133*, 150.

Watanabe, Y., Kawamoto, A., & Matsuda, K. (2002). Particle size distributions in functionally graded materials fabricated by the centrifugal solid-particle method. *Composites Science and Technology*, *62*(6), 881–888. doi:10.1016/S0266-3538(02)00023-4

Watanabe, Y., Sato, R., Matsuda, K., & Fukui, Y. (2004). Evaluation of particle size and particle shape distributions in Al-Al3Ni FGMs fabricated by a centrifugal in-situ method. *Science and Engineering of Composite Materials*, *11*(2-3), 185–200. doi:10.1515/SECM.2004.11.2-3.185

Wauthle, R., van der Stok, J., Amin Yavari, S., Van Humbeeck, J., Kruth, J. P., Zadpoor, A. A., Weinans, H., Mulier, M., & Schrooten, J. (2015). Additively manufactured porous tantalum implants. *Acta Biomaterialia*, *14*, 217–225. doi:10.1016/j.actbio.2014.12.003 PMID:25500631

Weigel, T., Schinkel, G., & Lendlein, A. (2006). Design and preparation of polymeric scaffolds for tissue engineering. *Expert Review of Medical Devices*, *3*(6), 835–851. doi:10.1586/17434440.3.6.835 PMID:17280547

Weinberger, P. (2014). 'Niels Bohr and the Dawn of Quantum Theory'. *Philosophical Magazine*, *94*(27), 3072–3087. doi:10.1080/14786435.2014.951710

Wei, S., Li, Y., Li, K., Kang, A., Zhang, S., Feng, T., Zhang, H., & Zhong, C. (2021). Functional amyloid-chitin hybrid ink coupled with flexible fabrication approaches for diverse macro and micro-structures. *Materials Today. Bio*, *13*, 100179. doi:10.1016/j.mtbio.2021.100179 PMID:34938993

Wichterle, O., & Lim, D. (1960). Hydrophilic gels for biological use. *Nature*, *185*(4706), 117–118. doi:10.1038/185117a0

Wik, T. S., Foss, O. A., Havik, S., Persen, L., Aamodt, A., & Witsø, E. (2010). Periprosthetic fracture caused by stress shielding after implantation of a femoral condyle endoprosthesis in a transfemoral amputee—A case report. *Acta Orthopaedica*, *81*(6), 765–767. doi:10.3109/17453674.2010.533937 PMID:21067436

Wilson, E. B., Decius, J. C., & Cross, P. C. (1955). *Molecular Vibrations: The Theory of Infrared and Raman Vibrational Spectra*. McGraw-Hill.

Wilson, E. B. Jr. (1939). A Method of Obtaining the Expanded Secular Equation for the Vibration Frequencies of a Molecule. *The Journal of Chemical Physics*, *7*(11), 1047–1052. doi:10.1063/1.1750363

Windhagen, H., Radtke, K., Weizbauer, A., Diekmann, J., Noll, Y., Kreimeyer, U., Schavan, R., Stukenborg-Colsman, C., & Waizy, H. (2013). Biodegradable magnesium-based screw clinically equivalent to titanium screw in hallux valgus surgery: Short term results of the first prospective, randomized, controlled clinical pilot study. *Biomedical Engineering Online*, *12*(1), 62. doi:10.1186/1475-925X-12-62 PMID:23819489

Wong, J. Y., & Bronzino, J. D. (2007). *Biomaterials*. CRC press.

Wong, W. L. E., & Gupta, M. (2015). Using microwave energy to synthesize light weight/ energy saving magnesium based materials: A review. *Technologies*, *3*(1), 1–18. doi:10.3390/ technologies3010001

Xiao Hui, L., Cheng, Z., Laurent, L. G., & ChunYing, C. (2010). "Smart" nanomaterials for cancer therapy. *Science China, 53*(11). doi:10.1007/s11426-010-4122-9

Xie, C. Z. (2015). Recent Advances in Shape Memory Polymers: Structure, Mechanism, Functionality, Modeling and Applications. *Progress in Polymer Science*, *49-50*, 79–120.

Xie, G., Zhang, K., Guo, B., Liu, Q., Fang, L., & Gong, J. R. (2013). Graphene-Based Materials for Hydrogen Generation from Light-Driven Water Splitting. *Advanced Materials*, *25*(28), 3820–3839. doi:10.1002/adma.201301207 PMID:23813606

Xiong, X., Campo, A. d., & Cui, J. (2019). Photoresponsive Polymers. Smart Polymers and Their Applications. Elsevier. doi:10.1016/B978-0-08-102416-4.00004-1

Xiong, Y., Tang, H., Wang, X., Zhao, Y., Fu, Q., Yang, J., & Xu, D. (2017). Significantly enhanced thermal conductivity of indium arsenide nanowires via sulfur passivation. *Scientific Reports*, *7*(1), 13252. doi:10.103841598-017-13792-4 PMID:29038573

Xu, Z. D., Xin, J. B., Fan, R. X., Wang, K., & Yang, J. (2018). A simple approach to fabricate CN/ MoS2 nanocomposite and its application as a lubricant additive. *Dig. J. Nanomater. Biostruct, 13*, 731-741. https://chalcogen.ro/731_XuZD.pdf

Xu, F. M., Zhu, S. J., Zhao, J., Qi, M., Wang, F. G., Li, S. X., & Wang, Z. G. (2004). Effect of stress ratio on fatigue crack propagation in a functionally graded metal matrix composite. *Composites Science and Technology*, *64*(12), 1795–1803. doi:10.1016/j.compscitech.2004.01.009

Xu, X., Cao, D., Yang, H., & He, M. (2018). Application of piezoelectric transducer in energy harvesting in pavement. *International Journal of Pavement Research and Technology*, *11*(4), 388–395. doi:10.1016/j.ijprt.2017.09.011

Yadav, P., Yadav, H., Shah, V. G., Shah, G., & Dhaka, G. (2015). Biomedical biopolymers, their origin and evolution in biomedical sciences: A systematic review. *Journal of Clinical and Diagnostic Research : JCDR*, *9*(9), ZE21. doi:10.7860/JCDR/2015/13907.6565 PMID:26501034

Yadav, S., Sadique, M. A., Kaushik, A., Ranjan, P., Khan, R., & Srivastava, A. K. (2022). Borophene as an emerging 2D flatland for biomedical applications: Current challenges and future prospects. *Journal of Materials Chemistry. B, Materials for Biology and Medicine*, *10*(8), 1146–1175. doi:10.1039/D1TB02277F PMID:35107476

Yamato, M., Konno, C., Utsumi, M., Kikuchi, A., & Okano, T. (2002). Thermally responsive polymer-grafted surfaces facilitate patterned cell seeding and co-culture. *Biomaterials*, *23*(2), 561–567. doi:10.1016/S0142-9612(01)00138-7 PMID:11761176

Yang, J., Tang, H., Zhao, Y., Zhang, Y., Li, J., Ni, Z., Chen, Y., & Xu, D. (2015). Thermal conductivity of zinc blende and wurtzite CdSe nanostructures. *Nanoscale*, *7*(38), 16071–16078. doi:10.1039/C5NR04117A PMID:26372172

Yang, K., & Xing, B. (2007). Desorption of polycyclic aromatic hydrocarbons from carbon nanomaterials in water. *Environmental Pollution*, *145*(2), 529–537. doi:10.1016/j.envpol.2006.04.020 PMID:16777283

Yang, Y., He, C., Dianyu, E., Yang, W., Qi, F., Xie, D., & Shuai, C. (2020). Mg bone implant: Features, developments and perspectives. *Materials & Design*, *185*, 108259. doi:10.1016/j.matdes.2019.108259

Yang, Y., Zeng, W., Huang, P., Zeng, X., & Mei, L. (2021). Smart materials for drug delivery and cancer therapy. *VIEW*, *2*(2), 20200042. doi:10.1002/VIW.20200042

Yao, B. D., Chan, Y. F., & Wang, N. (2002). Formation of ZnO nanostructures by a simple way of thermal evaporation. *Applied Physics Letters*, *81*(4), 757–759. doi:10.1063/1.1495878

Yao, K. S. (2019). Piezoelectric materials for biomedical applications: A review. *Materials Science and Engineering C*, 104.

Yao, S. H., Kao, W. H., Su, Y. L., & Liu, T. H. (2004). On the tribology and micro-drilling performance of TiN/AlN nanolayer coatings. *Materials Science and Engineering A*, *386*(1-2), 149–155. doi:10.1016/S0921-5093(04)00919-0

Yaraghi, N. A., & Kisailus, D. (2018). Biomimetic Structural Materials: Inspiration from Design and Assembly. *Annual Review of Physical Chemistry*, *69*(1), 23–57. doi:10.1146/annurev-physchem-040215-112621 PMID:29237136

Yates, M. R., & Barlow, C. Y. (2013). Life cycle assessments of biodegradable, commercial biopolymers—A critical review. *Resources, Conservation and Recycling*, *78*, 54–66. doi:10.1016/j.resconrec.2013.06.010

Yazdimamaghani, M., Razavi, M., Vashaee, D., & Tayebi, L. (2014). Development and degradation behavior of magnesium scaffolds coated with polycaprolactone for bone tissue engineering. *Materials Letters*, *132*, 106–110. doi:10.1016/j.matlet.2014.06.036

Ye, H. (2003). An overview of the development of Al-Si-alloy based material for engine applications. *Journal of Materials Engineering and Performance*, *12*(3), 288–297. doi:10.1361/105994903770343132

Yin, H., Zhao, S., Wan, J., Tang, H., Chang, L., He, L., Zhao, H., Gao, Y., & Tang, Z. (2013). Three-Dimensional Graphene/Metal Oxide Nanoparticle Hybrids for High-Performance Capacitive Deionization of Saline Water. *Advanced Materials, 25*(43), 6270–6276. Portico. . doi:10.1002/adma.201302223

Yong, Y., Cui, H., Zhou, Q., Su, X., Kuang, Y., & Li, X. (2018, June). 2019. 'Journal of Physics and Chemistry of Solids Ultrathin Nanowire Based on Icosahedral W @ Au 12 and Application as NO Gas Sensor'. *Journal of Physics and Chemistry of Solids, 127*, 68–75. doi:10.1016/j.jpcs.2018.12.008

You, F., Chen, X., Cooper, D. M. L., Chang, T., & Eames, B. F. (2018). Homogeneous hydroxyapatite/alginate composite hydrogel promotes calcified cartilage matrix deposition with potential for three-dimensional bioprinting. *Biofabrication, 11*(1), 015015. doi:10.1088/1758-5090/aaf44a PMID:30524110

You, J. O., Almeda, D., Ye, G. J., & Auguste, D. T. (2010). Bioresponsive matrices in drug delivery. *Journal of Biological Engineering, 4*(1), 1–13. doi:10.1186/1754-1611-4-15 PMID:21114841

Yu, H., Shi, R., Zhao, Y., Waterhouse, G. I. N., Wu, L. Z., Tung, C. H., & Zhang, T. (2016, September 14). Smart Utilization of Carbon Dots in Semiconductor Photocatalysis. *Advanced Materials, 28*(43), 9454–9477. doi:10.1002/adma.201602581 PMID:27623955

Yun, S., & Kim, J. (2009, August). Covalently bonded multi-walled carbon nanotubes-cellulose electro-active paper actuator. *Sensors and Actuators. A, Physical, 154*(1), 73–78. doi:10.1016/j.sna.2009.06.011

Yun, S., & Kim, J. (2011, February). Mechanical, electrical, piezoelectric and electro-active behavior of aligned multi-walled carbon nanotube/cellulose composites. *Carbon, 49*(2), 518–527. doi:10.1016/j.carbon.2010.09.051

Yun, S., Kim, J., & Ounaies, Z. (2006, April 25). Single-walled carbon nanotube/polyaniline coated cellulose based electro-active paper (EAPap) as hybrid actuator. *Smart Materials and Structures, 15*(3), N61–N65. doi:10.1088/0964-1726/15/3/N02

Yu, Y. I. (2016). Smart windows based on thermochromic materials: A review. *Solar Energy Materials and Solar Cells, 157*, 2–23.

Zadpoor, A. A. (2019). Additively manufactured porous metallic biomaterials. *Journal of Materials Chemistry. B, Materials for Biology and Medicine, 7*(26), 4088–4117. doi:10.1039/C9TB00420C PMID:31701985

Zadpoor, A. A., & Malda, J. (2017). Additive Manufacturing of Biomaterials, Tissues, and Organs. *Annals of Biomedical Engineering, 45*(1), 1–11. doi:10.100710439-016-1719-y PMID:27632024

Zafar, F., & Iqbal, A. (2016). Indium Phosphide Nanowires and Their Applications in Optoelectronic Devices. *Proceedings of the Royal Society A: Mathematical, Physical and Engineering Sciences*. Royal Society of London. 10.1098/rspa.2015.0804

Zafar, S., & Sharma, A. K. (2017). Microstructure and mechanical properties of microwave post-processed Ni coating. *Journal of Materials Engineering and Performance*, *26*(3), 1382–1390. doi:10.100711665-017-2540-y

Zeng, X. (2019). Recent progress in thermochromic materials for smart windows: A review. *Solar Energy Materials and Solar Cells*, *197*, 219–239.

Zhai, S., Jiang, W., Wei, L., Karahan, H. E., Yuan, Y., Ng, A. K., & Chen, Y. (2015). All-carbon solid-state yarn supercapacitors from activated carbon and carbon fibers for smart textiles. *Materials Horizons*, *2*(6), 598–605. doi:10.1039/C5MH00108K

Zhai, W., Li, Y., Cheng, Z., Sun, L., Li, F., & Li, J. (2020). Investigation on the forming force and surface quality during ultrasonic-assisted incremental sheet forming process. *International Journal of Advanced Manufacturing Technology*, *106*(7-8), 2703–2719. doi:10.100700170-019-04870-0

Zhang, B. M. (2006). *Centrifugal Casting* [M]. Machinery Industry Press.

Zhang, C., Lv, W., Tao, Y., & Yang, Q. H. (2015). Towards superior volumetric performance: Design and preparation of novel carbon materials for energy storage. *Energy & Environmental Science*, *8*(5), 1390–1403. doi:10.1039/C5EE00389J

Zhang, G., Zhao, P., Lin, L., Qin, L., Huan, Z., Leeflang, S., Zadpoor, A. A., Zhou, J., & Wu, L. (2021). Surface-treated 3D printed Ti-6Al-4V scaffolds with enhanced bone regeneration performance: An *in vivo* study. *Annals of Translational Medicine*, *9*(1), 39. doi:10.21037/atm-20-3829 PMID:33553332

Zhang, J., Jiang, Y., Shang, Z., Zhao, B., Jiao, M., Liu, W., Cheng, M., Zhai, B., Guo, Y., Liu, B., Shi, X., & Ma, B. (2021). Biodegradable metals for bone defect repair: A systematic review and meta-analysis based on animal studies. *Bioactive Materials*, *6*(11), 4027–4052. doi:10.1016/j.bioactmat.2021.03.035 PMID:33997491

Zhang, J., Wang, Y., Yang, B., & Zhou, B. (1999). Effects of Si content on the microstructure and tensile strength of an in situ Al/Mg_2Si composite. *Journal of Materials Research*, *14*(01), 68–74. doi:10.1557/JMR.1999.0012

Zhang, S., Zhang, X., Zhao, C., Li, J., Song, Y., Xie, C., Tao, H., Zhang, Y., He, Y., Jiang, Y., & Bian, Y. (2010). Research on an Mg–Zn alloy as a degradable biomaterial. *Acta Biomaterialia*, *6*(2), 626–640. doi:10.1016/j.actbio.2009.06.028 PMID:19545650

Zhang, X. S. (2020). Recent developments in electrochromic materials and devices. *Journal of Materials Chemistry. C, Materials for Optical and Electronic Devices*, *8*(36), 12423–12444.

Zhang, X. W. (2018). Progress in the development of piezoelectric materials for acoustic transducers: A review. *Progress in Materials Science*, *97*, 187–259.

Zhang, Y. L. (2016). A review of recent developments in sensing and monitoring of concrete structures using smart materials. *Sensors (Basel)*, *16*(10).

Zhang, Y., Jin, Z., Su, Y., & Wang, G. (2019). Charge separation and electron transfer routes modulated with Co-Mo-P over g-C3N4 photocatalyst. *Molecular Catalysis*, *462*, 46–55. doi:10.1016/j.mcat.2018.10.009

Zhang, Z., Demir, K. G., & Gu, G. X. (2019, March 19). Developments in 4D-printing: A review on current smart materials, technologies, and applications. *International Journal of Smart and Nano Materials*, *10*(3), 205–224. doi:10.1080/19475411.2019.1591541

Zhan, M., Liu, J., Wang, D., Chen, X., Zhang, L., & Wang, S. (2021). Optimized Neural Network Prediction Model of Shape Memory Alloy and Its Application for Structural Vibration Control. *Materials (Basel)*, *14*(21), 6593. doi:10.3390/ma14216593 PMID:34772118

Zhao, L. D. (2011). Thermoelectric materials: From history, a glance to future. *Materials Today*, *14*(7-8), 322–332.

Zhao, Y., Li, L., Liu, S., Wang, J., Xu, J., Shi, Y., Chen, K., Cabarrocas, P. R., & Yu, L. (2020). 'Germanium Quantum Dot Infrared Photodetectors Addressed by Self-Aligned Silicon Nanowire Electrodes'. *Nanotechnology*, *31*(14), 145602. doi:10.1088/1361-6528/ab647e PMID:31860876

Zheng, Y. F., Gu, X. N., & Witte, F. (2014). Biodegradable metals. *Materials Science and Engineering R Reports*, *77*, 1–34. doi:10.1016/j.mser.2014.01.001

Zheng, Y., Huang, K., You, X., Huang, B., Wu, J., & Gu, Z. (2017, November). Hybrid hydrogels with high strength and biocompatibility for bone regeneration. *International Journal of Biological Macromolecules*, *104*, 1143–1149. doi:10.1016/j.ijbiomac.2017.07.017 PMID:28687387

Zheng, Z., Jur, J., & Cheng, W. (2021). Smart materials and devices for electronic textiles. *MRS Bulletin*, *46*(6), 488–490. doi:10.155743577-021-00120-5

Zhong, Z., Yang, C., & Lieber, C. (2007). *Silicon Nanowires and Nanowire Heterostructures. Nanosilicon: new photonic and electronic material at the transition between solid and molecule*. Elsevier Ltd. doi:. doi:10.1016/B978-0-08-044528-1.50006-3

Zhu, J. N., Zhou, T. T., Zha, M., Li, C., Li, J. H., Wang, C., Gao, C.-L., Wang, H.-Y., & Jiang, Q. C. (2018). Microstructure and wear behaviour of Al-20Mg2Si alloy with combined Zr and Sb additions. *Journal of Alloys and Compounds*, *767*, 1109–1116. doi:10.1016/j.jallcom.2018.07.032

Zhu, L., Wang, Y., Hu, F., & Song, H. (2015). Structural and friction characteristics of g-C3N4/PVDF composites. *Applied Surface Science*, *345*, 349–354. doi:10.1016/j.apsusc.2015.03.197

Znaidi, L. (2010). Sol–gel-deposited ZnO thin films: A review. *Materials Science and Engineering B*, *174*(1-3), 18–30. doi:10.1016/j.mseb.2010.07.001

Zuo, L., Tang, X., & Zhang, P. S. (2015). *Electricity generating shock absorbers*. Google Patents.

Zweibel, K., Moskowitz, P., & Fthenakis, V. (1998). *Thin-Film Cadmium Telluride Photovoltaics: ES&H Issues, Solutions, and Perspectives*. OSTI. https://www.osti.gov/bridge

About the Contributors

Ashish Kumar Srivastava has completed his Ph.D. from Indian Institute of Technology (ISM) Dhanbad, Jharkhand, India in February 2018 and currently working as a professor in the department of mechanical engineering, G.L. Bajaj Institute of Technology and Management, Greater Noida, U.P., India. He is a good learner and listener and contributed greatly to society through academics and research. Dr. Srivastava has taught many engineering subjects at UG and PG level. Few subjects like additive manufacturing, engineering mechanics, strength of material, machine design, are the key subject of interest. As a researcher, he has published more than 100 research papers in various journal of repute, indexed in SCI/SCIE/SCOPUS. He has also presented more than 20 research papers in various international and national conferences in India and Abroad. Apart from this he has successfully completed 2 research funded projects under the collaborative research scheme of TEQIP-III and applied more than 5 research projects for funding in different govt. agencies. Recently he has Published 2 Indian patents and filled 2 more. Dr. Srivastava has guided 8 M.Tech Students and guiding 6 Ph.D. scholars.

* * *

Anamika Barman is a Ph.D. scholar, IIT Kharagpur.

Ranjit Barua is a Senior Researchers at Centre for Healthcare Science and Technology, IIEST-SHIBPUR. Core research area Bio-Mechanical System, 3D Bioprinting, Tissue Engineering.

Awani Bhushan has a Ph.D in Machine Design, Mechanical Engineering Indian Institute of Technology (Banaras Hindu University) Varanasi Completed: 2019 Thesis Title: Study of Weibull Design Criterion to Characterize the Fracture Behavior of Bimodular Nuclear Graphite Components. Supervisor: Prof. S.K. Panda, IIT (BHU) Varanasi.

Samir Das is Prime Minister Research Fellow Biomaterials and Tissue Engineering Lab School of Medical Science and Technology Indian Institute of Technology Kharagpur-721302 India

Sudipto Datta is a National Post Doctoral Fellow

Satish Geeri received Ph.D from Andhra University College of Engineering (A) in the year 2018. He is working as Associate Professor in the department of Mechanical Engineering, Pragati Engineering College (A). His research interest includes Nano Composites and their characterization.

Sunkulp Goel presently working as an Associate Professor, at Nanjing University of Science and Technology China, His research area is in the field of Nanoscience and Technology.

J. Mangaiyarkkarasi received her M.Phil. degree from Alagappa University, Karaikudi, Tamil Nadu, India. She completed her Ph.D. degree in 2017. She is currently working as an Associate Professor of Physics in NMSS Vellaichamy Nadar College, Nagamalai, Madurai- 625 019, Tamil Nadu, India. Her major research interests include synthesis of ceramic materials, powder X-ray analysis and electron density studies. She has published 9 research papers in the refereed International journals. She has prepared study materials (UG& PG Physics) for Madurai Kamaraj University Distance Education Course. She has 23 years of teaching experience.

Subash Chandra Ram is working as an associate Professor at Tula's Insitute Dehradun since March 2022 and he has published around 20 plus research publications in reputed Journals. The research area is the Mechanical and Wear behavior of Cast composites and Alloys.

Subhasis Roy is working as Assistant Professor at Department of Chemical Engineering, University of Calcutta, India. Dr. Roy received Ph.D. from IIT Kharagpur, India. He worked as a Postdoctoral Researcher at Sungkyunkwan University, South Korea, Research Scientist at IISER, Kolkata, and worked as a visiting researcher Raman Fellow, Ohio University, USA. Dr. Roy has published more than 65 scientific research papers in peer-reviewed journals and 22 conference proceedings. He has 18 book chapters to his credit and acted as editor for 2 books and filed several patents. Dr. Roy was awarded DST-Fast Track Young Scientist, TARE award by SERB and Mission innovation DBT-DST joint funding. He holds membership of many professional societies such as Royal Society of Chemistry (MRSC), National Academy of Sciences, India (MNASc), Fellow of Indian Chemical Society (FICS),

Linnean Society of London (FLS)and reviewers as well as editorial board member of many international journals. His expertise is in waste management and recycling, perovskite solar cells, multiferroic composite thin films, photocatalysts for water splitting, fuel cells, and hydrogen energy.

Preeti Singh Bahadur obtained her Ph.D Degree in Physics from Barkatullah University, Bhopal, M.P., India and is currently working as a Associate Professor in the Department of Applied Sciences, Amity University, Greater Noida, U.P., India. She has published many research papers in various reputed journals, indexed in SCI/SCOPUS and other indexing. She has presented more than 20 research papers in various international and national conferences.

Yusuf Waidi is a Research Scholar.

Index

N

P

R

S

T

V

Z

www.ingramcontent.com/pod-product-compliance
Lightning Source LLC
LaVergne TN
LVHW080632011026
849160LV00014B/1187